T0358362

INVERSE PROBLEMS

Tikhonov Theory and Algorithms

SERIES ON APPLIED MATHEMATICS

Editor-in-Chief: Zhong-ci Shi

---

*Published**

Vol. 8 Quaternary Codes
by Z.-X. Wan

Vol. 9 Finite Element Methods for Integrodifferential Equations
by C. M. Chen and T. M. Shih

Vol. 10 Statistical Quality Control — A Loss Minimization Approach
by D. Trietsch

Vol. 11 The Mathematical Theory of Nonblocking Switching Networks
by F. K. Hwang

Vol. 12 Combinatorial Group Testing and Its Applications (2nd Edition)
by D.-Z. Du and F. K. Hwang

Vol. 13 Inverse Problems for Electrical Networks
by E. B. Curtis and J. A. Morrow

Vol. 14 Combinatorial and Global Optimization
eds. P. M. Pardalos, A. Migdalas and R. E. Burkard

Vol. 15 The Mathematical Theory of Nonblocking Switching Networks
(2nd Edition)
by F. K. Hwang

Vol. 16 Ordinary Differential Equations with Applications
by S. B. Hsu

Vol. 17 Block Designs: Analysis, Combinatorics and Applications
by D. Raghavarao and L. V. Padgett

Vol. 18 Pooling Designs and Nonadaptive Group Testing — Important Tools
for DNA Sequencing
by D.-Z. Du and F. K. Hwang

Vol. 19 Partitions: Optimality and Clustering
Vol. I: Single-parameter
by F. K. Hwang and U. G. Rothblum

Vol. 20 Partitions: Optimality and Clustering
Vol. II: Multi-Parameter
by F. K. Hwang, U. G. Rothblum and H. B. Chen

Vol. 21 Ordinary Differential Equations with Applications (2nd Edition)
by S. B. Hsu

Vol. 22 Inverse Problems: Tikhonov Theory and Algorithms
by Kazufumi Ito and Bangti Jin

*For the complete list of the volumes in this series, please visit
http://www.worldscientific.com/series/sam

Series on Applied Mathematics Volume 22

INVERSE PROBLEMS

Tikhonov Theory and Algorithms

Kazufumi Ito
North Carolina State University, USA
Bangti Jin
University of California, Riverside, USA

NEW JERSEY • LONDON • SINGAPORE • BEIJING • SHANGHAI • HONG KONG • TAIPEI • CHENNAI

Published by

World Scientific Publishing Co. Pte. Ltd.
5 Toh Tuck Link, Singapore 596224
USA office: 27 Warren Street, Suite 401-402, Hackensack, NJ 07601
UK office: 57 Shelton Street, Covent Garden, London WC2H 9HE

Library of Congress Cataloging-in-Publication Data
Ito, Kazufumi, author.
Inverse problems : Tikhonov theory and algorithms / by Kazufumi Ito (North Carolina State University, USA) & Bangti Jin (University of California, Riverside, USA).
pages cm -- (Series on applied mathematics ; vol. 22)
Includes bibliographical references and index.
ISBN 978-9814596190 (hardcover : alk. paper)
1. Inverse problems (Differential equations)--Numerical solutions. I. Jin, Bangti, author. II. Title.
QA371.I88 2014
515'.357--dc23

2014013310

British Library Cataloguing-in-Publication Data
A catalogue record for this book is available from the British Library.

Copyright © 2015 by World Scientific Publishing Co. Pte. Ltd.

All rights reserved. This book, or parts thereof, may not be reproduced in any form or by any means, electronic or mechanical, including photocopying, recording or any information storage and retrieval system now known or to be invented, without written permission from the publisher.

For photocopying of material in this volume, please pay a copying fee through the Copyright Clearance Center, Inc., 222 Rosewood Drive, Danvers, MA 01923, USA. In this case permission to photocopy is not required from the publisher.

Printed in Singapore

Preface

Due to the pioneering works of many prominent mathematicians, including A. N. Tikhonov and M. A. Lavrentiev, the concept of inverse/ill-posed problems has been widely accepted and received much attention in mathematical sciences as well as applied disciplines, e.g., heat transfer, medical imaging and geophysics. Inverse theory has played an extremely important role in many scientific developments and technological innovations.

Amongst numerous existing approaches to numerically treat ill-posed inverse problems, Tikhonov regularization is the most powerful and versatile general-purposed method. Recently, Tikhonov regularization with nonsmooth penalties has demonstrated great potentials in many practical applications. The use of nonsmooth regularization can improve significantly the reconstruction quality. Their Bayesian counterparts also start to attract considerable attention. However, it also brings great challenges to its mathematical analysis and efficient numerical implementation.

The primary goal of this monograph is to blend up-to-date mathematical theory with state-of-art numerical algorithms for Tikhonov regularization. The main focus lies in nonsmooth regularization and their convergence analysis, parameter choice rules, nonlinear problems, efficient algorithms, direct inversion methods and Bayesian inversion. A clear understanding of these different facets of Tikhonov regularization, or more generally nonsmooth models for inverse problems, is expected to greatly broaden the scope of the approach and to promote further developments, in particular in the search of better model/methods. However, a seamless integration of these facets is still rare due to its relatively recent origin.

The presentation focuses on two components of applied inverse theory: mathematical theory (linear and nonlinear Tikhonov theory) and numerical algorithms (including nonsmooth optimization algorithms, direct inversion methods and Bayesian inversion). We discuss nonsmooth regularization in the context of classical regularization theory, especially consistency, convergence rates, and parameter choice rules. These theoretical developments cover both linear and nonlinear inverse problems. The nonsmoothness of the emerging models poses significant challenges to their efficient and accurate numerical solution. We describe a number of efficient algorithms for relevant nonsmooth optimization problems, e.g., augmented Lagrangian method and semismooth Newton method. The sparsity regularization is treated in great detail. In the application of these algorithms, often a good initial guess is very beneficial, which for a class of inverse problems can be obtained by direct inversion methods. Further, we describe the Bayesian framework, which quantifies the uncertainties associated with one particular solution, the Tikhonov solution, and provides the mechanism for choosing regularization parameters and selecting the proper regularization model. We shall describe the implementation details of relevant computational techniques.

The topic of Tikhonov regularization is very broad, and surely we are not able to cover it in one single volume. The choice of materials is strongly biased by our limited knowledge. In particular, we do not intend to present the theoretical results in their most general form, but to illustrate the main ideas. However, pointers to relevant references are provided throughout.

The book is intended for senior undergraduate students and beginning graduate students. The prerequisite includes basic partial differential equations and functional analysis. However, experienced researchers and practitioners in inverse problems may also find it useful.

The book project was started during the research stay of the second author at University of Bremen, as an Alexandre von Humboldt postdoctoral fellow, and largely developed the stay of the second author at Texas A&M University and his visits at North Carolina State University and The Chinese University of Hong Kong. The generous support of the Alexandre von Humboldt foundation and the hospitality of the hosts, Peter Maass, William Rundell, and Jun Zou, are gratefully acknowledged. The authors also benefitted a lot from the discussions with Dr. Tomoya Takeuchi of University of Tokyo.

Raleigh, NC and Riverside, CA

November 2013, *Kazufumi Ito* and *Bangti Jin*

Contents

Preface v

1. Introduction 1

2. Models in Inverse Problems 5

2.1 Introduction 5
2.2 Elliptic inverse problems 6
2.2.1 Cauchy problem 6
2.2.2 Inverse source problem 8
2.2.3 Inverse scattering problem 10
2.2.4 Inverse spectral problem 13
2.3 Tomography 15
2.3.1 Computerized tomography 16
2.3.2 Emission tomography 19
2.3.3 Electrical impedance tomography 20
2.3.4 Optical tomography 23
2.3.5 Photoacoustic tomography 25

3. Tikhonov Theory for Linear Problems 29

3.1 Well-posedness 31
3.2 Value function calculus 38
3.3 Basic estimates 44
3.3.1 Classical source condition 45
3.3.2 Higher-order source condition 52
3.4 A posteriori parameter choice rules 55
3.4.1 Discrepancy principle 55

3.4.2 Hanke-Raus rule 66
3.4.3 Quasi-optimality criterion 71
3.5 Augmented Tikhonov regularization 76
3.5.1 Augmented Tikhonov regularization 77
3.5.2 Variational characterization 80
3.5.3 Fixed point algorithm 88
3.6 Multi-parameter Tikhonov regularization 91
3.6.1 Balancing principle 92
3.6.2 Error estimates 95
3.6.3 Numerical algorithms 99

4. Tikhonov Theory for Nonlinear Inverse Problems 105

4.1 Well-posedness 106
4.2 Classical convergence rate analysis 112
4.2.1 A priori parameter choice 113
4.2.2 A posteriori parameter choice 117
4.2.3 Structural properties 124
4.3 A new convergence rate analysis 128
4.3.1 Necessary optimality condition 128
4.3.2 Source and nonlinearity conditions 129
4.3.3 Convergence rate analysis 134
4.4 A class of parameter identification problems 141
4.4.1 A general class of nonlinear inverse problems . . . 141
4.4.2 Bilinear problems 144
4.4.3 Three elliptic examples 145
4.5 Convergence rate analysis in Banach spaces 150
4.5.1 Extensions of the classical approach 150
4.5.2 Variational inequalities 152
4.6 Conditional stability 160

5. Nonsmooth Optimization 169

5.1 Existence and necessary optimality condition 172
5.1.1 Existence of minimizers 172
5.1.2 Necessary optimality 173
5.2 Nonsmooth optimization algorithms 177
5.2.1 Augmented Lagrangian method 177
5.2.2 Lagrange multiplier theory 181
5.2.3 Exact penalty method 184

5.2.4 Gauss-Newton method 188
5.2.5 Semismooth Newton Method 189
5.3 ℓ^p sparsity optimization 193
5.3.1 ℓ^0 optimization 194
5.3.2 ℓ^p $(0 < p < 1)$-optimization 195
5.3.3 Primal-dual active set method 197
5.4 Nonsmooth nonconvex optimization 200
5.4.1 Biconjugate function and relaxation 201
5.4.2 Semismooth Newton method 204
5.4.3 Constrained optimization 206

6. Direct Inversion Methods 209

6.1 Inverse scattering methods 209
6.1.1 The MUSIC algorithm 210
6.1.2 Linear sampling method 215
6.1.3 Direct sampling method 218
6.2 Point source identification 223
6.3 Numerical unique continuation 226
6.4 Gel'fand-Levitan-Marchenko transformation 228
6.4.1 Gel'fand-Levitan-Marchenko transformation . . . 228
6.4.2 Application to inverse Sturm-Liouville problem . . 232

7. Bayesian Inference 235

7.1 Fundamentals of Bayesian inference 237
7.2 Model selection 244
7.3 Markov chain Monte Carlo 250
7.3.1 Monte Carlo simulation 250
7.3.2 MCMC algorithms 253
7.3.3 Convergence analysis 258
7.3.4 Accelerating MCMC algorithms 261
7.4 Approximate inference 267
7.4.1 Kullback-Leibler divergence 268
7.4.2 Approximate inference algorithms 271

Appendix A Singular Value Decomposition 285

Appendix B Noise Models 289

Appendix C Exponential Families 295

Bibliography 299

Index 317

Chapter 1

Introduction

In this monograph we develop mathematical theory and solution methods for the model-based inverse problems. That is, we assume that the underlying mathematical models, which describe the map from the physical parameters and unknowns to observables, are known. The objective of the inverse problem is to construct a stable inverse map from observables to the unknowns. Thus, in order to formulate an inverse problem it is essential to select unknowns of physical relevance and analyze their physical properties and then to develop effective and accurate mathematical models for the forward map from the unknowns to observables. In practice, it is quite common that observables measure a partial information of the state variables and that the state variables and unknowns satisfy the binding constraints in terms of equations and inequalities. That is, observables only provide indirect information of unknowns.

The objective is to determine unknowns or its distribution from observables by constructing stable inverse maps. The mathematical analysis includes the uniqueness and stability of the inverse map, the development and analysis of reconstruction algorithms, and efficient and effective implementation. In order to construct a reconstruction algorithm it is necessary to de-convolute the convoluted forward map. To this end, we formulate the forward map in a function space framework and then develop a variational approach and a direct sampling method for the unknowns and use a statistical method based on the Bayes formula to analyze the distribution of unknowns. Both theoretical and computational aspects of inverse analysis are developed in the monograph. Throughout, illustrative examples will be given to demonstrate the feasibility and applicability of our specific analysis and formulation. The outline of our presentation is as follows.

A function space inverse problem has unknowns in function spaces and the constraints in the form of (nonlinear) partial differential and nonlocal equations for modeling many physical, bio-medical, chemical, social, and engineering processes. Unknowns may enter into the model in a very nonlinear manner. In Chapter 2 we describe several examples that motivate and illustrate our mathematical formulation and framework in various areas of inverse problems. For example, in the medium inverse problem such as electrical impedance tomography and inverse medium scattering, the unknown medium coefficient enters as a multiplication. These models will be used throughout the book to illustrate the underlying ideas of the theory and algorithms.

An essential issue of the inverse problem is about how to overcome the ill-posedness of the inverse map, i.e., a small change in the observables may result in a very large change in the constructed solution. Typical examples include the inverse of a compact linear operator and a matrix with many very small singular values. Many tomography problems can be formulated as solving a linear Volterra/Fredholm integral equation of the first kind for the unknown. For example, in the inverse medium scattering problem, we may use Born's approximation to formulate a linear integral equation for the medium coefficient. The inverse map can be either severely or mildly ill-posed, depending on the singularity strength of the kernel function. The inverse map is unbounded in either case and it is necessary to develop a robust generalized inverse map that allows a small variation in the problem data (e.g., right hand side and forward model) and computes an accurate regularized solution. In Chapters 3 and 4 we describe the Tikhonov regularization method and develop the theoretical and computational treatments of ill-posed inverse problems.

Specifically, we develop the Tikhonov value function calculus and the asymptotic error analysis for the Tikhonov regularized inverse map when the noise level of observables decreases to zero. The calculus can be used for analyzing several choice rules, including the discrepancy principle, Hanke-Raus rule and quasi-optimality criterion, for selecting the crucial Tikhonov regularization parameter, and for deriving a priori and a posteriori error estimates of the Tikhonov regularized solution.

It is very essential to incorporate all available prior information of unknowns into the mathematical formulation and reconstruction algorithms and also develop an effective selection method of the (Tikhonov) regularization parameters. We systematically use the Bayesian inference for this purpose and develop the augmented Tikhonov formulation. The formulation uses multiple prior distributions and Gamma distributions for the regularization parameters and an appropriate fidelity function based on the noise statistics. Based on a hierarchical Bayesian formulation we develop and analyze a general balancing principle for the selection of the regularization parameters. Further, we extend the augmented approach to the emerging topic of multi-parameter regularization, which enforces simultaneously multiple penalties in the hope of promoting multiple distinct features. We note that multiple/multiscale features can be observed in many applications, especially signal/image processing. A general convergence theory will be developed to partially justify their superior practical performance. The objective is to construct a robust but accurate inverse map for a class of real world inverse problems including inverse scattering, tomography problems, and image/signal processing.

Many distributed parameter identifications for partial differential equations are inherently nonlinear, even though the forward problem may be linear. The nonlinearity of the model calls for new ingredients for the mathematical analysis of related inverse problems. We extend the linear Tikhonov theory and analyze regularized solutions based on a generalized source condition. We shall recall the classical convergence theory, and develop a new approach from the viewpoint of optimization theory. In particular, the second-order optimality conditions provide a unified framework for deriving convergence rate results. The problem structure is highlighted for a general class of nonlinear parameter identification problems, and illustrated on concrete examples. The role of conditional stability in deriving convergence rates is also briefly discussed.

In Chapter 5 we discuss the optimization theory and algorithms for variational optimization, e.g., maximum likelihood estimate of unknowns based on the augmented Tikhonov formulation, optimal structural design, and optimal control problems. We focus on a class of nonsmooth optimization arising from the nonsmooth prior distribution, e.g., sparsity constraints. In order to enhance and improve the resolution of the reconstruction we use nonsmooth prior and the noise model in our Tikhonov formulation. For example, a Laplace distribution of function unknowns and its derivatives can be used. We derive the necessary optimality based on a generalized Lagrange multiplier theory for nonsmooth optimization. It is a coupled nonlinear system of equations for the primal variable and the Lagrange multiplier and results in a decomposition of coordinates. The nonsmoothness of the cost functional is treated using the complementarity condition on the Lagrange multiplier and the primal variable. Thus, Lagrange multiplier based gradient methods can be readily applied. Further, we develop a primal dual active set method based on the complementarity condition, and resulting algorithms are described in details and solve the inequality constraints, sparsity optimization, and a general class of nonsmooth optimizations. Also we develop a unified treatment of a class of nonconvex and nonsmooth optimizations, e.g., $L^0(\Omega)$ penalty, based on a generalized Lagrange multiplier theory.

For a class of inverse problems we can develop direct methods. For example, multiple signal classification (MUSIC) for estimating the frequency contents of a signal from the autocorrelation matrix has been used for determining point scatterers from the multi-static response matrix in inverse scattering. The direct sampling method is developed for the multipath scattering to probe medium inhomogeneities. For the inverse Sturm-Liouville problem, one develops an efficient iterative method based on the Gel'fand-Levitan-Marchenko transform. For the analytic extension of Cauchy data one can use the Taylor series expansion based on the Cauchy-Kowalevski theory. In Chapter 6 we present and analyze a class of direct methods for solving inverse problems. These methods can efficiently yield first estimates to certain nonlinear inverse problems, which can be either used as an initial guess for optimization algorithms in Chapter 5, or exploited to identify the region of interest and to shrink the computational domain.

In Chapter 7 we develop an effective use of Bayesian inference in the context of inverse problems, i.e., incorporation of model uncertainties, measurement noises and approximation errors in the posterior distribution, the evaluation of the effectiveness of the prior information, and the selection of regularization parameters and proper mathematical models. The Bayesian solution – the posterior distribution – encompasses an ensemble of inverse solutions that are consistent with the given data, and it can quantify the uncertainties associated with one particular solution, e.g., the Tikhonov minimizer, via credible intervals. We also discuss computational and theoretical issues for applying Markov chain Monte Carlo (MCMC) methods to inverse problems, including the Metropolis-Hastings algorithm and the Gibbs sampler. Advanced computational techniques for accelerating the MCMC methods, e.g., preconditioning, multilevel technique, and reduced-order modeling, are also discussed. Further, we discuss a class of deterministic approximate inference methods, which can deliver reasonable approximations within a small fraction of computational time needed for the MCMC methods.

Chapter 2

Models in Inverse Problems

2.1 Introduction

In this chapter, we describe several mathematical models for linear and nonlinear inverse problems arising in diverse practical applications. The examples focus on practical problems modeled by partial differential equations, where the available data consists of indirect measurements of the PDE solution. Illustrative examples showcasing the ill-posed nature of the inverse problem will also be presented. The main purpose of presenting these model problems is to show the diversity of inverse problems in practice and to describe their function space formulations.

Let us first recall the classical notion of well-posedness as formulated by the French mathematician Jacques Hadamard. A problem in mathematical physics is said to be well-posed if the following three requirements on existence, uniqueness and stability hold:

(a) There exists at least one solution;
(b) There is at most one solution;
(c) The solution depends continuously on the data.

The existence and uniqueness depend on the precise definition of a "solution", and stability is very much dependent on the topologies for measuring the solution and problem data. We note that mathematically, by suitably changing the topologies, an unstable problem might be rendered stable. However, such changes are not always plausible or possible for practical inverse problems for which the data is inevitably contaminated by noise due to imprecise data acquisition procedures. One prominent feature for practical inverse problems is the presence of data noise.

Given a problem in mathematical physics, if one of the three requirements fail to hold, then it is said to be ill-posed. For quite a long time, ill-posed problems were thought to be physically irrelevant and mathematically uninteresting. Nonetheless, it is now widely accepted that ill-posed problems arise naturally in almost all scientific and engineering disciplines, as we shall see below, and contribute significantly to scientific developments and technological innovations.

In this chapter, we present several model inverse problems, arising in elliptic (partial) differential equations and tomography, which serve as prototypical examples for linear and nonlinear inverse problems.

2.2 Elliptic inverse problems

In this part, we describe several inverse problems for a second-order elliptic differential equation

$$-\nabla \cdot (a(x)\nabla u) + \mathbf{b}(x) \cdot \nabla u + p(x)u = f(x) \quad \text{in } \Omega. \tag{2.1}$$

Here $\Omega \subset \mathbb{R}^{\mathrm{d}}$ ($\mathrm{d} = 1, 2, 3$) is an open domain with a boundary $\Gamma = \partial\Omega$. The equation (2.1) is equipped with suitable boundary conditions, e.g., Dirichlet or Neumann boundary conditions. The functions $a(x)$, $\mathbf{b}(x)$ and $p(x)$ are known as the conductivity/diffusivity, convection coefficient and the potential, respectively, and the function $f(x)$ is the source/sink term. In practice, the elliptic problem (2.1) can describe the stationary state of many physical processes, e.g., heat conduction, time-harmonic wave propagation, underground water flow and quasi-static electromagnetic processes.

There is an abundance of inverse problems related to equation (2.1), e.g., recovering one or several of the functions $a(x)$, $\mathbf{b}(x)$, $q(x)$ and $f(x)$, boundary conditions/coefficients, or other geometrical quantities from (often noisy) measurements of the solution u, which can be either in the interior of the domain Ω or on the boundary Γ. Below we describe five inverse problems: Cauchy problem, inverse source problem, inverse scattering problem, inverse spectral problem, and inverse conductivity problem. The last one will be described in Section 2.3.

2.2.1 *Cauchy problem*

The Cauchy problem for elliptic equations is fundamental to the study of many elliptic inverse problems, and has been intensively studied. One formulation is as follows. Let Γ_c and $\Gamma_i = \Gamma \setminus \Gamma_c$ be two disjointed parts of the boundary Γ, which refer to the experimentally accessible and inaccessible parts, respectively. Then the Cauchy problem reads: given the Cauchy data g and h on the boundary Γ_c, find u on the boundary Γ_i, i.e.,

$$\begin{cases} -\nabla \cdot (a(x)\nabla u) = 0 & \text{in } \Omega, \\ u = g & \text{on } \Gamma_c, \\ a\dfrac{\partial u}{\partial n} = h & \text{on } \Gamma_c. \end{cases}$$

This inverse problem itself arises also in many practical applications, e.g., thermal analysis of re-entry space shuttles/missiles [23], electrocardiography [121] and geophysical prospection. For example, in the analysis of space shuttles, one can measure the temperature and heat flux on the inner surface of the shuttle, and one is interested in the flux on the outer surface, which is not directly accessible.

The Cauchy problem is known to be severely ill-posed, and lacks a continuous dependence on the data [25]. There are numerous deep mathematical results on the Cauchy problem. We shall not delve into these results, but refer interested readers to the survey [2] for stability properties. In his essay of 1923 [122], Jacques Hadamard provided the following example, showing that the solution of the Cauchy problem for Laplace's equation does not depend continuously on the data.

Example 2.1. Let the domain $\Omega = \{(x_1, x_2) \in \mathbb{R}^2 : x_2 > 0\}$ be the upper half plane, and the boundary $\Gamma_c = \{(x_1, x_2) \in \mathbb{R}^2 : x_2 = 0\}$. Consider the solution $u = u_n$, $n = 1, 2, \ldots$, to the Cauchy problem

$$\begin{cases} \Delta u = 0 & \text{in } \Omega, \\ u = 0 & \text{on } \Gamma_c, \\ \dfrac{\partial u}{\partial n} = -n^{-1} \sin n x_1 & \text{on } \Gamma_c. \end{cases}$$

It can be verified directly that $u_n = n^{-2} \sin n x_1 \sinh n x_2$ is a solution to the problem, and further the uniqueness is a direct consequence of Holmgren's theorem for Laplace's equation. Hence it is the unique solution. Clearly, $\frac{\partial u_n}{\partial n}|_{\Gamma_c} \to 0$ uniformly as $n \to \infty$, whereas for any $x_2 > 0$, $u_n(x_1, x_2) = n^{-2} \sin n x_1 \sinh n x_2$ blows up as $n \to \infty$.

A closely related inverse problem is to recover the Robin coefficient $\gamma(x)$ on the boundary Γ_i from the Cauchy data g and h on the boundary Γ_c:

$$a\frac{\partial u}{\partial n} + \gamma u = 0 \quad \text{on } \Gamma_i.$$

It arises in the analysis of quenching process and nuclear reactors, and corrosion detection. In heat conduction, the Robin boundary condition describes a convective heat conduction across the interface Γ_i, and it is known as Newton's law for cooling. In this case the coefficient is also known as heat transfer coefficient, and it depends strongly on at least twelve variables or eight nondimensional groups [307]. Thus its accurate values are experimentally very expensive and inconvenient, if not impossible at all, to obtain. As a result, in thermal engineering, a constant value assumption on γ is often adopted, and a look-up table approach is common, which constrains the applicability of the model to simple situations. Hence, engineers seek to estimate the coefficient from measured temperature data. In corrosion detection, the Robin boundary condition occurs due to the roughening effect of corrosion as the thickness tends to zero and rapidity of the oscillations diverges, and the coefficient could represent the damage profile [192, 151]. It also arises naturally from a linearization of the nonlinear Stefan-Boltzmann law for heat conduction of radiation type. We refer interested readers to [175] for a numerical treatment via Tikhonov regularization and [53] for piecewise constant Robin coefficients.

2.2.2 *Inverse source problem*

A second classical linear inverse problem for equation (2.1) is to recover a source/sink term f, i.e.,

$$-\Delta u = f$$

from the Cauchy data (g, h) on the boundary Γ:

$$u = g \quad \text{and} \quad \frac{\partial u}{\partial n} = h \quad \text{on } \Gamma.$$

Exemplary applications include electroencephalography and electrocardiography. The former records brain's spontaneous electrical activities from electrodes placed on the scalp, whereas the latter determines the heart's electrical activity from the measured body-surface potential distribution. One possible clinical application of the inverse problem is the noninvasive localization of the accessory pathway tract in the Wolff-Parkinson-White syndrome. In the syndrome, the accessary path bridges atria and ventricles, resulting in a pre-excitation of the ventricles [121]. Hence the locus of the dipole "equivalent" sources during early pre-excitation can possibly serve as an indicator of the site of the accessory pathway.

Retrieving a general source term from the Cauchy data is not unique. Physically, this is well understood in electrocardiography, in view of the fact that the electric field that the source generates outside any closed surface completely enclosing them can be duplicated by equivalent single-layer or double-layer sources on the closed surface itself [121, 225]. Mathematically, it can be seen that by adding one compactly supported function, one obtains a different source without changing the Cauchy data. The uniqueness issue remains delicate even within the admissible set of characteristic functions. We illustrate the nonuniqueness with one example [86].

Example 2.2. Let Ω be an open bounded domain in $\mathbb{R}^{\mathrm{d}}$ with a boundary Γ. Let ω_i $(i = 1, 2)$ be two balls centered both at the origin O and with different radius r_i, respectively, and lie within the domain Ω, and choose the scalars λ_i such that $\lambda_1 r_1^{\mathrm{d}} = \lambda_2 r_2^{\mathrm{d}}$. Then we take the source $f_i = \lambda_i \chi_{\omega_i}$, where χ_S denotes the characteristic function of the set S, in the Laplace equation

$$\begin{cases} -\Delta u_i = f_i & \text{in } \Omega, \\ u_i = g & \text{on } \Gamma. \end{cases}$$

Then for any $v \in H(\Omega) = \{v \in H^2(\Omega) : \Delta v = 0\}$, we have by Green's identity that

$$\int_\Omega f_i v dx = \int_\Gamma g \frac{\partial v}{\partial n} ds - \int_\Gamma \frac{\partial u_i}{\partial n} v ds.$$

Meanwhile, by the mean value theorem for harmonic functions

$$\int_\Omega f_i v dx = |\omega_i| v(O).$$

By the construction of f_i and these two identities, we obtain

$$\int_\Gamma \left(\frac{\partial u_1}{\partial n} - \frac{\partial u_2}{\partial n} \right) v ds = 0 \quad \forall v \in H(\Omega).$$

Since for every $h \in H^{\frac{1}{2}}(\Gamma)$, there exists a harmonic function $v \in H(\Omega)$ such that $v = h$ on Γ and $H^{\frac{1}{2}}(\Gamma)$ is dense in $L^2(\Gamma)$, one can conclude that

$$\frac{\partial u_1}{\partial n} = \frac{\partial u_2}{\partial n} \quad \text{on } \Gamma,$$

i.e., the two sources yield identical Cauchy data.

For the purpose of numerical reconstruction, practitioners often content with minimum-norm sources or harmonic sources (i.e., $\Delta f = 0$). In some applications, the case of localized sources, as modeled by monopoles, dipoles or their combinations, is deemed more suitable. In the latter case, a unique recovery is ensured (as a consequence of Holmgren's theorem). Further, efficient direct algorithms for locating dipoles and monopoles have been developed; see Section 6.2. We refer to [121, 225] for an overview of regularization methods for electrocardiography inverse problem.

2.2.3 *Inverse scattering problem*

Inverse scattering is concerned with imaging obstacles and inhomogeneities via acoustic, electromagnetic and elastic waves with applications to a wide variety of fields, e.g., radar, sonar, geophysics, medical imaging (e.g., microwave tomography) and nondestructive evaluation. Microwave tomography provides one promising way to assess functional and pathological conditions of soft tissues, complementary to the more conventional computerized tomography and magnetic resonance imaging [275]. It is known that the dielectric properties of tissues with high (muscle) and low (fat and bone) water content are significantly different. The dielectric contrast between tissues forms its physical basis. Below we describe the case of acoustic waves, and refer to [73] for a comprehensive treatment.

We begin with the modeling of acoustic waves, where the medium can be air, water or human tissues. Generally, acoustic waves are considered as small perturbations in a gas or fluid. By linearizing the equations for fluid motion, we obtain the governing equation

$$\frac{1}{c^2}\frac{\partial^2 p}{\partial t^2} = \Delta p,$$

for the pressure $p = p(x,t)$, where $c = c(x)$ denotes the local speed of sound and the fluid velocity is proportional to ∇p. For time-harmonic acoustic waves of the form

$$p(x,t) = \Re\{u(x)e^{-\mathrm{i}\omega t}\}$$

with frequency $\omega > 0$, it follows that the complex valued space dependent part u satisfies the reduced wave equation

$$\Delta u + \frac{\omega^2}{c^2}u = 0.$$

In a homogeneous medium, the speed of sound c is constant and the equation becomes the Helmholtz equation

$$\Delta u + k^2 u = 0, \tag{2.2}$$

where the wave number k is given by $k = \omega/c$. A solution to the Helmholtz equation whose domain of definition contains the exterior of some sphere is called radiating if it satisfies the Sommerfeld radiation condition

$$\lim_{r\to\infty} r^{\frac{d-1}{2}}\left(\frac{\partial u^s}{\partial r} - iku^s\right) = 0, \tag{2.3}$$

where $r = |x|$ and the limit holds uniformly in all directions $x/|x|$.

We focus on the following two basic scattering scenarios, i.e., scattering by a bounded impenetrable obstacle and scattering by a penetrable inhomogeneous medium of compact support. First we note that for a vector $d \in \mathbb{S}^{\mathrm{d}-1}$, the function $e^{\mathrm{i}kx\cdot d}$ satisfies the Helmholtz equation (2.2) for all $x \in \mathbb{R}^{\mathrm{d}}$. It is called a plane wave, since $e^{\mathrm{i}(kx\cdot d-\omega t)}$ is constant on the planes $kx \cdot d = \text{const}$, where the wave fronts travel with a velocity c in the direction d. Throughout, we assume that the incident field u^i impinged on the scatterer/inhomogeneity is given by a plane wave $u^i(x) = e^{\mathrm{i}kx\cdot d}$.

Let $D \subset \mathbb{R}^{\mathrm{d}}$ be the space occupied by the obstacle. We assume that D is bounded, and its boundary ∂D is connected. Then the simplest obstacle scattering problem is to find the scattered field u^s satisfying (2.3) in the exterior $\mathbb{R}^{\mathrm{d}} \setminus D$ such that the total field $u = u^i + u^s$ satisfies the Helmholtz equation (2.2) in $\mathbb{R}^{\mathrm{d}} \setminus D$ and the Dirichlet boundary condition

$$u = 0 \quad \text{on } \partial D.$$

It corresponds to a sound-soft obstacle with the total pressure, i.e., the excess pressure over the static pressure, vanishing on the boundary. Alternative boundary conditions other than the Dirichlet one are also possible.

The simplest scattering problem for an inhomogeneous medium assumes that the speed of sound c is constant outside a bounded domain D. Then the total field $u = u^i + u^s$ satisfies

$$\Delta u + k^2 n^2 u = 0 \quad \text{in } \mathbb{R}^{\mathrm{d}} \tag{2.4}$$

and the scattered field u^s fulfills the Sommerfeld radiation condition (2.3), where the wave number k is given by $k = \omega/c_0$ and $n^2 = c_0^2/c^2$ is the refractive index, i.e., the ratio of the square of the sound speed c_0 in the homogeneous medium to that in the inhomogeneous one. The refractive index n^2 is always positive, with $n^2(x) = 1$ for $x \in \mathbb{R}^{\mathrm{d}} \setminus D$. Further, an absorbing medium can be modeled by adding an absorption term which leads to a refractive index n^2 with a positive imaginary part, i.e.,

$$n^2 = \frac{c_0^2}{c^2} + \mathrm{i}\frac{\gamma}{k},$$

where the absorption coefficient γ is possibly space dependent.

Now the direct scattering problem reads: given the incident wave $u^i = e^{\mathrm{i}kd\cdot x}$ and the physical properties of the scatterer, find the scattered field u^s and in particular its behavior at large distances from the scatterer, i.e., its far field behavior. Specifically, radiating solutions u^s have the asymptotic behavior

$$u^s(x) = \frac{e^{\mathrm{i}k|x|}}{|x|}\left(u_\infty(\hat{x}, d) + \frac{1}{|x|}\right) \quad \text{as } |x| \to \infty,$$

uniformly for all directions $\hat{x} = x/|x|$, where the function $u_\infty(\hat{x}, d)$, $\hat{x}, d \in \mathbb{S}^{\mathrm{d}-1}$, is known as the far field pattern of the scattered field u^s.

The inverse scattering problem is to determine either the sound-soft obstacle D or the refraction index n^2 from a knowledge of the far field pattern $u_\infty(\hat{x}, d)$ for $\hat{x}$ and d on the unit sphere $\mathbb{S}^{\mathrm{d}-1}$ (or a subset of $\mathbb{S}^{\mathrm{d}-1}$). In practice, near-field scattered data is also common. Then the inverse problem is to retrieve the shape of the scatterer Ω or the refractive index n^2 from noisy measurements of the scattered field u^s on a curve/surface Γ, corresponding to one or multiple incident fields (and/or multiple frequencies). Inverse obstacle scattering is an exemplary geometrical inverse problem, where the geometry of the scatterer (or qualitative information, e.g., the size, shape, locations, and the number of components) is sought for.

The inverse scattering problems as formulated above are highly nonlinear and ill-posed. There are many inverse scattering methods, which can be divided into two groups: indirect methods and direct methods. The former is usually iterative in nature, based on either Tikhonov regularization or iterative regularization. Such methods requires the existence of the Fréchet derivative of the solution operator, and its characterization (e.g., for Newton update) [198, 251]; see also [210] for a discussion on Tikhonov regularization. Generally, these methods are expensive due to the repeated evaluation of the forward operator and require a priori knowledge, e.g., the number of components. These issues can be overcome by direct inversion methods. Prominent direct methods include the linear sampling method, factorization method, multiple signal classification, and direct sampling method etc. We shall briefly survey these methods in Chapter 6. In general, indirect methods are efficient, but yield only information about the scatterer support, which might be sufficient in some practical applications, whereas indirect methods can yield distributed profiles with full details but at the expense of much increased computational efforts.

Analogous methods exist for the inverse medium scattering problem. Here reconstruction algorithms are generally based on an equivalent reformulation of (2.4), i.e., the Lippmann-Schwinger integral equation

$$u = u^i + k^2 \int_\Omega (n^2(y) - 1) G(x, y) u(y) dy, \tag{2.5}$$

where $G(x, y)$ is the fundamental solution for the open field, i.e.,

$$G(x, y) = \begin{cases} \frac{\mathrm{i}}{4} H_0^1(k|x-y|), & \mathrm{d} = 2, \\ \frac{1}{4\pi} \frac{e^{\mathrm{i}k|x-y|}}{|x-y|}, & \mathrm{d} = 3, \end{cases}$$

where H_0^1 refers to the zeroth-order Hankel function of the first kind. Meanwhile, direct methods apply also to the inverse medium scattering problem (2.4), with the goal of determining the support of the refractive index n^2-1. Further, we note that by ignoring multiple scattering, we arrive the following linearized model to (2.5):

$$u = u^i + k^2 \int_\Omega (n^2(y) - 1)G(x,y)u^i(y)dy,$$

which is obtained by approximating the total field u by the incident field u^i. It is known as Born's approximation in the literature, and has been customarily adopted in reconstruction algorithms.

2.2.4 *Inverse spectral problem*

Eigenvalues and eigenfunctions are fundamental to the understanding of many physical problems, especially the behavior of dynamical systems, e.g., beams and membranes. Here eigenvalues, often known as the natural frequencies or energy states, can be measured by observing the dynamical behavior of the system. Naturally, one expects eigenvalues and eigenfunctions can tell a lot about the underlying system, which gives rise to assorted inverse problems with spectral data.

Generally, the forward problem can be formulated as

$$\mathcal{L}u = \lambda u \quad \text{in } \Omega,$$

with suitable boundary condition on $\partial\Omega$, where $\mathcal{L}$ is an elliptic operator, and $\lambda \in \mathbb{C}$ and u are the eigenvalue and respective eigenfunction. The operator $\mathcal{L}$ can also be a discrete analogue of the continuous formulation, resulting from proper discretization via, e.g., finite difference method or finite element method. In the latter case, it amounts to matrix eigenvalue problem with a structured matrix, e.g., tridiagonal or circulant. The matrix formulation is common in structural analysis, e.g., vibration.

The inverse spectral problem is to recover the coefficients in the operator $\mathcal{L}$ or the geometry of the domain Ω from partial or multiple spectral data, where the spectral data refer to the knowledge of complete or partial information of the eigenvalues or eigenfunctions. In the discrete case, it is concerned with reconstructing a structured matrix from the prescribed spectral data. Below we describe two versions of inverse spectral problems, i.e., inverse Sturm-Liouville problem and isospectral problem.

The simplest elliptic differential operator $\mathcal{L}$ is given by $\mathcal{L}u = -u'' + qu$ over the unit interval $(0, 1)$, where q is a potential. Then the classical

Sturm-Liouville problem reads: given a potential q and nonnegative constants h and H, find the eigenvalues $\{\lambda_k\}$ and eigenfunctions $\{u_k\}$ such that

$$\begin{cases} -u'' + q(x)u = \lambda u \\ u'(0) - hu(0) = 0, \\ u'(1) + Hu(1) = 0. \end{cases}$$

The set of eigenvalues $\{\lambda_k\}$ are real and countable. The respective inverse problem, i.e., the inverse Sturm-Liouville problem, consists of recovering the potential $q(x)$, h and H from a knowledge of spectral data. The spectral data can take several different forms, and this gives rise to a whole family of related inverse problems. A first spectral data is one complete spectrum $\{\lambda_k\}_{k=1}^{\infty}$. It is well known that this is insufficient for the recovery of a general potential q, and thus some additional information must be provided. Several possible choices of extra data are listed below.

(i) The two-spectrum case. In addition to the spectrum $\{\lambda_k\}_{k=1}^{\infty}$, a second spectrum $\{\mu_k\}_{k=1}^{\infty}$ is provided, where H is replaced by $\tilde{H} \neq H$. Then the potential q, h, H and $\tilde{H}$ can be uniquely determined from the spectra $\{\lambda_k\}_{k=1}^{\infty}$ and $\{\mu_k\}_{k=1}^{\infty}$ [34, 214]. It is one of the earliest inverse problems studied mathematically, dating at least back to 1946 [34].

(ii) Spectral function data. Here one seeks to reconstruct the potential q from its spectral function, i.e., the eigenvalues $\{\lambda_k\}_{k=1}^{\infty}$ and the norming constants $\rho_k := \|u_k\|_{L^2(0,1)}^2/u_k(0)^2$ for a finite h and $\rho_k := \|u_k\|_{L^2(0,1)}^2/u_k'(0)^2$ when $h = \infty$. Then the spectral data $\{\lambda_k\}_{k=1}^{\infty}$ and $\{\rho_k\}_{k=1}^{\infty}$ uniquely determines the set $q(x)$, h, H, and $\tilde{H}$ [100].

(iii) The symmetric case. If it is known that q is symmetric about the midpoint of the interval, i.e., $q(x) = q(1-x)$, and the boundary condition obeys the symmetry condition $h = H$, then the knowledge of a single spectrum $\{\lambda_k\}_{k=1}^{\infty}$ uniquely determines q [34].

(iv) Partially known $q(x)$. If the potential q is given over at least one half of the interval, e.g., $1/2 \leq x \leq 1$, then again one spectrum $\{\lambda_k\}_{k=1}^{\infty}$ suffices to recover the potential q [138].

Apart from eigenvalues and norming constants, other spectral data is also possible. One such data is nodal points, i.e., locations of the zeros of the eigenfunctions. In the context of vibrating systems, the nodal position is the location where the system does not vibrate. The knowledge of the position of one node of each eigenfunction and the average of the potential q uniquely determines the potential [232].

Now we turn to the two-dimensional case: What does the complete spectrum (with multiplicity counted) tell us about the domain Ω. A special case leads to the famous question raised by Mark Kac [185], i.e., "Can one hears the shape of a drum?". Physically, the drum is considered as an elastic membrane whose boundary is clamped, and the sound it makes is the list of overtones. Then we need to infer information about the shape of the drumhead from the list of overtones. Mathematically, the problem can be formulated as the Dirichlet eigenvalue problem on the domain $\Omega \subset \mathbb{R}^2$:

$$\begin{aligned} -\Delta u &= \lambda u \quad \text{in } \Omega, \\ u &= 0 \quad \text{on } \partial\Omega. \end{aligned}$$

Then the inverse problem is: given the frequencies $\{\lambda_k\}$, can we tell the shape of the drum, i.e., the domain Ω? The problem was answered negatively in 1992 by Gordon, Webb and Wolpert [113], who constructed a pair of regions in the plane that have different shapes but identical eigenvalues. These regions are nonconvex polygons. So the answer to Kac's question is: for many shapes, one cannot hear the shape of the drum completely. However, some information can be still inferred, e.g., domain volume.

The numerical treatment of inverse spectral problems is generally delicate. Least squares type methods are often inefficient, and constructive algorithms (often originating from uniqueness proofs) are more efficient. We refer to [266] for an elegant approach for the inverse Sturm-Liouville problem, and [67] for a comprehensive treatment of inverse (matrix) eigenvalue problems. The multidimensional inverse spectral problems are numerically very challenging, and little is known. We will describe one approach for the inverse Sturm-Liouville problem in Chapter 6.

2.3 Tomography

In this part, we describe several tomographic imaging techniques, which are very popular in medical imaging and nondestructive evaluation. We begin with two classical tomography problems of integral geometry type, i.e., computerized tomography and emission tomography, and then turn to PDE-based imaging modalities, including electrical impedance tomography, optical tomography and photoacoustic tomography.

2.3.1 *Computerized tomography*

Computerized tomography (CT) is a medical imaging technique that uses computer-processed X-rays to produce images of specific areas of the body. It can provide information about the anatomical details of an organ: the map of the linear attenuation function is essentially the map of the density. The cross-sectional images are useful for diagnostic and therapeutic purposes. The physics behind CT is as follows. Suppose a narrow beam of X-ray photons passes through a path L. Then according to Beer's law, the observed beam density I is given by

$$\frac{I}{I_0} = e^{-\int_L \mu(x)dx},$$

where I_0 is the input intensity, and $\mu = \mu(x)$ is the attenuation coefficient. It depends on both the density of the material and the nuclear composition characterized by the atomic number. By taking negative logarithm on both sides, we get

$$\int_L \mu(x)dx = -\log\frac{I}{I_0}.$$

The inverse problem is to recover the attenuation coefficient μ from the measured fractional decrease in intensity.

Mathematically, the integral transform here is known as the Radon transform. It is named after Austrian mathematician Johann Radon, who in 1917 introduced the two-dimensional version and also provided a formula for the inverse transformation. Below we briefly describe the transform and its inverse in the two-dimensional case, and refer to [240] for the general d-dimensional case.

In the 2D case, the line L with a unit normal vector $\theta(\alpha) = (\cos\alpha, \sin\alpha)$ (i.e., α is the angle between the normal vector to L and the x_1-axis) and distance s to the origin is given by

$$L_{\alpha,s} = \{x \in \mathbb{R}^2 :\ x\cdot\theta = s\}.$$

Then the Radon transform of a function $f : \mathbb{R}^2 \to \mathbb{R}$ is a function defined on the set of lines

$$Rf(\alpha, s) = \int_{L_{\alpha,s}} f(x)ds(x).$$

The line $L_{\alpha,s}$ can be parameterized with respect to arc length t by

$$(x_1(t), x_2(t)) = s\theta(\alpha) + t\theta(\alpha)^\perp,$$

where $\theta^\perp = (\sin\alpha, -\cos\alpha)$. Then the transform can be rewritten as

$$Rf(\alpha, s) = \int_{-\infty}^{\infty} f(s\theta + t\theta^\perp)dt.$$

The Radon transform is closely related to the Fourier transform, which in d-dimension is defined by

$$\widehat{f}(\omega) = \frac{1}{(2\pi)^{\frac{d}{2}}} \int f(x)e^{-ix\cdot\omega}dx,$$

and the inverse transform is given by

$$\check{f}(x) = \frac{1}{(2\pi)^{\frac{d}{2}}} \int \widehat{f}(\omega)e^{ix\cdot\omega}d\omega.$$

To see the connection, we denote $R_\alpha f(s) = Rf(\alpha, s)$ since the Fourier transform makes sense only in the s variable. Then there holds [240]

$$\widehat{R_\alpha[f]}(\sigma) = \sqrt{2\pi}\widehat{f}(\sigma\theta(\alpha)).$$

Roughly speaking, the two-dimensional Fourier transform of f along the direction θ coincides with the Fourier transform of its Radon transform in the variable s. This connection allows one to show the unique invertibility of the transform on suitably chosen function spaces, and to derive analytic inversion formulas, e.g., the popular filtered backprojection method and its variants for practical reconstruction.

We conclude this part with the singular value decomposition (SVD), cf. Appendix A, of the Radon transform [78].

Example 2.3. In this example we compute the SVD of the Radon transform. We assume that f is square integrable and supported on the unit disc D centered at the origin. Then the Radon transform Rf is given by

$$(Rf)(\alpha, s) = \int_{-w(s)}^{w(s)} f(s\theta + t\theta^\perp)dt, \quad |s| \le 1,$$

where $w(s) = \sqrt{1-s^2}$. By the Cauchy-Schwarz inequality

$$|Rf(\alpha, s)|^2 \le 2w(s) \int_{-w(s)}^{w(s)} |f(s\theta + t\theta^\perp)|^2 dt,$$

which upon manipulation yields

$$\int_{-1}^{1} w^{-1}(s)|Rf(\alpha, s)|^2 ds \le 2\int_D |f(x)|^2 dx.$$

This naturally suggests the following weighted norm on the range Y of the Radon transform

$$\|g\|_Y^2 = \int_0^{2\pi} \int_{-1}^{1} w^{-1}(s)|g(\alpha, s)|^2 ds d\alpha.$$

Next we recall Chebyshev polynomials $U_m(s)$ of the second kind, defined by

$$U_m(s) = \frac{\sin(m+1)\arccos s}{\sin \arccos s}, \quad m = 0, 1, \ldots$$

which are orthogonal with respect to the weight $w(s)$:

$$\int_{-1}^{1} w(s) U_m(s) U_{m'}(s) ds = \frac{\pi}{2}\delta_{m,m'}.$$

Further, any function $g(\alpha, s)$ in Y can be represented in terms of $w(s)U_m(s)$ for fixed α. This suggests to consider the subspace Y_m of Y spanned by

$$g_m(\alpha, s) = \sqrt{\frac{2}{\pi}} w(s) U_m(s) u(\alpha), \quad m = 0, 1, \ldots$$

where $u(\alpha)$ is an arbitrary square integrable function of α. Clearly, $\|g_m\|_Y^2 = \int_0^{2\pi} |u(\alpha)|^2 d\alpha$. The next step is to show that RR^* maps Y_m into itself. It is easy to verify that

$$R^* g(x) = \int_0^{2\pi} g(\alpha, \theta \cdot x) w^{-1}(\theta \cdot x) d\alpha.$$

Hence for $g_m \in Y_m$, there holds

$$\begin{aligned} RR^* g_m(\alpha, s) &= \sqrt{\frac{2}{\pi}} \int_{-w(s)}^{w(s)} \int_0^{2\pi} U_m(\theta' \cdot (s\theta + t\theta^\perp)) u(\alpha') d\alpha' \\ &= \frac{4\pi}{m+1} \sqrt{\frac{2}{\pi}} w(s) U_m(s) \bar{u}(\alpha), \end{aligned}$$

with

$$\bar{u}(\alpha) = \frac{1}{2\pi} \frac{\sin(m+1)(\alpha - \alpha')}{\sin(\alpha - \alpha')} u(\alpha') d\alpha'. \tag{2.6}$$

Hence the operator RR^* transforms a function in Y_m into another function in Y_m, and further, the restriction of RR^* to the subspace Y_m is equivalent to the integral operator defined in (2.6). In view of the completeness of Chebyshev polynomials, we can find all eigenvalues and eigenfunctions of RR^*. Upon noting the identity

$$\frac{1}{2\pi} \frac{\sin(m+1)(\alpha - \alpha')}{\sin(\alpha - \alpha')} = \sum_{k=0}^{m} Y_{m-2k}(\alpha) Y_{m-2k}^*(\alpha'),$$

with $Y_l(\alpha) = \frac{1}{\sqrt{2\pi}} e^{-il\alpha}$ and by the orthonormality of $Y_l(\alpha)$, $Y_{m-2k}(\alpha)$ are the eigenfunctions of the integral operator associated with the eigenvalue 1. Next we introduce the functions

$$u_{m,k}(\alpha, s) = \sqrt{\frac{2}{\pi}} w(s) U_m(s) Y_{m-k}(\alpha) \quad k = 0, 1, \ldots, m.$$

Clearly, these functions are orthonormal in Y, and further,

$$RR^* u_{m,k} = \sigma_m^2 u_{m,k}, \quad k = 0, 1, \ldots, m,$$

with $\sigma_m = \sqrt{\frac{4\pi}{m+1}}$. It remains to show that the functions

$$v_{m,k}(x) = \frac{1}{\sigma_m} R^* u_{m,k}$$

constitute a complete set of orthonormal functions in $L^2(D)$, which follows from the fact that $v_{m,k}(x)$ can be expressed in terms of Jacobi polynomials [240]. Therefore, the singular values of the Radon transform decays to zero, and with the multiplicity counted, it decays at a rate $1/\sqrt{m+1}$, which is fairly mild, indicating that the inverse problem is only mildly ill-posed.

2.3.2 *Emission tomography*

In emission tomography one determines the distribution f of a radiopharmaceutical in the interior of an object by measuring the radiation outside the object in a tomographic fashion. Let μ be the attenuation distribution of the object, which one aims to determine. Then the intensity measured by a detector collimated to pick up only radiation along the line L is given by

$$I = \int_L f(x) e^{-\int_{L(x)} \mu(y) dy} dx, \tag{2.7}$$

where $L(x)$ is the line segment of L between x and the detector. This is the mathematical model for single particle emission computed tomography (SPECT). Thus SPECT gives rise to the attenuated ray transform

$$(P_\mu f)(\theta, x) = \int f(x + t\theta) e^{-\int_t^\infty \mu(x+\tau\theta) d\tau} dt, \quad x \in \theta^\perp, \quad \theta \in \mathbb{S}^{d-1}.$$

In positron emission tomography (PET), the sources eject particles pairwise in opposite directions, and they are detected in coincidence mode, i.e., only events with two particles arriving at opposite detectors at the same time are counted. In that case, (2.7) has to be replaced by

$$I = \int_L f(x) e^{-\int_{L_+(x)} \mu(y) dy - \int_{L_-(x)} \mu(y) dy} dx, \tag{2.8}$$

where $L_+(x)$ and $L_-(x)$ are two half-lines of L with end point x. Since the exponent adds up to the integral over L, we can write

$$I = e^{-\int_L \mu(y)dy} \int_L f(x)dx.$$

In PET, one is only interested in f, not μ. Usually one determines f from the measurements, assuming μ to be known or simply ignoring it.

Emission tomography is essentially stochastic in nature. In case of a small number of events, the stochastic aspect is pronounced. Thus besides the above models based on integral transforms, stochastic models have been popular in the applied community. These models are completely discrete. We describe a widely used model for PET due to Shepp and Vardi [276].

In the model, we subdivide the reconstruction region into pixels or voxels. The number of events in pixel (or voxel) j is a Poisson random variable ξ_j whose expectation $f_j = E[\xi_j]$ is a measure of the activity in pixel/voxel j. The vector $f = (f_1, \ldots, f_m)^t \in \mathbb{R}^m$ is the sought-for quantity. The measurement vector $g = (g_1, \ldots, g_n)^t \in \mathbb{R}^n$ is considered a realization of a Poisson random variable $\gamma = (\gamma_1, \ldots, \gamma_n)^t$, where γ_i is the number of events detected in detector i. The model is described by the matrix $A = [a_{ij}] \in \mathbb{R}^{n\times m}$, where the entry a_{ij} denotes the probability that an event in pixel/voxel j is detected in detector i. These probabilities are determined either theoretically or by measurements. We have

$$E[\gamma] = Af. \tag{2.9}$$

One conventional approach to estimate f is the maximum likelihood method, which consists in maximizing the likelihood function

$$L(f) = \prod \frac{(Af)_i^{g_i}}{g_i!} e^{-(Af)_i}.$$

Shepp and Vardi [276] devised an expectation maximization algorithm for efficiently finding the maximizer; see [298] for the convergence analysis and [119, 47] for further extensions to regularized variants.

2.3.3 *Electrical impedance tomography*

Electrical impedance tomography is a diffusive imaging modality for determining the electrical conductivity of an object from electrical measurements on the boundary [33]. The experimental setup is as follows. One first attaches a set of electrodes to the surface of the object, then injects an electrical current through these electrodes and measures the resulting electrical voltages on these electrodes. In practice, several input currents are applied, and the induced electrical potentials are measured. The goal is to determine the conductivity distribution from the noisy voltage data. We refer to Fig. 2.1 for a schematic illustration of one EIT system at University of Eastern Finland.

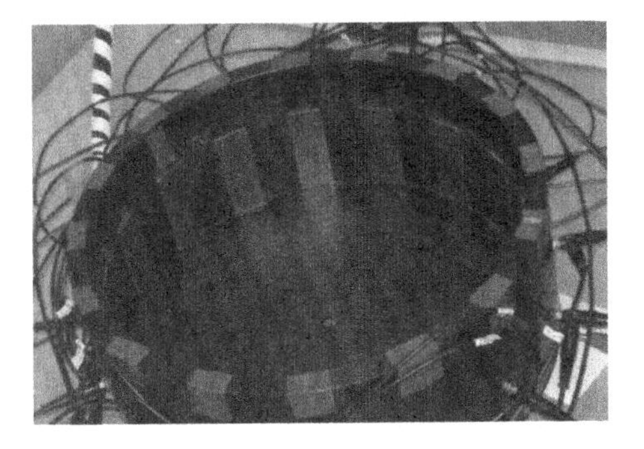

Fig. 2.1 Schematic illustration of EIT system.

There are several different mathematical models for the experiment. One popular model in medical imaging is the complete electrode model [278]. Let Ω be an open bounded domain, referring to the space occupied by the object, in $\mathbb{R}^{\mathrm{d}}$ ($\mathrm{d} = 2, 3$) and Γ be its boundary. The electrical potential u in the interior of the domain is governed by the following second-order elliptic differential equation

$$-\nabla \cdot (\sigma \nabla u) = 0 \quad \text{in } \Omega.$$

A careful modeling of boundary conditions is very important for accurately reproducing experimental data. Let $\{e_l\}_{l=1}^{L} \subset \Gamma$ be a set of L electrodes. We assume that each electrode e_l consists of an open and connected subset of the boundary Γ, and the electrodes are pairwise disjoint. Let $I_l \in \mathbb{R}$ be the current applied to the lth electrode e_l and denote by $I = (I_1, \ldots, I_L)^{\mathrm{t}}$ the input current pattern. Then we can describe the boundary conditions on the electrodes by

$$\begin{cases} u + z_l \sigma \dfrac{\partial u}{\partial n} = U_l & \text{on } e_l, \ l = 1, 2, \ldots, L, \\ \displaystyle\int_{e_l} \sigma \frac{\partial u}{\partial n} ds = I_l & \text{for } l = 1, 2, \ldots, L, \\ \sigma \dfrac{\partial u}{\partial n} = 0 & \text{on } \Gamma \backslash \cup_{l=1}^{L} e_L, \end{cases}$$

where $\{z_l\}$ are contact impedances of the electrodes.

The complex boundary conditions takes into account the following important physical characteristics of the experiment: (a) The electrodes are inherently discrete; (b) The electrode e_l is a perfect conductor, which implies that the potential along each electrode is constant: $u|_{e_l} = U_l$, and the current I_l sent to the lth electrode e_l is completely confined to e_l; (c) When a current is applied, a highly resistive layer forms at the electrode-electrolyte interface due to dermal moisture, with contact impedances $\{z_l\}_{l=1}^{L}$, which is known as the contact impedance effect in the literature. Ohm's law asserts that the potential at electrode e_l drops by $z_l \sigma \frac{\partial u}{\partial n}$. Experimental studies show that the model can achieve an accuracy comparable with the measurement precision [278].

The inverse problem consists of estimating the conductivity distribution σ from the measured voltages $U = (U_1, \ldots, U_L)^{\mathrm{t}} \in \mathbb{R}^L$ for multiple input currents. It has found applications in noninvasive imaging, e.g., detection of skin cancer and location of epileptic foci, and nondestructive testing, e.g., locating resistivity anomalies due to the presence of minerals [64].

In the idealistic situation, one assumes that the input current g is applied at every point on the boundary $\partial\Omega$, i.e.,

$$\sigma \frac{\partial u}{\partial n} = g \quad \text{on } \partial\Omega,$$

and further the potential u is measured everywhere on the boundary $\partial\Omega$, i.e., the Dirichlet trace $u = f$. If the measurement can be made for every possible input current g, then the data consists of the complete Neumann-to-Dirichlet map Λ_σ. This model is known as the continuum model, and it is very convenient for mathematical studies, i.e., uniqueness and stability. EIT is an example of inverse problems with operator-valued data. We note that the complete electrode model can be regarded as a Galerkin approximation of the continuum model [148].

We refer interested readers to [264, 68, 263, 171] for impedance imaging with Tikhonov regularization.

We end this section with the ill-posedness of the EIT problem.

Example 2.4. We consider a radially symmetric case. For the unit disk $\Omega = \{x \in \mathbb{R}^2 : |x| < 1\}$, consider the conductivity distribution

$$\sigma(x) = \begin{cases} \kappa, \ |x| < \rho, \\ 1, \ \rho < |x| < 1, \end{cases}$$

with a constant $\kappa > 0$ and $0 < \rho < 1$. Solving the forward problem explicitly using polar coordinates $x = re^{\mathrm{i}\xi}$ yields the spectral decomposition

$$\Lambda_\sigma : \left\{ \begin{array}{c} \frac{1}{\sqrt{\pi}} \cos k\xi \\ \frac{1}{\sqrt{\pi}} \sin k\xi \end{array} \right\} \mapsto \frac{1}{k} \frac{1+\mu\rho^{2k}}{1-\mu\rho^{2k}} \left\{ \begin{array}{c} \frac{1}{\sqrt{\pi}} \cos k\xi \\ \frac{1}{\sqrt{\pi}} \sin k\xi \end{array} \right\},$$

with $\mu = \frac{1-\kappa}{1+\kappa} \in (-1,1)$; cf. [250]. Hence asymptotically, the eigenvalue of the Neumann-to-Dirichlet map decays at a rate $O(k^{-1})$. Further, the smaller (exponentially decaying) is the radius ρ, the smaller perturbation on the Neumann boundary condition. This indicates that the inclusions far away from the boundary are more challenging to recover.

2.3.4 *Optical tomography*

Optical tomography is a relatively new imaging modality. It images the optical properties of the medium from measurements of near-infrared light on the surface of the object. In a typical experiment, a highly scattering medium is illuminated by a narrow collimated beam and the light that propagates through the medium is collected by an array of detectors. It has potential applications in, e.g., breast cancer detection, monitoring of infant brain tissue oxygenation level and functional brain activation studies. The inverse problem is to reconstruct optical properties (predominantly absorption and scattering distribution) of the medium from these boundary measurements. We refer to [14] for comprehensive surveys.

The mathematical formulation of the forward problem is dictated primarily by the spatial scale, ranging from the Maxwell equations at the microscale, to the radiative transport equation at the mesoscale, and to diffusion equations at the macroscale. Below we describe the radiative transport equation and its diffusion approximation, following [14].

Light propagation in tissues is usually described by the radiative transport equation. It is a one-speed approximation of the transport equation, and it assumes that the energy of the particles does not change in the collisions and that the refractive index is constant within the medium. Let $\Omega \subset \mathbb{R}^{\mathrm{d}}$, $\mathrm{d} = 2, 3$, be the physical domain with a boundary $\partial\Omega$, and $\hat{s} \in \mathbb{S}^{\mathrm{d}-1}$ denote the unit vector in the direction of interest. Then the frequency domain radiative transport equation is of the form

$$\begin{aligned} &\frac{\mathrm{i}\omega}{c}\phi(x,\hat{s}) + \hat{s}\cdot\nabla\phi(x,\hat{s}) + (\mu_s+\mu_a)\phi(x,\hat{s}) \\ &\qquad = \mu_s \int_{\mathbb{S}^{\mathrm{d}-1}} \Theta(\hat{s},\hat{s}')\phi(x,\hat{s}')d\hat{s}' + q(x,\hat{s}), \quad x \in \Omega, \end{aligned}$$

where c is the speed of light in the medium, ω is the angular modulation frequency of the input signal, and $\mu_a = \mu_a(x)$ and $\mu_s = \mu_s(x)$ are the absorption and scattering coefficients of the medium, respectively. Further, $\phi(x, \hat{s})$ is the radiance, $\Theta(\hat{s}, \hat{s}')$ is the scattering phase function and $q(x, \hat{s})$ is the source inside Ω. The function $\Theta(\hat{s}, \hat{s}')$ describes the probability that a photon with an initial direction $\hat{s}'$ will have a direction $\hat{s}$ after a scattering event. The most usual phase function $\Theta(\hat{s} \cdot \hat{s}')$ is the Henyey-Greenstein scattering function, given by

$$\Theta(\hat{s} \cdot \hat{s}') = \begin{cases} \dfrac{1}{2\pi} \dfrac{1 - g^2}{1 + g^2 - 2g\hat{s} \cdot \hat{s}'}, & \mathrm{d} = 2, \\ \dfrac{1}{4\pi} \dfrac{1 - g^2}{(1 + g^2 - 2g\hat{s} \cdot \hat{s}')^{3/2}}, & \mathrm{d} = 3. \end{cases}$$

The scattering shape parameter g, taking values in $(-1, 1)$, defines the shape of the probability density.

For the boundary condition, we assume that no photons travel in an inward direction at the boundary $\partial\Omega$ except at the source point $\epsilon_j \subset \partial\Omega$, thus

$$\phi(x, \hat{s}) = \begin{cases} \phi_0(x, \hat{s}), & x \in \epsilon_j, \quad \hat{s} \cdot n < 0, \\ 0, & x \in \partial\Omega \setminus \cup\epsilon_j, \quad \hat{s} \cdot n < 0, \end{cases} \tag{2.10}$$

where $\phi_0(x, \hat{s})$ is a boundary source term. This boundary condition implies that once a photon escapes the domain Ω it does not re-enter the domain. In optical tomography, the measurable quantity is the exitance $J^+(x)$ on the boundary $\partial\Omega$ of the domain Ω, which is given by

$$J^+(x) = \int_{\hat{s} \cdot n > 0} (\hat{s} \cdot n)\phi(x, \hat{s}) d\hat{s}, \quad x \in \partial\Omega.$$

The forward simulation of the radiative transport equation is fairly expensive, due to its involvement of the scattering term. Hence simplifying models are often adopted. Here we describe the popular diffusion approximation, which is a first-order spherical harmonic approximation to the radiative transport equation. Specifically, the radiance $\phi(x, \hat{s})$ is approximated by

$$\phi(x, \hat{s}) \approx \Phi(x) - \mathrm{d}\hat{s} \cdot (\kappa \nabla \Phi(x)),$$

where $\Phi(x)$ is the photon density defined by

$$\Phi(x) = \frac{1}{|\mathbb{S}^{\mathrm{d}-1}|} \int_{S^{\mathrm{d}-1}} \phi(x, \hat{s}) d\hat{s}.$$

Here $\kappa = (\mathrm{d}(\mu_a + \mu_s'))^{-1}$ is the diffusion coefficient, with $\mu_s' = (1 - g_1)\mu_s$ being the reduced scattering coefficient and g_1 being the mean of the cosine of the scattering angle. In the case of the Henyey-Greenstein scattering function, we have $g_1 = g$. The diffusion coefficient κ represents the length scale of an equivalent random walk step. By inserting the approximation and adopting similar approximations for the source term $q(x, \hat{s})$ and the scattering phase function $\Theta(\hat{s}, \hat{s}')$, we obtain

$$-\nabla \cdot (\kappa \nabla \Phi(x)) + \mu_a \Phi(x) + \tfrac{\mathrm{i}\omega}{c}\Phi(x) = q_0(x),$$

where $q_0(x)$ is the source inside Ω. This represents the governing equation of the diffusion approximation.

The boundary condition (2.10) cannot be expressed using variables of the diffusion approximation directly. Instead it is often replaced by an approximation that the total inward directed photon current is zero. Further, to take into account the mismatch between the refractive indices of the medium and surrounding medium, a Robin type boundary condition is often adopted. Then the boundary condition can be written as

$$\Phi(x) + \frac{1}{2\gamma_{\mathrm{d}}}\kappa\xi\frac{\partial\Phi(x)}{\partial n} = \begin{cases} \dfrac{I_s}{\gamma_{\mathrm{d}}}, \ x \in \epsilon_i, \\ 0, \ x \in \partial\Omega \setminus \cup\epsilon_i, \end{cases}$$

where I_s is a diffuse boundary current at the source position $\epsilon_j \subset \partial\Omega$, γ_{d} is a constant with $\gamma_2 = 1/\pi$ and $\gamma_3 = 1/4$, and the parameter ξ determines the internal reflection at the boundary $\partial\Omega$. In the case of matched refractive index, $\xi = 1$. Further, the exitance $J^+(x)$ is given by

$$J^+(x) = -\kappa\frac{\partial\Phi(x)}{\partial n} = \frac{2\gamma_{\mathrm{d}}}{\xi}\Phi(x).$$

The standard inverse problem in optical tomography is to recover intrinsic optical parameters, i.e., the absorption coefficient μ_a and scattering coefficient μ_s, from boundary measurements of the transmitted and/or reflected light. We refer to [85] and [281] for an analysis of Tikhonov regularization formulations for the diffusion approximation and the radiative transport equation, respectively.

2.3.5 *Photoacoustic tomography*

Photoacoustic tomography (PAT), also known as thermoacoustic or optoacoustic tomography, is a rapidly emerging technique that holds great potentials for biomedical imaging. It exploits the thermoacoustic effect for signal generation, first discovered by Alexander Graham Bell in 1880, and seeks to combine the high electromagnetic contrast of tissue with the high spatial resolution of ultrasonic methods. It has several distinct features. Because the optical absorption properties of a tissue is highly correlated to its molecular constitution, PAT images can reveal the pathological condition of the tissue, and hence facilitate a wide range of diagnostic tasks. Further, when employed with optical contrast agents, it has the potential to facilitate high-resolution molecular imaging of deep structures, which cannot be easily achieved with pure optical methods. Below we describe the mathematical model following [215, 304].

In PAT, a laser or microwave source is used to irradiate an object, and the thermoacoustic effect results in the emission of acoustic signals, indicated by the pressure field $p(x,t)$, which can be measured by using wide-band ultrasonic transducers located on a measurement aperture. The objective of PAT is to estimate spatially varying electromagnetic absorption properties of the tissue from the measured acoustic signals. The photoacoustic wavefield $p(x,t)$ in an inviscid and lossless medium is governed by a wave equation

$$\left(\nabla^2 - \frac{1}{c^2}\frac{\partial^2}{\partial t^2}\right) p(x,t) = -\frac{\beta}{\kappa c^2}\frac{\partial^2 T(x,t)}{\partial t^2},$$

where $T(x,t)$ denotes the temperature rise within the object. The quantities β, κ and c denote the thermal coefficient of volume expansion, isothermal compressibility and speed of sound, respectively.

When the temporal width of the exciting electromagnetic pulse is sufficiently short, the pressure wavefield is produced before significant heat conduction can take place. This occurs when the temporal width τ of the exciting electromagnetic pulse satisfies $\tau < \frac{d_c^2}{4\alpha_{th}}$, where d_c and α_{th} denote the characteristic dimension of the heated region and the thermal diffusivity, respectively. Then the temperature function $T(x,t)$ satisfies

$$\rho C_V \frac{\partial T(x,t)}{\partial t} = H(x,t),$$

where ρ and C_V denote the mass density and specific heat capacity of the medium at constant volume, respectively. The heating function $H(x,t)$ describes the energy per unit volume per unit time that is deposited in the medium by the exciting electromagnetic pulse. The amount of heat generated by tissue is proportional to the strength of the radiation. Consequently, one obtains the simplified photoacoustic wave equation

$$\left(\nabla^2 - \frac{1}{c^2}\frac{\partial^2}{\partial t^2}\right) p(x,t) = -\frac{\beta}{C_p}\frac{\partial H(x,t)}{\partial t},$$

where $C_p = \rho c^2 \kappa C_V$ denotes the specific heat capacity of the medium at constant pressure. Sometimes, it is convenient to work with the velocity potential $\phi(x,t)$, i.e., $p(x,t) = -\rho\frac{\partial\phi(x,t)}{\partial t}$. Then we have

$$\left(\nabla^2 - \frac{1}{c^2}\frac{\partial^2}{\partial t^2}\right)\phi(x,t) = \frac{\beta}{\rho C_p}H(x,t).$$

In practice, it is appropriate to consider the following separable form of the heating function

$$H(x,t) = A(x)I(t),$$

where $A(x)$ is the absorbed energy density and $I(t)$ denotes the temporal profile of the illuminating pulse. When the exciting electromagnetic pulse duration τ is short enough, i.e., $\tau < \frac{d_c}{c}$, all the thermal energy has been deposited by the electromagnetic pulse before the mass density or volume of the medium has had time to change. Then one can approximate $I(t)$ by a Dirac delta function $I(t) \approx \delta(t)$. Hence, the absorbed energy density $A(x)$ is related to the induced pressure wavefield $p(x,t)$ at $t = 0$ as

$$p(x, t = 0) = \Gamma A(x),$$

where Γ is the dimensionless Grueneisen parameter. The goal of PAT is to determine $A(x)$, or equivalently $p(x, t = 0)$ from measurements of $p(x,t)$ acquired on a measurement aperture. Mathematically, it is equivalent to the initial value problem

$$\left(\nabla^2 - \frac{1}{c^2}\frac{\partial^2}{\partial t^2}\right)p(x,t) = 0,$$

subject to

$$p(x,t)|_{t=0} = \Gamma A(x) \quad \text{and} \quad \frac{\partial p(x,t)}{\partial t}|_{t=0} = 0.$$

When the object possesses homogeneous acoustic properties that match a uniform and lossless background medium, and the duration of the irradiating pulse is negligible, the pressure wavefield $p(x_0,t)$ recorded at transducer location x_0 can be expressed as

$$p(x_0,t) = \frac{\beta}{4\pi C_p}\int A(x)\frac{d}{dt}\frac{\delta(t - \frac{|x_0-x|}{c_0})}{|x_0 - x|}dx, \tag{2.11}$$

where c_0 is the speed of sound in the object and background medium. The function $A(x)$ is compactly supported, bounded and nonnegative. Equation (2.11) represents the canonical imaging model for PAT. The inverse problem is then to estimate $A(x)$ from the knowledge of $p(x_0,t)$.

Equation (2.11) can be expressed in an alternative but mathematically equivalent form as

$$g(x_0, t) = \int A(x)\delta\left(t - \frac{|x_0 - x|}{c_0}\right) dx, \tag{2.12}$$

where the integrated data function $g(x_0, t)$ is defined as

$$g(x_0, t) \equiv \frac{4\pi C_p c_0}{\beta} t \int_0^t p(x_0, t')dt'.$$

Note that $g(x_0, t)$ represents a scaled version of the acoustic velocity potential $\phi(x_0, t)$. The reformulation (2.12) represents a spherical Radon transform, and indicates that the integrated data function describes integrals over concentric spherical surfaces of radii $c_0 t$ that are centered at the receiving transducer location x_0. Equations (2.11) and (2.12) form the basis for deriving exact reconstruction formulas for special geometries, e.g., cylindrical, spherical or planar surfaces; see [204] for a comprehensive overview.

Chapter 3

Tikhonov Theory for Linear Problems

Inverse problems suffer from instability, which poses significant challenges to their stable and accurate numerical solution. Therefore, specialized techniques are required. Since the ground-breaking work of the Russian mathematician A. N. Tikhonov [286–288], regularization, especially Tikhonov regularization, has been established as one of the most powerful and popular techniques for solving inverse problems.

In this chapter, we discuss Tikhonov regularization for linear inverse problems

$$Ku = g^\dagger, \tag{3.1}$$

where $K : X \to Y$ is a bounded linear operator, and the spaces X and Y are Banach spaces. In practice, we have at hand only noisy data g^δ, whose accuracy with respect to the exact data $g^\dagger = Ku^\dagger$ ($u^\dagger$ is the true solution) is quantified in some error metric ϕ, which measures the model output $g^\dagger$ relative to the measurement data g^δ. We will denote the accuracy by $\phi(u, g^\delta)$ to indicate its dependence on the data g^δ, and mostly we are concerned with the choice

$$\phi(u, g^\delta) = \|Ku - g^\delta\|^p.$$

We refer to Table 3.1 for a few common choices, and Appendix B for their statistical motivation.

In Tikhonov regularization, we solve a nearby *well-posed* optimization problem of the form

$$\min_{u \in C} \left\{ J_\alpha(u) = \phi(u, g^\delta) + \alpha\psi(u) \right\}, \tag{3.2}$$

and take its minimizer, denoted by u_α^δ, as a solution. The functional J_α is called the Tikhonov functional. It consists of two terms, the fidelity term $\phi(u, g^\delta)$ and the regularization term $\psi(u)$. Roughly, the former measures the proximity of the model output Ku to the observational data g^δ and hence incorporates the information in the data g^δ, whereas the latter encodes a priori information, e.g., smoothness, sparsity, monotonicity, and other structural properties of the unknown. The nonnegative scalar α is known as the regularization parameter, and it controls the relative weight given to the two terms. The choice of the parameter α is essential for successfully applying Tikhonov regularization to practical problems. The set $\mathcal{C} \subset X$ is convex and closed, and it reflects constraint on the sought-for solution, e.g., positivity and other physical constraints.

An appropriate choice of the functionals ϕ and ψ depends very much on specific applications, or more precisely, on the noise model and prior information, respectively. A list of some common choices for the fidelity ϕ and regularization ψ is shown in Tables 3.1 and 3.2, respectively. The fidelity ϕ and penalty ψ are often derived in a Bayesian setting, since Tikhonov minimizer is identical with the maximum a posteriori estimate of the posterior distribution; see Chapter 7 for related discussions.

Table 3.1 Common data fidelity functionals $\phi(u, g^\delta)$. In the table, Ω is bounded domain. For Poisson and speckle noises, the fidelity functionals are not nonnegative, since a constant depending on the data g^δ is omitted.

noise model	$\phi(u, g^\delta)$
additive Gaussian	$\|Ku - g^\delta\|^2_{L^2(\Omega)}$
additive impulsive	$\|Ku - g^\delta\|_{L^1(\Omega)}$
Poisson	$\int_\Omega (Ku - g^\delta \log Ku) dx$
speckle noise	$\int_\Omega (\log Ku + \frac{g^\delta}{Ku}) dx$
Huber	$\|L_\delta(Ku - g^\delta)\|_{L^1(\Omega)}$, $L_\delta(t) = \begin{cases} \frac{1}{2}t^2, & \|t\| \leq \delta, \\ \delta(t - \frac{\delta}{2}), & \|t\| > \delta. \end{cases}$

Table 3.2 Several common regularization functionals $\psi(u)$. In the table, Ω is an open bounded domain.

prior model	$\psi(u)$
generalized Gaussian	$\|u\|^p_{L^p(\Omega)}$, $1 \leq \nu \leq 2$
BV or TV	$\|u\|_{BV(\Omega)}$ or $\|u\|_{\mathrm{TV}(\Omega)}$
Sobolev	$\|u\|^p_{W^{1,p}(\Omega)}$
elastic-net	$\|u\|_{\ell^1} + \frac{\gamma}{2}\|u\|^2_{\ell^2}$
sparsity	$\|u\|^p_{\ell^p}$, $0 \leq p < 2$

The rest of the chapter is organized as follows. First we discuss the fundamental question of the existence and stability of a Tikhonov minimizer

u_α^δ, and consistency, i.e., the convergence of the minimizer u_α^δ as the noise level δ tends to zero (with the parameter α suitably chosen). These are necessary for the well-posedness of the Tikhonov formulation. Then we discuss the value function calculus due to A. N. Tikhonov, which are important for designing and analyzing Tikhonov models. Next we discuss convergence rates, which is concerned with the quality of the approximation u_α^δ relative to a true solution $u^\dagger$, under different conditions on the "true" solution $u^\dagger$. The regularization parameter α is essential for the performance of the Tikhonov method, and we discuss several choice rules in Sections 3.4; see also Section 3.5 for related discussions. Finally, we discuss an extension of the formulation J_α to the case of multiple penalties, to accommodate multiple distinct features, as are often observed in many practical problems.

3.1 Well-posedness

Now we discuss the well-posedness of the Tikhonov model, i.e., existence, stability and consistency of a minimizer. For simplicity, we restrict our discussion to the case that the fidelity ϕ is norm powered, i.e.,

$$\phi(u, g^\delta) = \|Ku - g^\delta\|^p.$$

The results below can be extended more general cases.

Throughout, we make the following assumption on ϕ and ψ.

Assumption 3.1. Let X be reflexive, and the set $\mathcal{C}$ be convex and closed. Let the nonnegative functionals $\phi(u, y^\delta) = \|Ku - g^\delta\|^p$ and $\psi(u)$ satisfy:

(a) The functional J_α is coercive, i.e., for any sequence $\{u_n\}$ such that the functional value $\{J_\alpha(u_n)\}$ is uniformly bounded, then the sequence $\{u_n\}$ is uniformly bounded in X.
(b) The functional ψ is sequentially weak lower semicontinuous.
(c) The operator K is bounded.

Remark 3.1. The assumption that X is reflexive can be relaxed to nonreflexive. Accordingly the weak lower semicontinuity in (b) should be replaced with weak $*$ lower semicontinuity. Also the coercivity on J_α can be replaced with a compactness assumption of the constraint set $\mathcal{C}$.

Remark 3.2. The boundedness of the operator K implies that K is weakly continuous, i.e., $u_n \to u$ weakly in X implies $Ku_n \to Ku$ weakly in Y.

We recall also the H-property of a functional ψ.

Definition 3.1. A functional $\psi(u)$ is said to have the H-property on the space X if any sequence $\{u_n\} \subset X$ that satisfies that the conditions $u_n \to u$ weakly for some $u \in X$ and $\psi(u_n) \to \psi(u)$ imply that u_n converges to u in X.

Remark 3.3. Norms on Hilbert spaces, $L^p(\Omega)$ spaces and Sobolev spaces $W^{m,p}(\Omega)$ with $1 < p < \infty$ and $m \geq 1$ satisfy the H-property.

Now we state an existence result.

Theorem 3.1. *Let Assumption 3.1 hold. Then for every $\alpha > 0$, there exists at least one minimizer u_α^δ to the functional J_α defined by* (3.2).

Proof. Since the functionals ϕ and ψ are nonnegative, there exists a minimizing sequence $\{u_n\} \subset \mathcal{C}$, such that

$$\lim_{n\to\infty} \|Ku_n - g^\delta\|^p + \alpha\psi(u_n) = \inf_{u\in\mathcal{C}} \|Ku - g^\delta\|^p + \alpha\psi(u) := \eta.$$

Hence the sequence of functional values $\{J_\alpha(u_n)\}$ is uniformly bounded. By Assumption 3.1(a), there exists a subsequence of $\{u_n\}$, also denoted by $\{u_n\}$, and some $u^* \in X$ such that $u_n \to u^*$ weakly. By the closedness and convexity of the set $\mathcal{C}$, we have $u^* \in \mathcal{C}$. Now by the lower semi-continuity of ψ from Assumption 3.1(b) and the weak lower semicontinuity of norms, we deduce

$$\begin{aligned}
\|Ku^* - g^\delta\|^p + \alpha\psi(u^*) &\leq \liminf_{n\to\infty} \|Ku_n - g^\delta\|^p + \alpha \liminf_{n\to\infty} \psi(u_n) \\
&\leq \liminf_{n\to\infty} (\|Ku_n - g^\delta\|^p + \alpha\psi(u_n)) = \eta,
\end{aligned}$$

i.e., u^* is a minimizer to J_α. □

Remark 3.4. The existence result holds if the functional $\phi(u, g^\delta)$ is weakly lower semicontinuous with respect to u, for any fixed $g^\delta \in Y$.

Now we illustrate Theorem 3.1 with several examples.

Example 3.1. This example is concerned with the L^1-TV model for restoring images subjected to blurring and impulsive noise, i.e.,

$$J_\alpha(u) = \|Ku - g^\delta\|_{L^1(\Omega)} + \alpha|u|_{\mathrm{TV}(\Omega)},$$

where $\Omega \subset \mathbb{R}^2$ is an open bounded domain, and the blurring operator K is continuous with respect to $L^p(\Omega)$, $1 < p < 2$ and $1 \notin \mathrm{Null}(K)$. The L^1 fidelity term makes the model robust with respect to outliers in the data, and the penalty term preserves the edges in the images [265]. Here the space $X = \mathrm{BV}(\Omega)$ and $Y = L^1(\Omega)$. The space $\mathrm{BV}(\Omega)$ of functions of bounded variation is given by

$$\mathrm{BV}(\Omega) = \{v \in L^1(\Omega) : \|v\|_{L^1(\Omega)} + |v|_{\mathrm{TV}(\Omega)} < \infty\},$$

where the total variation seminorm $|v|_{\mathrm{TV}(\Omega)}$ is defined by

$$|v|_{\mathrm{TV}(\Omega)} = \sup_{\substack{w \in C_0^1(\Omega;\mathbb{R}^{\mathrm{d}}) \\ \|w\|_{L^\infty(\Omega)} \leq 1}} \int v \mathrm{div}\phi dx.$$

By the compact embedding of the space $\mathrm{BV}(\Omega)$ into $L^p(\Omega)$, $1 < p < 2$ [16], and the continuity of K in $L^p(\Omega)$, the operator K is weakly continuous. The assumption $1 \notin \mathrm{Null}(K)$ implies that functional is coercive on the space $X = \mathrm{BV}(\Omega)$. Further, the regularization functional $\psi(u) = |u|_{\mathrm{TV}}$ is BV weak $*$ lower semicontinuous. Therefore, by Theorem 3.1, there exists at least one minimizer to J_α.

Example 3.2. In this example, we consider the sparsity regularization

$$J_\alpha(u) = \tfrac{1}{2}\|Ku - g^\delta\|^2 + \alpha\|u\|_{\ell^p}^p,$$

with $p \in (0, 1]$, where $\ell^p = \{u \in \ell^2 : \sum_k |u_k|^p < \infty\}$, and endowed with $\|u\|_{\ell^p}^p = \sum_k |u_k|^p$. For $0 < p < 1$, $\|\cdot\|_{\ell^p}$ is a quasi-norm. The operator K is bounded from ℓ^2 to ℓ^2. To show the existence, we use a reparametrization $u = \gamma(v)$, $v \in \ell^2$, introduced in [311], where

$$u_i = \gamma(v)_i = |v_i|^{2/p}\mathrm{sgn}(v_i).$$

Then $v_i = |u_i|^{p/2-1}u_i$, and $\gamma : \ell^2 \to \ell^p$ is an isomorphism, with $|\gamma(v)|_p = |v|_2^{2/p}$. Then it follows that the problem is equivalent to

$$J_\alpha(v) = \tfrac{1}{2}\|K\gamma(v) - g^\delta\|^2 + \alpha\|v\|^2.$$

Next we claim that $\gamma : \ell^2 \to \ell^2$ is sequentially weakly continuous, i.e. $v^n \to \bar{v}$ weakly in ℓ^2 implies that $\gamma(v^n) \to \gamma(\bar{v})$ weakly in ℓ^2. Let $r = \frac{2}{p}+1 \in [3, \infty)$, and r^* be the conjugate exponent, given by $r^* = \frac{p}{2} + 1 \in (1, 3/2]$. Then γ is the duality mapping from ℓ^r to ℓ^{r^*}, i.e.,

$$(\gamma(v), v)_{\ell^{r^*},\ell^r} = \|\gamma(v)\|_{\ell^{r^*}}\|v\|_{\ell^r}, \quad |\gamma(v)|_{\ell^{r^*}} = |v|_{\ell^r}.$$

If $v^n \to \bar{v}$ weakly in ℓ^2, then $v^n \to \bar{v}$ weakly in ℓ^r. Since the duality mapping $\gamma : \ell^r \to \ell^{r^*}$ is sequentially weakly continuous ([69], pp. 97) we have $\gamma(v^n) \to \gamma(\bar{v})$ weakly in ℓ^{r^*}. Using that $r^* \leq 2$, this implies that $\gamma(v^n) \to \gamma(\bar{v})$ weakly in ℓ^2. Then the existence follows from the arguments in Theorem 3.1.

The Tikhonov model J_α is stable, i.e., the minimizer u_α^δ depends continuously on the data g^δ, which is stated next.

Theorem 3.2. *Let Assumption 3.1 hold. Let the sequence* $\{g_n\} \subset Y$ *be convergent to* $g^\delta \in Y$*, and* u_n *be a minimizer to the functional* J_α *with* g_n *in place of* g^δ*. Then the sequence* $\{u_n\}$ *contains a subsequence converging to a minimizer to* J_α*. Further, if the minimizer to* J_α *is unique, then the whole sequence converges. Moreover, if the functional* ψ *satisfies the H-property, then the convergence is actually strong.*

Proof. Let u_α^δ be a minimizer to J_α. The minimizing property of u_n yields

$$\|Ku_n - g_n\|^p + \alpha\psi(u_n) \leq \|Ku_\alpha^\delta - g_n\|^p + \alpha\psi(u_\alpha^\delta).$$

Hence the sequences $\{\|Ku_n - g_n\|^p\}$ and $\{\psi(u_n)\}$ are uniformly bounded. Further, in view of the triangle inequality,

$$\|Ku_n - g^\delta\|^p \leq 2^{p-1}(\|Ku_n - g_n\|^p + \|g_n - g^\delta\|^p),$$

the sequence $\{\|Ku_n - g^\delta\|\}$ is also uniformly bounded. Then, by Assumption 3.1(a), the sequence $\{u_n\}$ is uniformly bounded in X, and there exists a subsequence, also denoted by $\{u_n\}$, converging weakly in X to some $u^* \in \mathcal{C}$. By the weak continuity of K and the convergence of g_n to g^δ in Y, we have $Ku_n - g_n \to Ku^* - g^\delta$ weakly. Now the weak lower semicontinuity of norms and Assumption 3.1(b) imply that

$$\begin{aligned}
\|Ku^* - g^\delta\|^p + \alpha\psi(u^*) &\leq \liminf_{n\to\infty} \|Ku_n - g_n\|^p + \alpha \liminf_{n\to\infty} \psi(u_n) \\
&\leq \liminf_{n\to\infty} (\|Ku_n - g_n\|^p + \alpha\psi(u_n)) \\
&\leq \liminf_{n\to\infty} \|Ku_\alpha^\delta - g_n\|^p + \alpha\psi(u_\alpha^\delta) \\
&= \|Ku_\alpha^\delta - g^\delta\|^p + \alpha\psi(u_\alpha^\delta).
\end{aligned} \tag{3.3}$$

Hence, the limit u^* is a minimizer to the functional J_α. If the minimizer u_α^δ is unique, then $u^* = u_\alpha^\delta$, and every subsequence contains a subsubsequence that converges to u^* weakly. Therefore, the whole sequence converges weakly.

Now we show that the functional value $\psi(u_n)$ converges to $\psi(u^*)$. Assume the contrary, i.e., $\limsup_{n\to\infty} \psi(u_n) > \liminf_{n\to\infty} \psi(u_n) \geq \psi(u^*)$. Then there exists a subsequence $\{u_{n_k}\}$ of $\{u_n\}$ such that

$$c := \limsup_{k\to\infty} \psi(u_{n_k}) > \psi(u^*).$$

Meanwhile, by taking $u_\alpha^\delta = u^*$ in (3.3) yields

$$\lim_{k\to\infty} \|Ku_{n_k} - g_{n_k}\|^p + \alpha\psi(u_{n_k}) = \|Ku^* - g^\delta\|^p + \alpha\psi(u^*).$$

This two identities together yield

$$\lim_{k\to\infty} \|Ku_{n_k} - g_{n_k}\|^p = \|Ku^* - g^\delta\|^p + \alpha(\psi(u^*) - c) < \|Ku^* - g^\delta\|^p,$$

which is in contradiction with the weak lower semicontinuity of norms, i.e.,

$$\|Ku^* - g^\delta\| \le \liminf_{k\to\infty} \|Ku_{n_k} - g_{n_k}\|.$$

Hence $\lim_{n\to\infty} \psi(u_n) = \psi(u^*)$. This together with the H-property yields the third assertion. □

Next we turn to the behavior of the minimizer u_α^δ, as the noise level δ goes to zero. A fundamental analysis on the Tikhonov regularization method is whether the approximate solution u_α^δ converges to the true solution, i.e., the ψ-minimizing solution $u^\dagger$ defined below, as the noise level δ tends to zero. This is referred to as the consistency in the literature.

Definition 3.2. An element $u^\dagger \in X$ is called a ψ-minimizing solution to the problem $Ku = g^\dagger$ if it satisfies

$$Ku^\dagger = g^\dagger \quad \text{and} \quad \psi(u^\dagger) \le \psi(u) \quad \forall u \in \{u \in \mathcal{C} : \ Ku = g^\dagger\}.$$

The existence of a ψ-minimizing solution follows from Assumption 3.1.

Theorem 3.3. *Let Assumption 3.1 hold. Then there exists at least one ψ-minimizing solution to* (3.1).

Proof. Suppose that there does not exist a ψ-minimizing solution in $\mathcal{C}$. Then there exists a sequence $\{u_n\} \subset \mathcal{C}$ of solutions to (3.1) such that $\psi(u_n) \to c$ and

$$c < \psi(u), \quad \forall u \in \{u \in \mathcal{C} : K(u) = g^\dagger\}.$$

Hence, the functional $J_\alpha(u_n)$ (for any fixed α) with $g^\dagger$ in place of g^δ is uniformly bounded, and by the coercivity of J_α, cf. Assumption 3.1(i), the sequence $\{u_n\}$ contains a subsequence, also denoted by $\{u_n\}$, and some $u^* \in \mathcal{C}$ such that $u_n \to u^*$ weakly. Then $Ku^* = g^\dagger$, and by Assumption 3.1(ii), there holds $\psi(u^*) = \lim_{n\to\infty} \psi(u_n) = c$. This contradicts the assumption, and completes the proof. □

Theorem 3.4. *Let Assumption 3.1 hold, and $\{g^{\delta_n}\}_n \subset Y$ be a sequence of noisy data, with $\delta_n = \|g^\dagger - g^\delta\| \to 0$. Then the sequence of minimizers $\{u_{\alpha_n}^{\delta_n}\}$ has a subsequence converging weakly to a ψ-minimizing solution $u^\dagger$, if the regularization parameter $\alpha_n \equiv \alpha(\delta_n)$ satisfies*

$$\lim_{n\to\infty} \frac{\delta_n^p}{\alpha_n} = 0 \quad \text{and} \quad \lim_{n\to\infty} \alpha_n = 0.$$

Further, if the ψ-minimizing solution $u^\dagger$ is unique, then the whole sequence converges weakly. Moreover, if the functional ψ satisfies the H-property, then the convergence is actually strong.

Proof. By the minimizing property of the minimizer $u_{\alpha_n}^{\delta_n}$, we have

$$\begin{aligned}\|Ku_{\alpha_n}^{\delta_n} - g^{\delta_n}\|^p + \alpha_n\psi(u_{\alpha_n}^{\delta_n}) &\le \|Ku^\dagger - g^{\delta_n}\|^p + \alpha_n\psi(u^\dagger)\\ &\le \delta_n^p + \alpha_n\psi(u^\dagger).\end{aligned}$$

By the choice of the sequence $\{\alpha_n\}$, the sequences $\{\|u_{\alpha_n}^{\delta_n} - g^{\delta_n}\|^p\}$ and $\{\psi(u_{\alpha_n}^{\delta_n})\}$ are both uniformly bounded. Further,

$$\|Ku_{\alpha_n}^{\delta_n} - g^\dagger\|^p \le 2^{p-1}(\|Ku_{\alpha_n}^{\delta_n} - g^{\delta_n}\|^p + \|g^{\delta_n} - g^\dagger\|^p).$$

Thus the sequence $\{\|Ku_{\alpha_n}^{\delta_n} - g^\dagger\|^p\}$ is also uniformly bounded. By Assumption 3.1(a), the sequence $\{u_{\alpha_n}^{\delta_n}\}_n$ is uniformly bounded, and thus contains a subsequence, also denoted by $\{u_{\alpha_n}^{\delta_n}\}$, converging to some u^* weakly. By the convexity and closedness of the set $\mathcal{C}$, $u^* \in \mathcal{C}$.

By the weak lower semicontinuity of norms we deduce

$$\begin{aligned}\|Ku^* - g^\dagger\|^p &\le \liminf_{n\to\infty} \|Ku_{\alpha_n}^{\delta_n} - g^{\delta_n}\|^p\\ &\le \limsup_{n\to\infty} \|Ku_{\alpha_n}^{\delta_n} - g^{\delta_n}\|^p + \alpha_n\psi(u_{\alpha_n}^{\delta_n}) = 0,\end{aligned}$$

i.e., $Ku^* = g$. Next we show that u^* is a ψ-minimizing solution. It follows from the minimizing property of $u_{\alpha_n}^{\delta_n}$ and weak lower semicontinuity of the functional ψ, cf. Assumption 3.1(b), that

$$\begin{aligned}\psi(u^*) &\le \liminf_{n\to\infty} \psi(u_{\alpha_n}^{\delta_n}) \le \limsup_{n\to\infty} \psi(u_{\alpha_n}^{\delta_n})\\ &\le \lim_{n\to\infty} \left(\frac{\delta_n^p}{\alpha_n} + \psi(u^\dagger)\right) = \psi(u^\dagger).\end{aligned}$$

Now the first assertion follows directly from this and the fact that $u^\dagger$ is a ψ-minimizing solution.

If the ψ-minimizing solution $u^\dagger$ is unique, then $u^* = u^\dagger$. Clearly, every subsequence of $\{u_{\alpha_n}^{\delta_n}\}$ contains a subsubsequence converging weakly to $u^\dagger$, and thus the whole sequence converges weakly. This shows the second assertion. By taking $u^\dagger = u^*$ in the preceding inequality yields $\lim_{n\to\infty} \psi(u_{\alpha_n}^{\delta_n}) = \psi(u^*)$. This together with the H-property of ψ yields the third assertion. □

Definition 3.3. A normed vector space X is called uniformly convex if for every $\epsilon > 0$ there is some $\delta > 0$ so that for any two vectors with $\|u\| = 1$ and $\|v\| = 1$, the condition $\|u - v\| \geq \epsilon$ implies that $\left\|\frac{u+v}{2}\right\| \leq 1 - \delta$.

Hilbert spaces and $L^p(\Omega)$ spaces, $1 < p < \infty$, are examples of uniformly convex spaces. If the space X is uniformly convex, and $\psi(u) = \|u\|_X^p$, $1 < p < \infty$, then the ψ-minimizing solution $u^\dagger$ can be characterized by $\xi = \partial\psi(u^\dagger) \in \overline{\text{Range}(K^*)}$. Here the subdifferential $\partial\psi(u)$ for a convex functional ψ on X denotes the set of all subgradients $\xi \in X^*$ such that

$$\psi(\tilde{u}) \geq \psi(u) + \langle \xi, \tilde{u} - u \rangle \qquad \forall \tilde{u} \in X.$$

Proposition 3.1. *If X is uniformly convex, $\psi(u) = \|u\|_X^p$, $1 < p < \infty$, and $\mathcal{C} = X$. Then there exists a ψ-minimizing solution $u^\dagger$, and it satisfies $\partial\psi(u^\dagger) \in \overline{\text{Range}(K^*)}$.*

Proof. The set $S = \{u \in X : Ku = g^\dagger\} \subset X$ is nonempty and closed since $g^\dagger \in \text{Range}(K)$ and K is a continuous linear operator. Hence the uniform convexity of the space X guarantees the existence and uniqueness of a ψ-minimizing solution $u^\dagger$. Further, there holds for any $\xi \in \partial\psi(u^\dagger)$

$$\langle \xi, u^\dagger \rangle \leq \langle \xi, u \rangle, \quad \forall u \in S.$$

Let v be an arbitrary element in $\text{Null}(K)$. Then $u^\dagger \pm v \in S$, and thus

$$\langle \xi, u^\dagger \rangle \leq \langle \xi, u^\dagger \pm v \rangle = \langle \xi, u^\dagger \rangle \pm \langle \xi, v \rangle.$$

It follows that $\langle \xi, v \rangle = 0$. Hence $\xi \in \text{Null}(K)^\perp = \overline{\text{Range}(K^*)}$. □

Last, we derive a result characterizing the solution u_α^δ for the case Y is a Hilbert space, $\phi(u, g^\delta) = \frac{1}{2}\|Ku - g^\delta\|^2$ and $\psi(u) = \frac{1}{p}\|u\|_X^p$ with $p \geq 1$, and in the absence of convex constraint $\mathcal{C}$. Below $\|\cdot\|_{X^*}$ denotes the dual norm of the dual space X^*.

Theorem 3.5. *Let Y be a Hilbert space, $\phi(u, g^\delta) = \frac{1}{2}\|Ku - g^\delta\|^2$, $\psi(u) = \frac{1}{p}\|u\|_X^p$, and $\|K^* g^\delta\|_{X^*} \neq 0$. Then the following statements hold.*

(i) If $p > 1$, then for any $\alpha > 0$, there holds

$$\langle K^*(Ku_\alpha^\delta - g^\delta), u_\alpha^\delta \rangle = -p\alpha\psi(u_\alpha^\delta),$$
$$\|K^*(g^\delta - Ku_\alpha^\delta)\|_{X^*} = \alpha\|u_\alpha^\delta\|_X^{p-1}.$$

(ii) If $p = 1$ and $\alpha^* = \|K^* g^\delta\|_{X^*}$, then for $\alpha \geq \alpha^*$, $u_\alpha^\delta = 0$ and for $\alpha < \alpha^*$, there hold

$$\langle K^*(Ku_\alpha^\delta - g^\delta), u_\alpha^\delta \rangle = -\alpha\psi(u_\alpha^\delta),$$
$$\|K^*(g^\delta - Ku_\alpha^\delta)\|_{X^*} = \alpha.$$

In either case, if the two relations hold, then u_α^δ is a minimizer of J_α.

Proof. First we show (i). Note that the optimality condition for u_α^δ reads

$$K^*(Ku_\alpha^\delta - g^\delta) + \alpha\xi_\alpha^\delta = 0,$$

where $\xi_\alpha^\delta \in \partial\psi(u_\alpha^\delta)$. Then $\partial\psi(u_\alpha^\delta) = I_p(u_\alpha^\delta)$, the duality mapping of gauge $g(t) = t^{p-1}$ [69] . It satisfies

$$\|I_p(u_\alpha^\delta)\|_{X^*} = \|u_\alpha^\delta\|_X^{p-1} \quad \text{and} \quad \langle I_p(u_\alpha^\delta), u_\alpha^\delta\rangle = \|u\|_X^p.$$

Consequently, the desired relations follow directly. Conversely, the two relations imply directly that [69]

$$\alpha^{-1}K^*(g^\delta - Ku_\alpha^\delta) = I_p(u_\alpha^\delta) \in \partial\psi(u_\alpha^\delta),$$

which is a sufficient condition for the optimality of u_α^δ by noting the convexity of the functional J_α.

Now we turn to (ii). First we note that the optimality condition for u_α^δ reads

$$K^*(Ku_\alpha^\delta - g^\delta) + \alpha\xi_\alpha^\delta = 0,$$

where $\xi_\alpha^\delta \in \partial\psi(u_\alpha^\delta)$. By [69] , we have $\partial\psi(u) = \{u^* \in X^* : \langle u^*, u\rangle = \|u\|_X, \|u^*\|_{X^*} = 1\}$ for $u \neq 0$. Therefore, at $u_\alpha^\delta \neq 0$, the two relations obviously hold, and the converse follows as before. If $u_\alpha^\delta = 0$, the extremal property implies that for $\forall h \in X, \varepsilon > 0$,

$$\tfrac{1}{2}\|K(\varepsilon h) - g^\delta\|^2 + \alpha\psi(\varepsilon h) \geq \tfrac{1}{2}\|g^\delta\|^2.$$

Letting $\varepsilon \to 0^+$ yields $\alpha\psi(h) - \langle Kh, g^\delta\rangle \geq 0$, Since the inequality holds for arbitrarily $h \in X$ and $\|K^*g^\delta\|_{X^*} \neq 0$, we have $\alpha \geq \alpha^* = \|K^*y^\delta\|_{X^*}$. The converse follows by reversing the arguments. □

Remark 3.5. In case of $\psi(u)$ being a seminorm (or its power) on the space X, i.e., $\psi(u) = \frac{1}{p}|u|_X^p$ with $p \geq 1$, at a minimum u_α^δ, the identity $\langle K^*(Ku_\alpha^\delta - g^\delta), u_\alpha^\delta\rangle = -p\alpha\psi(u_\alpha^\delta)$ remains valid. However, an explicit characterization of the critical value α^* is missing.

3.2 Value function calculus

In this section we present the value function calculus due to A. N. Tikhonov: it was already used in the classical book by Tikhonov and Arsenin [289], and further studied in [292]. Value function plays an important role in the study of Tikhonov regularization, especially in analyzing parameter choice rules, e.g., discrepancy principle and balancing principle, and their numerical implementations. Our description follows [154].

For the model $J_\alpha(u)$, the value function $F(\alpha) : \mathbb{R}^+ \to \mathbb{R}$ is defined by

$$F(\alpha) = \inf_{u \in \mathcal{C}} J_\alpha(u). \tag{3.4}$$

We will discuss its important properties, especially differentiability.

A first result shows the monotonicity and concavity of F.

Theorem 3.6. *The value function $F(\alpha)$ is nondecreasing and concave.*

Proof. Given a $\tilde{\alpha} < \alpha$, for any $u \in X$, by the nonnegativity of the functional $\psi(u)$, we have

$$F(\tilde{\alpha}) \leq J_{\tilde{\alpha}}(u) = \phi(u, g^\delta) + \tilde{\alpha}\psi(u) \leq \phi(u, g^\delta) + \alpha\psi(u).$$

Taking the infimum with respect to $u \in \mathcal{C}$ yields $F(\tilde{\alpha}) \leq F(\alpha)$. This shows the monotonicity of the value function $F(\alpha)$.

Next we show the concavity of the function F. Let $\alpha_1 > 0$ and $\alpha_2 > 0$ be given, and set $\alpha_t = (1-t)\alpha_1 + t\alpha_2$ for any $t \in [0, 1]$. Then

$$\begin{aligned} F((1-t)\alpha_1 + t\alpha_2) &= \inf_{u \in \mathcal{C}} J_{\alpha_t}(u) \\ &= \inf_{u \in \mathcal{C}} \left\{ \phi(u, g^\delta) + ((1-t)\alpha_1 + t\alpha_2)\psi(u) \right\} \\ &\geq (1-t) \inf_{u \in \mathcal{C}} J_{\alpha_1}(u) + t \inf_{u \in \mathcal{C}} J_{\alpha_2}(u) \\ &= (1-t)F(\alpha_1) + tF(\alpha_2). \end{aligned}$$

Therefore, the function $F(\alpha)$ is concave. □

A direct consequence of concavity is continuity.

Corollary 3.1. *The function $F(\alpha)$ is continuous everywhere.*

Next we examine the differentiability of the value function F. To this end, recall first the definition of one-sided derivatives (Dini derivatives) $D^\pm F$ of F defined by

$$\begin{aligned} D^- F(\alpha) &= \lim_{h \to 0^+} \frac{F(\alpha) - F(\alpha - h)}{h}, \\ D^+ F(\alpha) &= \lim_{h \to 0^+} \frac{F(\alpha + h) - F(\alpha)}{h}. \end{aligned}$$

The concavity and monotonicity of the function F in Theorem 3.6 ensures the existence of one-sided derivatives $D^\pm F(\alpha)$.

Lemma 3.1. *The one-sided derivatives $D^- F(\alpha)$ and $D^+ F(\alpha)$ of the value function $F(\alpha)$ exist for any $\alpha > 0$, and moreover $D^\pm F(\alpha) \geq 0$.*

Proof. For a given $\alpha > 0$, take any $0 < h_1 < h_2 < \alpha$ and set $t = 1 - \frac{h_1}{h_2} < 1$. Then $\alpha - h_1 = t\alpha + (1-t)(\alpha - h_2)$. Now by the concavity of F, we have

$$\begin{aligned} F(\alpha - h_1) &\geq tF(\alpha) + (1-t)F(\alpha - h_2) \\ &= \left(1 - \frac{h_1}{h_2}\right) F(\alpha) + \frac{h_1}{h_2} F(\alpha - h_2). \end{aligned}$$

Rearranging the terms gives

$$\frac{F(\alpha) - F(\alpha - h_1)}{h_1} \leq \frac{F(\alpha) - F(\alpha - h_2)}{h_2}.$$

Hence the sequence $\{\frac{F(\alpha)-F(\alpha-h)}{h}\}_{h>0}$ is monotonically decreasing as h tends to zero and bounded from below by zero, and the limit $\lim_{h\to 0^+} \frac{F(\alpha)-F(\alpha-h)}{h}$ exists, i.e., the left-sided derivative $D^-F(\alpha)$ exists. The existence of the right-sided derivative $D^+F(\alpha)$ follows analogously. □

As a consequence of the definition of one-sided derivatives, we have:

Corollary 3.2. *The one-sided derivatives $D^-F(\alpha)$ and $D^+F(\alpha)$ are left- and right continuous, respectively.*

Remark 3.6. The preceding results on $F(\alpha)$ do not require the existence of a minimizer u_α^δ to the Tikhonov functional J_α, and are valid for any space. In particular, it is also true for nonlinear operators.

Now we characterize the value function F by $\bar{\phi}$ and $\bar{\psi}$ defined by

$$\bar{\phi}(\alpha) := \phi(u_\alpha^\delta, g^\delta) \quad \text{and} \quad \bar{\psi}(\alpha) := \psi(u_\alpha^\delta).$$

Here and below, let Assumption 3.1 hold, and denote the set of minimizers to J_α by $S(\alpha)$, i.e.,

$$S(\alpha) = \arg\min_{u\in\mathcal{C}} J_\alpha(u).$$

The set $S(\alpha)$ might contain multiple elements, i.e., distinct $u_\alpha^\delta,\ \tilde{u}_\alpha^\delta \in S(\alpha)$ such that

$$F(\alpha) = \phi(u_\alpha^\delta, g^\delta) + \alpha\psi(u_\alpha^\delta) = \phi(\tilde{u}_\alpha^\delta, g^\delta) + \alpha\psi(\tilde{u}_\alpha^\delta),$$

while $\phi(u_\alpha^\delta, g^\delta) < \phi(\tilde{u}_\alpha^\delta, g^\delta)$ and $\psi(u_\alpha^\delta) > \psi(\tilde{u}_\alpha^\delta)$. That is, the functions $\bar{\phi}$ and $\bar{\psi}$ can be set valued. We always use any but fixed representation $\bar{\phi}(\alpha) = \phi(u_\alpha^\delta, g^\delta)$ and $\bar{\psi}(\alpha) = \psi(u_\alpha^\delta)$ for some $u_\alpha^\delta \in S(\alpha)$.

We illustrate the discontinuity with the L^1-TV model.

Example 3.3. Consider the following total variation image denoising model with the L^1 fidelity, i.e.,

$$J_\alpha(u) = \|u - g^\delta\|_{L^1(\Omega)} + \alpha|u|_{\mathrm{TV}(\Omega)},$$

and the observed image $g^\delta = \chi_{B_r(0)}(x)$, where $B_r(0)$ is a disk centered at the origin and radius r. One can deduce that for each α, every minimizer has to be of the form $c\chi_{B_r(0)}(x)$ for some $c \in [0,1]$ [56]. Hence one only needs to minimize

$$2\pi\alpha rc + \pi r^2|1-c|$$

over $c \in [0,1]$. Hence the solution set $S(\alpha)$ is given by

$$S(\alpha) = \begin{cases} 0, & \alpha > r/2, \\ \{c\chi_{B_r(0)}, c \in [0,1]\}, & \alpha = r/2, \\ \chi_{B_r(0)}, & \alpha < r/2. \end{cases}$$

Hence the solution is unique for all except one value of the parameter $\alpha = r/2$. Clearly, at this α value, the functions $\bar{\phi}$ and $\bar{\psi}$ are discontinuous.

The continuity of the functions $\bar{\phi}$ and $\bar{\psi}$ at α can be guaranteed if the Tikhonov functional J_α has a unique minimizer.

Lemma 3.2. *Let Assumption 3.1(a) and (b) hold, and $\phi(u, g^\delta)$ be weakly lower semicontinuous in u. If J_α, $\alpha > 0$, has a unique minimizer u_α^δ, then the functions $\bar{\phi}$ and $\bar{\psi}$ are continuous at α.*

Proof. Let $\{\alpha_n\}$ be any sequence converging to α. Then by the minimizing property of $u_n \equiv u_{\alpha_n}^\delta$ we have

$$\phi(u_n, g^\delta) + \alpha_n\psi(u_n) \leq \phi(u_\alpha^\delta, g^\delta) + \alpha_n\psi(u_\alpha^\delta).$$

Therefore the sequences $\{\phi(u_n, g^\delta)\}$ and $\{\psi(u_n)\}$ are uniformly bounded. By Assumption 3.1(a), the sequence $\{u_n\}$ is uniformly bounded, and there exists a subsequence, also denoted by $\{u_n\}$, such that $u_n \to u^*$ weakly for some $u^* \in \mathcal{C}$. By the weak lower semicontinuity of ϕ and ψ we deduce

$$\begin{aligned} \phi(u^*, g^\delta) + \alpha\psi(u^*) &\leq \liminf_{n\to\infty} \phi(u_n, g^\delta) + \liminf_{n\to\infty} \alpha_n\psi(u_n) \\ &\leq \liminf(\phi(u_n, g^\delta) + \alpha_n\psi(u_n)) \\ &\leq \liminf_{n\to\infty}(\phi(u_\alpha^\delta, g^\delta) + \alpha_n\psi(u_\alpha^\delta)) \\ &= \phi(u_\alpha^\delta, g^\delta) + \alpha\psi(u_\alpha^\delta), \end{aligned}$$

i.e., u^* is a minimizer to the Tikhonov functional J_α, and by the uniqueness assumption, $u^* = u_\alpha^\delta$. A standard argument yields that the whole sequence converges weakly to u_α^δ. Now we can repeat the proof of Theorem 3.2 to show $\lim_{n\to\infty}\psi(u_n) = \psi(u_\alpha^\delta)$ and $\lim_{n\to\infty}\phi(u_n, g^\delta) = \phi(u_\alpha^\delta, g^\delta)$, i.e., $\lim_{n\to\infty}\bar{\psi}(\alpha_n) = \bar{\psi}(\alpha)$ and $\lim_{n\to\infty}\bar{\phi}(\alpha_n) = \bar{\phi}(\alpha)$. □

In any case, the functions $\bar{\phi}$ and $\bar{\psi}$ are monotone in α.

Lemma 3.3. *Given $\alpha_1, \alpha_2 > 0$, if $S(\alpha_1)$ and $S(\alpha_2)$ are both nonempty, then for any representations $\bar{\phi}$ and $\bar{\psi}$, there hold*

$$(\bar{\psi}(\alpha_1) - \bar{\psi}(\alpha_2))(\alpha_1 - \alpha_2) \leq 0,$$
$$(\bar{\phi}(\alpha_1) - \bar{\phi}(\alpha_2))(\alpha_1 - \alpha_2) \geq 0.$$

Proof. The minimizing property of any $u_{\alpha_1}^\delta \in S(\alpha_1)$ and $u_{\alpha_2}^\delta \in S(\alpha_2)$ gives

$$\bar{\phi}(\alpha_1) + \alpha_1\bar{\psi}(\alpha_1) \leq \bar{\phi}(\alpha_2) + \alpha_1\bar{\psi}(\alpha_2),$$
$$\bar{\phi}(\alpha_2) + \alpha_2\bar{\psi}(\alpha_2) \leq \bar{\phi}(\alpha_1) + \alpha_2\bar{\psi}(\alpha_1).$$

Adding these two inequalities yields the first inequality. Dividing the first and the second equations by α_1 and α_2, respectively, and then adding them yields the second inequality. □

Now the one-sided derivatives $D^{\pm}F(\alpha)$ can be made more precise.

Lemma 3.4. *Suppose that for a given $\alpha > 0$ the set $S(\alpha)$ is nonempty. Then there hold*

$$D^+F(\alpha) \leq \bar{\psi}(\alpha) \leq D^-F(\alpha),$$
$$F(\alpha) - \alpha D^-F(\alpha) \leq \bar{\phi}(\alpha) \leq F(\alpha) - \alpha D^+F(\alpha).$$

Proof. For any $\tilde{\alpha}$ such that $0 < \tilde{\alpha} < \alpha$, we have

$$F(\tilde{\alpha}) = \inf_{u \in C} J_{\tilde{\alpha}}(u) \leq J_{\tilde{\alpha}}(u_\alpha^\delta) = \bar{\phi}(\alpha) + \tilde{\alpha}\bar{\psi}(\alpha).$$

Therefore, we have

$$\begin{aligned} F(\alpha) - F(\tilde{\alpha}) &\geq \bar{\phi}(\alpha) + \alpha\bar{\psi}(\alpha) - \bar{\phi}(\alpha) - \tilde{\alpha}\bar{\psi}(\alpha) \\ &= (\alpha - \tilde{\alpha})\bar{\psi}(\alpha). \end{aligned}$$

Thus we obtain

$$\frac{F(\alpha) - F(\tilde{\alpha})}{\alpha - \tilde{\alpha}} \geq \bar{\psi}(\alpha).$$

By passing to limit $\tilde{\alpha} \to \alpha$, it follows that $D^-F(\alpha) \geq \bar{\psi}(\alpha)$. The inequality $D^+F(\alpha) \leq \bar{\psi}(\alpha)$ follows analogously. The remaining assertion follows from these two inequalities and the definition of the value function F. □

The next result is an immediate consequence of the above two lemmas, and it represents the fundamental formula for the value function calculus.

Lemma 3.5. *Let the value function F be defined in* (3.4).

(a) *If $F'(\alpha)$ exists at $\alpha > 0$, then*

$$\bar{\psi}(\alpha) = F'(\alpha) \quad \text{and} \quad \bar{\phi}(\alpha) = F(\alpha) - \alpha F'(\alpha).$$

(b) *There exists a countable set $Q \subset \mathbb{R}^+$ such that for any $\alpha \in \mathbb{R}^+ \setminus Q$, F is differentiable, and $\bar{\phi}(\alpha)$ and $\bar{\psi}(\alpha)$ are continuous and*

$$\bar{\psi}(\alpha) = F'(\alpha) \quad \text{and} \quad \bar{\phi}(\alpha) = F(\alpha) - \alpha F'(\alpha).$$

Proof. The assertion (a) is a direct consequence of Lemma 3.4. Assertion (b) follows from the monotonicity of $\bar{\psi}(\alpha)$ in α from Lemma 3.3, and thus there exist at most countable discontinuity points. □

By Lemma 3.3, any representations $\bar{\phi}(\alpha)$ and $\bar{\psi}(\alpha)$ define measures $d\bar{\phi}$ and $d\bar{\psi}$ [91]. Moreover, the following relation holds.

Theorem 3.7. *The measure $d\bar{\phi}$ is absolutely continuous with respect to the measure $d\bar{\psi}$, and the Radon-Nikodym derivative is given by*

$$\frac{d\bar{\phi}}{d\bar{\psi}}(\alpha) = -\alpha. \tag{3.5}$$

In particular, if either $\bar{\phi}(\alpha)$ or $\bar{\psi}(\alpha)$ is differentiable, then the other is also differentiable and $\bar{\phi}'(\alpha) + \alpha\bar{\psi}'(\alpha) = 0$.

Proof. For any $\alpha, \tilde{\alpha} > 0$, the minimizing property of $u_\alpha^\delta \in S(\alpha)$ and $u_{\tilde{\alpha}}^\delta \in S(\tilde{\alpha})$ gives

$$\bar{\phi}(\alpha) + \alpha\bar{\psi}(\alpha) \le \bar{\phi}(\tilde{\alpha}) + \alpha\bar{\psi}(\tilde{\alpha}),$$
$$\bar{\phi}(\tilde{\alpha}) + \tilde{\alpha}\bar{\psi}(\tilde{\alpha}) \le \bar{\phi}(\alpha) + \tilde{\alpha}\bar{\psi}(\alpha).$$

By rearranging these two relations, we get

$$-\tilde{\alpha}(\bar{\psi}(\alpha) - \bar{\psi}(\tilde{\alpha})) \le \bar{\phi}(\alpha) - \bar{\phi}(\tilde{\alpha}) \le -\alpha(\bar{\psi}(\alpha) - \bar{\psi}(\tilde{\alpha})).$$

This shows the first assertion. The rest is obvious from Lemma 3.5. □

Lemma 3.3 indicates the functions $\bar{\phi}$ and $\bar{\psi}$ are monotone. The next result provides one simple condition for ensuring their strict monotonicity.

Lemma 3.6. *Let the functional J_α satisfy $S(\alpha_1) \cap S(\alpha_2) = \emptyset$ for all distinct $\alpha_1, \alpha_2 > 0$. Then the functions $\bar{\phi}(\alpha)$ and $\bar{\psi}(\alpha)$ are strictly monotone.*

Proof. We proceed by means of contradiction. Assume that there exist two distinct values α_1 and α_2 such that $\bar{\phi}(\alpha_1) = \bar{\phi}(\alpha_2)$. By the assumption $S(\alpha_1) \cap S(\alpha_2) = \emptyset$, we have

$$\bar{\phi}(\alpha_1) + \alpha_1\bar{\psi}(\alpha_1) < \bar{\phi}(\alpha_2) + \alpha_1\bar{\psi}(\alpha_2),$$

which consequently implies that $\bar{\psi}(\alpha_1) < \bar{\psi}(\alpha_2)$. However, reversing the role of α_1 and α_2 gives a contradicting inequality $\bar{\psi}(\alpha_2) < \bar{\psi}(\alpha_1)$. This shows the assertion on $\bar{\phi}$. The assertion on $\bar{\psi}$ follows analogously. □

By imposing differentiability on ϕ and ψ, we have the next lemma ensuring the condition $S(\alpha_1) \cap S(\alpha_2) = \emptyset$ for distinct α_1 and α_2 in the absence of constraint [292].

Lemma 3.7. *Let $\alpha^* = \inf\{\alpha : \psi(u_\alpha^\delta) = \inf_{u\in X} \psi(u)\}$, ϕ and ψ be continuously Frechét differentiable, and ψ be convex. If $\alpha^* < \infty$, then the identity $S(\alpha_1) \cap S(\alpha_2) \neq \emptyset$ for distinct α_1 and α_2 holds if and only if $\alpha_1, \alpha_2 \geq \alpha^*$.*

Proof. The optimality condition reads

$$J'_\alpha(u_\alpha^\delta) = \phi'(u_\alpha^\delta, g^\delta) + \alpha\psi'(u_\alpha^\delta) = 0.$$

Letting $\alpha_2 > \alpha_1 > 0$ and noting $u_{\alpha_1}^\delta = u_{\alpha_2}^\delta$ for some $u_{\alpha_1}^\delta \in S(\alpha_1)$ and $u_{\alpha_2}^\delta \in S(\alpha_2)$ yield $(\alpha_2 - \alpha_1)\psi'(u_{\alpha_1}^\delta) = 0$. However, ψ is convex, it is necessary and sufficient that

$$u_{\alpha_1}^\delta = u_{\alpha_2}^\delta \in \arg\min_{u\in X} \psi(u),$$

and hence that $\psi(u_{\alpha_1}^\delta) = \psi(u_{\alpha_2}^\delta) = \min_{u\in X} \psi(u)$. By Lemma 3.3, $\psi(u_\alpha^\delta)$ is monotone, and thus the last equality is possible only if $\alpha_1, \alpha_2 \geq \alpha^*$. □

Remark 3.7. Lemma 3.7 implies that for any distinct $\alpha_1, \alpha_2 \in (0, \alpha^*)$, we have $S(\alpha_1) \cap S(\alpha_2) = \emptyset$.

A similar result holds for nonsmooth models. We illustrate this on the L^2-TV and L^2-ℓ^1 models.

Corollary 3.3. *Let $\phi(u, g^\delta) = \frac{1}{2}\|Ku - g^\delta\|^2$, $\psi(u) = \frac{1}{p}\|u\|^p$ with $p \geq 1$ and $\|K^*g^\delta\|_{X^*} > 0$. Let $\alpha^* = \|K^*g^\delta\|$ if $p = 1$, and $\alpha^* = \infty$, if $p > 1$. Then for $\alpha < \alpha^*$, the functions $\bar{\phi}(\alpha)$ and $\bar{\psi}(\alpha)$ are strictly monotone.*

Proof. By Theorem 3.5, the minimizers $u_{\alpha_1}^\delta$ and $u_{\alpha_2}^\delta$ satisfy

$$\langle K^*(g^\delta - Ku_{\alpha_i}^\delta), u_{\alpha_i}^\delta \rangle = \alpha_i \|u_{\alpha_i}^\delta\|_X^p, \quad i = 1, 2.$$

This implies $S(\alpha_1) \cap S(\alpha_2) = \emptyset$, and the desired assertion follows from Lemma 3.6. □

3.3 Basic estimates

In Section 3.1, we have established basic properties, i.e. existence, stability and consistency, of the minimizer u_α^δ to the functional J_α. Now we derive basic estimates of the minimizer under *source conditions* on the true solution $u^\dagger$. For the ease of exposition, we focus our discussion on the case Y being a Hilbert space, together with a quadratic fidelity, i.e.,

$$\phi(u, g^\delta) = \tfrac{1}{2}\|Ku - g^\delta\|^2.$$

In particular, we are interested in the quality of the approximation u_α^δ, which represents a central topic in regularization theory. We discuss two typical approaches separately, i.e., classical source condition and higher-order source condition.

3.3.1 *Classical source condition*

We begin with the most classical source condition: there exists a representer $w \in Y$ and an element $\xi^\dagger \in \partial\psi(u^\dagger)$ such that

$$\begin{aligned} K^* w + \mu^\dagger &= \xi^\dagger \in \partial\psi(u^\dagger), \\ \langle \mu^\dagger, u - u^\dagger \rangle &\geq 0, \quad \forall u \in \mathcal{C}, \end{aligned} \tag{3.6}$$

where $\mu^\dagger$ is the Lagrangian multiplier associated the constraint $\mathcal{C}$.

From the viewpoint of optimization theory, the source condition (3.6) can be regarded as a necessary optimality condition for the constrained optimization problem:

$$\min_{u \in \mathcal{C}} \psi(u) \quad \text{subject to } Ku = g^\dagger.$$

To see this, we recall the necessary optimality condition for u_α, the minimizer corresponding to $g^\dagger$, which reads

$$\begin{aligned} \alpha\xi_\alpha + K^*(Ku_\alpha - g^\dagger) - \alpha\mu_\alpha = 0, \\ \langle \mu_\alpha, u - u_\alpha \rangle \geq 0, \quad \forall u \in \mathcal{C}, \end{aligned}$$

where $\xi_\alpha \in \partial\psi(u_\alpha)$, and μ_α is the Lagrangian multiplier for the constraint $\mathcal{C}$. Hence,

$$\begin{aligned} \xi_\alpha &= K^* \frac{g^\dagger - Ku_\alpha}{\alpha} + \mu_\alpha, \\ \langle \mu_\alpha, u - u_\alpha \rangle &\geq 0, \quad \forall u \in \mathcal{C}. \end{aligned}$$

Under minor conditions on ψ, we have $u_\alpha \to u^\dagger$, and hence $\xi_\alpha \to \xi^\dagger \in \partial\psi(u^\dagger)$. If, further, $\frac{y^\dagger - Ku_\alpha}{\alpha} \to w$ weakly (weak $*$) and $\mu_\alpha \to \mu^\dagger$ weakly (weak $*$), then we arrive at the source condition (3.6). The weak ($*$) convergence does not hold a priori, due to nonclosedness of the operator K^*.

Remark 3.8. In view of Proposition 3.1, in the absence of constraint $\mathcal{C}$, the source condition $\xi^\dagger \in \mathrm{range}(K^*)$ is indeed stronger than the necessary optimality condition $\xi^\dagger \in \overline{\mathrm{range}(K^*)}$ for the ψ-minimizing solution $u^\dagger$.

The rest of this part is devoted to derive various estimates under the source condition (3.6). The estimates are measured in Bregman distance $d_\xi(\tilde{u}, u)$ between $\tilde{u}$ and u with respect to $\xi \in \partial\psi(u)$ given by

$$d_\xi(\tilde{u}, u) = \psi(\tilde{u}) - \psi(u) - \langle \xi, \tilde{u} - u \rangle.$$

Bregman distance was introduced in 1967 by L. M. Bregman [40], and its role in convergence rate analysis was first shown in the seminal paper [50] (see also [258] for further discussions). By definition, it is always nonnegative. However, the Bregman distance is not symmetric and does not satisfy the triangle inequality, and hence it is not a distance in the usual sense. Further, to handle the constraint $\mathcal{C}$, we introduce a generalized Bregman distance

$$d_{\xi,\mu}(\tilde{u}, u) = \psi(\tilde{u}) - \psi(u) - \langle \xi, \tilde{u} - u \rangle + \langle \mu, \tilde{u} - u \rangle,$$

where μ is a Lagrangian multiplier associated with $\mathcal{C}$, i.e.,

$$\langle \mu, \tilde{u} - u \rangle \geq 0 \quad \forall \tilde{u} \in \mathcal{C}.$$

In the absence of the constraint $\mathcal{C}$, the generalized Bregman distance $d_{\xi,\mu}(\tilde{u}, u)$ coincides with the Bregman distance $d_\xi(\tilde{u}, u)$.

Example 3.4. If $\psi(u) = \|u\|^2$, then $\partial\psi(u) = \{2u\}$, and the Bregman distance $d_\xi(\tilde{u}, u)$ is given by

$$d_\xi(\tilde{u}, u) = \|\tilde{u}\|^2 - \|u\|^2 - \langle 2u, \tilde{u} - u \rangle = \|\tilde{u} - u\|^2.$$

Remark 3.9. In case of constrained Tikhonov regularization, we may have $w = 0$. Further, if the set $\{\mu^\dagger \neq 0\}$ has a positive measure, then the term $\langle \mu^\dagger, u - u^\dagger \rangle$ provides a strictly positive contribution. These are possible beneficial consequences due to the presence of constraint.

Remark 3.10. For the constrained Tikhonov regularization a Hilbert space setting, Neubauer [242, 243] established an $O(\delta^{1/2})$ convergence rate under a projected source condition, and also discussed its finite dimensional discretization. Chavent and Kunisch [58] revisited the constrained Tikhonov regularization from the viewpoint of optimization theory; see also [152] for further refinement along the line, which will be presented in Chapter 4. Recently, Flemming and Hofmann [95] studied the connections between the projected source condition and variational inequalities, an alternative approach to convergence rate analysis; see Section 4.5.

Theorem 3.8. *If the source condition* (3.6) *holds, then*

$$d_{\xi^\dagger,\mu^\dagger}(u_\alpha^\delta, u^\dagger) \leq \tfrac{1}{2}\left(\tfrac{\delta}{\sqrt{\alpha}} + \sqrt{\alpha}\|w\|\right)^2,$$
$$\|Ku_\alpha^\delta - g^\delta\| \leq \delta + 2\alpha\|w\|.$$

Proof. By the minimizing properties of x_α^δ, we have

$$\tfrac{1}{2}\|Ku_\alpha^\delta - g^\delta\|^2 + \alpha\psi(u_\alpha^\delta) \leq \tfrac{1}{2}\|Ku^\dagger - g^\delta\|^2 + \alpha\psi(u^\dagger),$$

which together with the source condition (3.6) gives

$$\tfrac{1}{2}\|Ku_\alpha^\delta - g^\delta\|^2 + \alpha d_{\xi^\dagger,\mu^\dagger}(u_\alpha^\delta, u^\dagger) \leq \tfrac{1}{2}\delta^2 - \alpha\langle w, K(u_\alpha^\delta - u^\dagger)\rangle.$$

By completing square, the identities $Ku^\dagger = g^\dagger$ and $\|g^\delta - g^\dagger\| = \delta$, and the Cauchy-Schwarz inequality, we get

$$\begin{aligned}\tfrac{1}{2}\|Ku_\alpha^\delta - g^\delta + \alpha w\|^2 &+ \alpha d_{\xi^\dagger,\mu^\dagger}(u_\alpha^\delta, u^\dagger) \\ &\leq \tfrac{1}{2}\delta^2 + \alpha\|w\|\delta + \tfrac{1}{2}\alpha^2\|w\|^2 = \tfrac{1}{2}(\delta + \alpha\|w\|)^2.\end{aligned}$$

The assertion follows directly from this and the triangle inequality. □

Corollary 3.4. *Let the source condition* (3.6) *hold. Then for the choice* $\alpha = \delta/\|w\|$, *there holds*

$$d_{\xi^\dagger,\mu^\dagger}(u_\alpha^\delta, u^\dagger) \leq 2\|w\|\delta \quad \textit{and} \quad \|Ku_\alpha^\delta - g^\delta\| \leq 3\delta.$$

Remark 3.11. In case that the functional ψ satisfies extra conditions, the estimate in the Bregman distance also provides an estimate in norm. In particular, for q-convex functionals, i.e.,

$$c\|\tilde{u} - u\|_H^q \leq d_\xi(\tilde{u}, u),$$

the estimate in Corollary 3.4 provides a norm estimate. Examples of q-convex functional including $\psi(u) = \frac{1}{p}\|u\|^p$, where X is a uniformly convex Banach space and embeds continuously into H, and $1 < p \leq q$ [309]. All ℓ^q, $q < q \leq 2$, are uniformly convex and continuously embeds into ℓ^2 [32].

To gain further insights into the regularization process, one can also decompose the error, i.e., the Bregman distance $d_{\xi^\dagger,\mu^\dagger}(u_\alpha^\delta, u^\dagger)$ from u_α^δ to $u^\dagger$, into the *approximation error* and the *propagation error*, which refer to the distance from u_α to $u^\dagger$ and that from u_α^δ to u_α, respectively. Theoretically, the behavior of the approximation error $u_\alpha - u^\dagger$ contains information about how difficult it is to approximate the unknown solution $u^\dagger$ and provides hints on what conditions on $u^\dagger$ may be helpful. The behavior of the propagation error $u_\alpha^\delta - u_\alpha$ shows how the data noise influences the reconstruction accuracy. Next we provide estimates for these different terms, which will be exploited below for deriving a posteriori error estimates for heuristic rules.

Theorem 3.9. *Let the source condition (3.6) hold. Then the approximation error and its discrepancy satisfy*

$$d_{\xi^\dagger,\mu^\dagger}(u_\alpha,u^\dagger) \leq \frac{\|w\|^2}{2}\alpha \quad \text{and} \quad \|Ku_\alpha - g^\dagger\| \leq 2\|w\|\alpha.$$

With the choice $\xi_\alpha = -K^*(Ku_\alpha - g^\dagger)/\alpha + \mu_\alpha$, μ_α being the Lagrangian multiplier for the constraint $\mathcal{C}$, the propagation error and its discrepancy satisfy

$$d_{\xi_\alpha,\mu_\alpha}(u_\alpha^\delta,u_\alpha) \leq \frac{\delta^2}{2\alpha} \quad \text{and} \quad \|K(u_\alpha^\delta - u_\alpha)\| \leq 2\delta.$$

Proof. The minimizing property of u_α yields

$$\tfrac{1}{2}\|Ku_\alpha - g^\dagger\|^2 + \alpha\psi(u_\alpha) \leq \alpha\psi(u^\dagger).$$

Rearranging the terms and noting $\xi^\dagger \in \partial\psi(u^\dagger)$ and $\xi^\dagger = K^*w + \mu^\dagger$ yield

$$\tfrac{1}{2}\|Ku_\alpha - g^\dagger\|^2 + \alpha d_{\xi^\dagger,\mu^\dagger}(u_\alpha,u^\dagger) \leq -\alpha\langle w, Ku_\alpha - g^\dagger\rangle. \tag{3.7}$$

The non-negativity of $d_{\xi^\dagger,\mu^\dagger}(u_\alpha,u^\dagger)$ and the Cauchy-Schwarz inequality imply

$$\tfrac{1}{2}\|Ku_\alpha - g^\dagger\|^2 \leq \alpha\|w\|\|Ku_\alpha - g^\dagger\|.$$

Appealing again to inequality (3.7) and using the Cauchy-Schwarz and Young's inequalities, we arrive at

$$\tfrac{1}{2}\|Ku_\alpha - g^\dagger\|^2 + \alpha d_{\xi^\dagger,\mu^\dagger}(u_\alpha,u^\dagger) \leq \tfrac{\alpha^2\|w\|^2}{2} + \tfrac{1}{2}\|Ku_\alpha - g^\dagger\|^2.$$

This shows the first assertion.

Next we use the minimizing property of u_α^δ to get

$$\tfrac{1}{2}\|Ku_\alpha^\delta - g^\delta\|^2 + \alpha d_{\xi_\alpha}(u_\alpha^\delta,u_\alpha) \leq \tfrac{1}{2}\|Ku_\alpha - g^\delta\|^2 - \alpha\langle\xi_\alpha, u_\alpha^\delta - u_\alpha\rangle.$$

Using the optimality of u_α, we have

$$\begin{aligned}&\xi_\alpha = -K^*(Ku_\alpha - g^\dagger)/\alpha + \mu_\alpha \in \partial\psi(u_\alpha),\\ &\langle\mu_\alpha, u - u_\alpha\rangle \geq 0, \quad \forall u \in \mathcal{C}.\end{aligned}$$

Plugging in the expression for ξ_α and rearranging the formula give

$$\tfrac{1}{2}\|K(u_\alpha^\delta - u_\alpha)\|^2 + \alpha d_{\xi_\alpha,\mu_\alpha}(u_\alpha^\delta,u_\alpha) \leq -\langle g^\dagger - g^\delta, K(u_\alpha^\delta - u_\alpha)\rangle. \tag{3.8}$$

Now the non-negativity of $d_{\xi_\alpha,\mu_\alpha}(u_\alpha^\delta,u_\alpha)$ and the Cauchy-Schwarz inequality yield $\|K(u_\alpha - u_\alpha^\delta)\| \leq 2\delta$. Next by inequality (3.8) and the Cauchy-Schwarz and Young's inequalities, we obtain

$$\begin{aligned}&\tfrac{1}{2}\|K(u_\alpha^\delta - u_\alpha)\|^2 + \alpha d_{\xi_\alpha,\mu_\alpha}(u_\alpha^\delta,u_\alpha)\\ &\qquad\qquad \leq \delta\|K(u_\alpha^\delta - u_\alpha)\| \leq \tfrac{\delta^2}{2} + \tfrac{1}{2}\|K(u_\alpha^\delta - u_\alpha)\|^2,\end{aligned}$$

which concludes the proof. □

As was noted earlier, the Bregman distance $d_{\xi,\mu}(\tilde{u},u)$ in general does not satisfy a triangle inequality. Nonetheless, the total error $d_{\xi^\dagger,\mu^\dagger}(u_\alpha^\delta,u^\dagger)$ behaves like the sum of the approximation error $d_{\xi^\dagger,\mu^\dagger}(u_\alpha,u^\dagger)$ and the propagation error $d_{\xi_\alpha,\mu_\alpha}(u_\alpha^\delta,u_\alpha)$. Indeed there holds for $a,b\geq 0$ that $(a+b)^2/2\leq a^2+b^2\leq (a+b)^2$ and hence, we see that the total error does behave like the sum of $d_{\xi^\dagger,\mu^\dagger}(u_\alpha,u^\dagger)$ and $d_{\xi_\alpha,\mu_\alpha}(u_\alpha^\delta,u_\alpha)$.

The connection between the total error in the Bregman distance and the approximation and propagation errors can be made a bit more precise using the following Pythagoras identity for the Bregman distance.

Lemma 3.8. *Let $\xi^\dagger\in\partial\psi(u^\dagger)$ and $\xi\in\partial\psi(u)$. Then there holds for any $\tilde{u}$*

$$d_{\xi^\dagger}(\tilde{u},u^\dagger)=d_\xi(\tilde{u},u)+d_{\xi^\dagger}(u,u^\dagger)+\langle\xi^\dagger-\xi,u-\tilde{u}\rangle.$$

Proof. Straightforward calculation yields

$$\begin{aligned}d_{\xi^\dagger}(\tilde{u},u^\dagger)&=\psi(\tilde{u})-\psi(u^\dagger)-\langle\xi^\dagger,\tilde{u}-u^\dagger\rangle\\&=d_\xi(\tilde{u},u)+\psi(u)-\psi(u^\dagger)-\langle\xi^\dagger,\tilde{u}-u^\dagger\rangle+\langle\xi,\tilde{u}-u\rangle\\&=d_\xi(\tilde{u},u)+d_{\xi^\dagger}(u,u^\dagger)+\langle\xi^\dagger-\xi,u-\tilde{u}\rangle.\end{aligned}$$

This shows the identity. □

The following useful result is a consequence of Lemma 3.8.

Corollary 3.5. *Let the source condition* (3.6) *be fulfilled. Then for the choice $\xi_\alpha=-K^*(Ku_\alpha-g^\dagger)/\alpha+\mu_\alpha$, μ_α being the Lagrangian multiplier for the constraint $\mathcal{C}$, there holds*

$$|d_{\xi^\dagger,\mu^\dagger}(u_\alpha^\delta,u^\dagger)-(d_{\xi_\alpha,\mu_\alpha}(u_\alpha^\delta,u_\alpha)+d_{\xi^\dagger,\mu^\dagger}(u_\alpha,u^\dagger))|\leq 6\|w\|\delta.$$

Proof. Taking $u=u_\alpha$, $\tilde{u}=u_\alpha^\delta$, $\xi^\dagger=K^*w+\mu^\dagger$ and $\xi=\xi_\alpha=-K^*(Ku_\alpha-g^\dagger)/\alpha+\mu_\alpha$, in the Pythagoras identity (cf. Lemma 3.8), we deduce

$$d_{\xi^\dagger}(u_\alpha^\delta,u^\dagger)=d_{\xi_\alpha}(u_\alpha^\delta,u_\alpha)+d_{\xi^\dagger}(u_\alpha,u^\dagger)+\langle\xi^\dagger-\xi_\alpha,u_\alpha-u_\alpha^\delta\rangle.$$

Now for the last term, we have

$$\begin{aligned}&\langle\xi^\dagger-\xi_\alpha,u_\alpha-u_\alpha^\delta\rangle\\&=\langle K^*w+\mu^\dagger+K^*(Ku_\alpha-g^\dagger)/\alpha-\mu_\alpha,u_\alpha-u_\alpha^\delta\rangle\\&=\langle w+(Ku_\alpha-g^\dagger)/\alpha,K(u_\alpha-u_\alpha^\delta)\rangle+\langle\mu^\dagger-\mu_\alpha,u_\alpha-u_\alpha^\delta\rangle,\end{aligned}$$

and

$$\langle\mu^\dagger-\mu_\alpha,u_\alpha-u_\alpha^\delta\rangle=\langle\mu^\dagger,u_\alpha-u^\dagger\rangle-\langle\mu^\dagger,u_\alpha^\delta-u^\dagger\rangle+\langle\mu_\alpha,u_\alpha^\delta-u_\alpha\rangle.$$

Consequently, we arrive at

$$d_{\xi^\dagger,\mu^\dagger}(u_\alpha^\delta, u^\dagger) = d_{\xi_\alpha,\mu_\alpha}(u_\alpha^\delta, u_\alpha) + d_{\xi^\dagger,u^\dagger}(u_\alpha, u^\dagger) \\ + \langle w + (Ku_\alpha - g^\dagger)/\alpha, K(u_\alpha - u_\alpha^\delta)\rangle.$$

Now the Cauchy-Schwarz inequality and Theorem 3.9 give

$$|\langle w + (Ku_\alpha - g^\dagger)/\alpha, K(u_\alpha - u_\alpha^\delta)\rangle| \\ \leq (\|w\| + \|Ku_\alpha - g^\dagger\|/\alpha)\|K(u_\alpha^\delta - u_\alpha)\| \leq 6\|w\|\delta.$$

This concludes the proof. □

Hence the total error $d_{\xi^\dagger,\mu^\dagger}(u_\alpha^\delta, u^\dagger)$ differs from the sum of approximation and propagation errors only by a term of order δ. In general, the difference can be either positive or negative.

Last, we give an estimate for the generalized Bregman distance between two regularized solutions for the same data g^δ but different regularization parameters. This estimate underlies the quasi-optimality criterion in Section 3.4.

Proposition 3.2. *Let $\xi_\alpha^\delta = -K^*(Ku_\alpha^\delta - g^\delta)/\alpha + \mu_\alpha^\delta$, with μ_α^δ being the Lagrangian multiplier for the constraint $\mathcal{C}$. For $q \in]0,1[$, there holds*

$$d_{\xi_\alpha^\delta,\mu_\alpha^\delta}(u_{q\alpha}^\delta, u_\alpha^\delta) \leq \frac{(1-q)^2\|Ku_\alpha^\delta - g^\delta\|^2}{2\alpha q}.$$

Moreover, if the source condition (3.6) *holds, then*

$$\|K(u_{q\alpha}^\delta - u_\alpha^\delta)\| \leq 2(1-q)(\delta + 2\alpha\|w\|).$$

Proof. The minimizing property of $u_{q\alpha}^\delta$ implies

$$\tfrac{1}{2}\|Ku_{q\alpha}^\delta - g^\delta\|^2 + q\alpha\psi(u_{q\alpha}^\delta) \leq \tfrac{1}{2}\|Ku_\alpha^\delta - g^\delta\|^2 + q\alpha\psi(u_\alpha^\delta).$$

Rearranging the terms gives

$$\tfrac{1}{2}\|Ku_{q\alpha}^\delta - g^\delta\|^2 + q\alpha d_{\xi_\alpha^\delta,\mu_\alpha^\delta}(u_{q\alpha}^\delta, u_\alpha^\delta) \\ \leq \tfrac{1}{2}\|Ku_\alpha^\delta - g^\delta\|^2 + q\langle Ku_\alpha^\delta - g^\delta, K(u_{q\alpha}^\delta - u_\alpha^\delta)\rangle,$$

which leads to

$$q\alpha d_{\xi_\alpha^\delta,\mu_\alpha^\delta}(u_{q\alpha}^\delta, u_\alpha^\delta) \leq -(1-q)\langle Ku_\alpha^\delta - g^\delta, K(u_{q\alpha}^\delta - u_\alpha^\delta)\rangle - \tfrac{1}{2}\|K(u_{q\alpha}^\delta - u_\alpha^\delta)\|^2.$$

Appealing again to the Cauchy-Schwarz and Young's inequalities gives the first assertion. Using the Cauchy-Schwarz inequality in

$$\tfrac{1}{2}\|K(u_{q\alpha}^\delta - u_\alpha^\delta)\|^2 \leq -(1-q)\langle Ku_\alpha^\delta - g^\delta, K(u_{q\alpha}^\delta - u_\alpha^\delta)\rangle,$$

and Theorem 3.8 shows the remaining assertion. □

Last, we note that the solution u_α^δ can also be measured using the symmetric Bregman distance $d^s(\tilde{u}, u)$ defined by

$$\begin{aligned} d^s(\tilde{u}, u) &= d_{\xi,\mu}(\tilde{u}, u) + d_{\tilde{\xi},\tilde{\mu}}(u, \tilde{u}) \\ &= \langle \tilde{\xi} - \xi - \tilde{\mu} + \mu, \tilde{u} - u \rangle, \end{aligned}$$

where $\xi \in \partial\psi(u)$ and $\tilde{\xi} \in \psi(\tilde{u})$, and μ and $\tilde{\mu}$ are the respective Lagrange multiplier for the constraint $\mathcal{C}$. The symmetric Bregman distance is always nonnegative, and it depends on the choice of the subgradients ξ and $\tilde{\xi}$. Then we can state a basic estimate due to [27].

Theorem 3.10. *Let the source condition* (3.6) *hold. Then for the choice* $\xi_\alpha = -K^*(Ku_\alpha - g^\dagger)/\alpha + \mu_\alpha$, μ_α *being the Lagrangian multiplier for the constraint* $\mathcal{C}$, *there holds*

$$\begin{aligned} \tfrac{1}{2}\|Ku_\alpha^\delta - g^\delta\|^2 + \tfrac{1}{2}\|K(u_\alpha^\delta - u^\dagger)\|^2 + \alpha d^s(u_\alpha^\delta, u^\dagger) \\ \leq \tfrac{1}{2}\|Ku^\dagger - g^\delta\|^2 - \alpha\langle w, K(u_\alpha^\delta - u^\dagger)\rangle. \end{aligned}$$

Proof. The optimality condition for u_α^δ implies

$$\xi_\alpha^\delta = K^*(g^\delta - Ku_\alpha^\delta)/\alpha + \mu_\alpha^\delta \in \partial\psi(u_\alpha^\delta),$$

with μ_α^δ being the Lagrange multiplier for the constraint $\mathcal{C}$, i.e.,

$$\langle \mu_\alpha^\delta, u - u_\alpha^\delta \rangle \geq 0 \quad \forall u \in \mathcal{C}.$$

Consequently, we have

$$\begin{aligned} \langle K^*(Ku_\alpha^\delta - g^\delta), u_\alpha^\delta - u^\dagger\rangle + \alpha\langle(\xi_\alpha^\delta - \mu_\alpha^\delta) - (\xi^\dagger - \mu^\dagger), u_\alpha^\delta - u^\dagger\rangle \\ = -\alpha\langle(\xi^\dagger - \mu^\dagger), u_\alpha^\delta - u^\dagger\rangle, \end{aligned}$$

which together with the source condition (3.6) yields

$$\langle Ku_\alpha^\delta - g^\delta, K(u_\alpha^\delta - u^\dagger)\rangle + \alpha d^s(u_\alpha^\delta, u^\dagger) = -\alpha\langle w, K(u_\alpha^\delta - u^\dagger)\rangle.$$

Now from the identity

$$\begin{aligned} \tfrac{1}{2}\|Ku^\dagger - g^\delta\|^2 - \tfrac{1}{2}\|Ku_\alpha^\delta - g^\delta\|^2 - \langle g^\delta - Ku_\alpha^\delta, K(u_\alpha^\delta - u^\dagger)\rangle \\ = \tfrac{1}{2}\|Ku_\alpha^\delta - Ku^\dagger\|^2 \geq 0, \end{aligned}$$

the desired assertion follows immediately. □

Corollary 3.6. *Let the source condition* (3.6) *hold. Then for the choice* $\alpha = \delta/\|w\|$, *there holds*

$$d^s(u_\alpha^\delta, u^\dagger) \leq \delta\|w\| \quad \textit{and} \quad \|Ku_\alpha^\delta - g^\delta\| \leq \sqrt{2}\delta.$$

Proof. It follows from Theorem 3.10 and Young's inequality that

$$d^s(u_\alpha^\delta, u^\dagger) \leq \tfrac{1}{2}\left(\tfrac{\delta^2}{\alpha} + \alpha\|w\|^2\right) \quad \text{and} \quad \|Ku_\alpha^\delta - g^\delta\| \leq \sqrt{\delta^2 + \alpha^2\|w\|^2}.$$

The rest follows directly from these estimates and the choice of α. □

Remark 3.12. Other basic estimates can also be derived in terms of the symmetric Bregman distance $d^s(u, \tilde{u})$. In particular, this implies that the (a posteriori) error estimates in Section 3.4 might also be measured in $d^s(u, \tilde{u})$.

3.3.2 *Higher-order source condition*

One can obtain higher-order convergence rates by imposing stronger source conditions. One such condition reads

$$K^*Kw = \xi^\dagger \in \partial\psi(u^\dagger). \tag{3.9}$$

This condition was first considered in [258] in the context of the Banach space, which generalizes the classical approach and its proof [88]; see also [245, 131, 132]. An alternative approach is via a dual formulation of the Tikhonov functional [116]. We shall describe the results from [258] when there is no convex constraint $\mathcal{C}$.

Theorem 3.11. *Let condition* (3.9) *hold. Then for* $\alpha > 0$, *there holds*

$$d_{\xi^\dagger}(u_\alpha, u^\dagger) \leq d_{\xi^\dagger}(u^\dagger - \alpha w, u^\dagger),$$

$$\|K(u_\alpha - u^\dagger)\| \leq \alpha\|Kw\| + \sqrt{2\alpha d_{\xi^\dagger}(u^\dagger - \alpha w, u^\dagger)}.$$

Proof. It follows from the minimizing property of u_α that

$$\underbrace{\tfrac{1}{2}\left[\|Ku_\alpha - g^\dagger\|^2 - \|Ku - g^\dagger\|^2\right]}_{I} + \alpha\underbrace{\left(\psi(u_\alpha) - \psi(u)\right)}_{II} \leq 0.$$

By expanding the terms, the term I can be rewritten as

$$\begin{aligned} I &= \|Ku_\alpha\|^2 - \|Ku\|^2 - 2\langle K(u_\alpha - u), g^\dagger\rangle \\ &= \|Ku_\alpha\|^2 - \|Ku\|^2 - 2\langle K(u_\alpha - u), Ku^\dagger\rangle. \end{aligned}$$

Appealing to the source condition (3.9), the term II can be rewritten into

$$\begin{aligned} II &= d_{\xi^\dagger}(u_\alpha, u^\dagger) - d_{\xi^\dagger}(u, u^\dagger) + \langle K^*Kw, u_\alpha - u\rangle \\ &= d_{\xi^\dagger}(u_\alpha, u^\dagger) - d_{\xi^\dagger}(u, u^\dagger) + \langle K(u_\alpha - u), Kw\rangle. \end{aligned}$$

Further, by completing the square, we deduce

$$\tfrac{1}{2}I + \alpha\langle K(u_\alpha - u), Kw\rangle = \tfrac{1}{2}\|K(u_\alpha - u^\dagger + \alpha w)\|^2 - \tfrac{1}{2}\|K(u - u^\dagger + \alpha w)\|^2.$$

Combining these identities together yields

$$\begin{aligned} &\tfrac{1}{2}\|K(u_\alpha - u^\dagger + \alpha w)\|^2 + \alpha d_{\xi^\dagger}(u_\alpha, u^\dagger) \\ &\qquad \leq \tfrac{1}{2}\|K(u - u^\dagger + \alpha w)\|^2 + \alpha d_{\xi^\dagger}(u, u^\dagger). \end{aligned}$$

This last inequality is valid for any $u \in X$, in particular $u = u^\dagger - \alpha w$. Hence, we have

$$\tfrac{1}{2}\|K(u_\alpha - u^\dagger + \alpha w)\|^2 + \alpha d_{\xi^\dagger}(u_\alpha, u^\dagger) \leq \alpha d_{\xi^\dagger}(u^\dagger - \alpha w, u^\dagger).$$

The first estimate is an immediate consequence of this inequality. The second estimate follows from the triangle inequality

$$\begin{aligned} \|K(u_\alpha - u^\dagger)\| &\leq \alpha\|Kw\| + \|K(u_\alpha - u^\dagger + \alpha w)\| \\ &\leq \alpha\|Kw\| + \sqrt{2\alpha d_{\xi^\dagger}(u^\dagger - \alpha w, u^\dagger)}. \end{aligned}$$

This completes the proof. □

The following result gives an estimate in case of noisy data.

Theorem 3.12. *Let condition* (3.9) *hold. Then for* $\alpha > 0$, *there holds*

$$d_{\xi^\dagger}(u_\alpha^\delta, u^\dagger) \leq d_{\xi^\dagger}(u^\dagger - \alpha w, u^\dagger) + \frac{\delta^2}{\alpha} + \frac{\delta}{\alpha}\sqrt{\delta^2 + \alpha d_{\xi^\dagger}(u^\dagger - \alpha w, u^\dagger)},$$

$$\|K(u_\alpha^\delta - u^\dagger)\| \leq \alpha\|Kw\| + \delta + \sqrt{\delta^2 + \alpha d_{\xi^\dagger}(u^\dagger - \alpha w, u^\dagger)}.$$

Proof. It follows from the minimizing property of u_α^δ that

$$\underbrace{\tfrac{1}{2}\left[\|Ku_\alpha^\delta - g^\delta\|^2 - \|Ku - g^\delta\|^2\right]}_{I} + \alpha\underbrace{\left(\psi(u_\alpha^\delta) - \psi(u)\right)}_{II} \leq 0.$$

Now by the relation $Ku^\dagger = g^\dagger$, the term I can be rewritten as

$$\begin{aligned} I &= \|Ku_\alpha^\delta\|^2 - \|Ku\|^2 - 2\langle K(u_\alpha^\delta - u), g^\delta\rangle \\ &= \|Ku_\alpha^\delta\|^2 - \|Ku\|^2 - 2\langle K(u_\alpha^\delta - u), Ku^\dagger\rangle - 2\langle K(u_\alpha^\delta - u), g^\delta - g^\dagger\rangle. \end{aligned}$$

Appealing to the source condition (3.9), we deduce

$$II = d_{\xi^\dagger}(u_\alpha^\delta, u^\dagger) - d_{\xi^\dagger}(u, u^\dagger) + \langle K(u_\alpha^\delta - u), Kw\rangle.$$

Now by completing the square, we deduce

$$\begin{aligned} \tfrac{1}{2}I + \alpha\langle K(u_\alpha^\delta - u), Kw\rangle &= \tfrac{1}{2}\|K(u_\alpha^\delta - u^\dagger + \alpha w)\|^2 \\ &\quad - \tfrac{1}{2}\|K(u - u^\dagger + \alpha w)\|^2 - \langle K(u_\alpha^\delta - u), g^\dagger - g^\delta\rangle. \end{aligned}$$

Therefore, one has for any $u \in X$ that

$$T(u_\alpha^\delta) \leq T(u),$$

where the functional $T(u)$ is defined by

$$T(u) = \tfrac{1}{2}\|K(u - u^\dagger + \alpha w)\|^2 + \alpha d_{\xi^\dagger}(u, u^\dagger) - \langle Ku, g^\delta - g^\dagger\rangle.$$

Upon setting $u = u^\dagger - \alpha w$, one obtains

$$\begin{aligned} &\tfrac{1}{2}\|K(u_\alpha^\delta - u^\dagger + \alpha w)\|^2 + \alpha d_{\xi^\dagger}(u_\alpha^\delta, u^\dagger) \\ &\qquad \leq \langle K(u_\alpha^\delta - u^\dagger + \alpha w), g^\delta - g^\dagger\rangle + \alpha d_{\xi^\dagger}(u^\dagger - \alpha w, u^\dagger). \end{aligned}$$

Let $\gamma = \|K(u_\alpha^\delta - u^\dagger - \alpha w)\|^2$. Then by neglecting the first and then the second term in the left-hand side one gets respectively,

$$\alpha d_{\xi^\dagger}(u_\alpha^\delta, u^\dagger) \leq \gamma\delta + \alpha d_{\xi^\dagger}(u^\dagger - \alpha w, u^\dagger)$$

and

$$\gamma^2 \leq 2\gamma\delta + 2\alpha d_{\xi^\dagger}(u^\dagger - \alpha w, u^\dagger),$$

which implies

$$\gamma \leq \delta + \sqrt{\delta^2 + \alpha d_{\xi^\dagger}(u^\dagger - \alpha w, u^\dagger)}.$$

Consequently, the first estimate holds. The second estimate follows from this and the triangle inequality

$$\|K(u_\alpha^\delta - u^\dagger)\| \leq \gamma + \alpha\|Kw\|.$$

This completes the proof of the theorem. □

A direct consequence of Theorems 3.11 and 3.12 is as follows.

Corollary 3.7. *Let condition (3.9) hold. If the functional ψ is twice differentiable in a neighborhood of $u^\dagger$ and there exists an $M > 0$ such that $\langle \psi''(u)v, v\rangle \leq M\|v\|^2$ for any u in the neighborhood and for any $v \in X$, then*

$$
\begin{aligned}
d_{\xi^\dagger}(u_\alpha, u^\dagger) &= O(\alpha^2), \\
d_{\xi^\dagger}(u_\alpha^\delta, u^\dagger) &= O(\delta^{\frac{4}{3}}) \quad \text{if } \alpha \sim \delta^{\frac{2}{3}}.
\end{aligned}
$$

Proof. By Theorems 3.11 and 3.12, it suffices to estimate $d_{\xi^\dagger}(u^\dagger - \alpha w, u^\dagger)$. By Taylor's expansion around $u^\dagger$ and the identity $\xi^\dagger = \psi'(u^\dagger)$, one has for small α

$$
\begin{aligned}
d_{\xi^\dagger}(u^\dagger - \alpha w, u^\dagger) &= \psi(u^\dagger - \alpha w) - \psi(u^\dagger) - \langle \psi'(u^\dagger), -\alpha w\rangle \\
&= \langle \psi''(\zeta_\alpha)(-\alpha w), -\alpha w\rangle \\
&= \alpha^2 \langle \psi''(\zeta_\alpha) w, w\rangle \leq M\|w\|^2\alpha^2,
\end{aligned}
$$

where ζ_α belongs to the segment $[u^\dagger, u^\dagger - \alpha w]$. The desired assertion follows from the choice of α. □

Remark 3.13. In the presence of a convex constraint $\mathcal{C}$, it is unclear whether the estimates in Theorems 3.11 and 3.12 are still valid, since the crucial choice $u = u^\dagger - \alpha w$ does not necessarily belong to $\mathcal{C}$ and thus may be infeasible.

There are several alternative approaches to convergence rates, e.g., approximate source condition [130, 133], variational inequalities [141], and conditional stability [65]. We will discuss variational inequalities and conditional stability in Chapter 4.

Remark 3.14. The a priori parameter choice, i.e., α is a function of the noise level δ, has been customarily used for choosing the regularization parameter. However, it is not always convenient for practical applications, since the optimal choice would require the knowledge of the source representer w, which is generally inaccessible. Therefore, it is of interest to derive a posteriori parameter choice rules, which uses only the given data g^δ and possibly also the noise level δ, for selecting the regularization parameter α. In the following sections, we describe several a posteriori choice rules, e.g., discrepancy principle and balancing principle.

3.4 A posteriori parameter choice rules

A proper choice of the regularization parameter α lies at the very heart of practical applications of the Tikhonov model J_α. However, it is a highly nontrivial issue. In this part, we discuss three a posteriori parameter choice rules, including discrepancy principle, Hanke-Raus rule and quasi-optimality criterion. In comparison with the a priori choices discussed in Section 3.3, they do not require a knowledge of the source representer w, and are readily applicable to practical problems. Throughout, we focus on the Tikhonov model J_α discussed in Section 3.3, i.e., Y is a Hilbert space, and $\phi(u, g^\delta) = \frac{1}{2}\|Ku - g^\delta\|^2$.

3.4.1 *Discrepancy principle*

Amongst existing a posteriori choice rules, the discrepancy principle due to Morozov [238] or its variants remain the most popular choice, provided that an estimate of the noise level δ is available. This is attributed to its solid theoretical underpinnings and easy numerical implementation; see [31, 173, 9, 7] for some studies in the context of nonsmooth regularization. The principle determines the parameter $\alpha = \alpha(\delta)$ such that

$$\|Ku_\alpha^\delta - g^\delta\| = c_m\delta, \tag{3.10}$$

where $c_m \geq 1$ is a fixed constant. A slightly relaxed version takes the form

$$c_{m,1}\delta \leq \|Ku_\alpha^\delta - g^\delta\| \leq c_{m,2}\delta,$$

for some constants $1 \leq c_{m,1} < c_{m,2}$. The rationale of the principle is that one cannot expect the inverse solution u_α^δ to be more accurate than the data g^δ in terms of the residual. Note that the inverse solution u_α^δ is only implicitly defined through the functional J_α, and thus in practice the principle requires solving a highly nonlinear and potentially nonsmooth equation (3.10) of α. In this part, we discuss its theoretical justifications and efficient numerical realizations.

Remark 3.15. The discrepancy principle can be characterized using the value function $F(\alpha)$. To see this, we define a function $\Upsilon(\alpha)$ by

$$\Upsilon(\alpha) = \frac{F(\alpha) - \delta^2/2}{\alpha^{\frac{1}{1-\gamma}}},$$

where the constant $\gamma \in [0, 1)$. A critical point α of Υ satisfies

$$\gamma\alpha\bar{\psi}(\alpha) + \bar{\phi}(\alpha) = \tfrac{\delta^2}{2},$$

which for $\gamma = 0$ reduces to the discrepancy principle (3.10). In general, the rule $\Upsilon(\alpha)$ can be regarded as a damped discrepancy principle.

First we state the existence of a solution $\alpha(\delta)$ to equation (3.10).

Theorem 3.13. *If the function* $\|Ku_\alpha^\delta - g^\delta\|$ *is continuous with respect to* α, *and* $\lim_{\alpha\to+\infty} \|Ku_\alpha^\delta - g^\delta\| > c_m\delta$ *and* $\lim_{\alpha\to 0^+} \|Ku_\alpha^\delta - g^\delta\| < c_m\delta$, *then there exists at least one positive solution* $\alpha(\delta)$ *to equation* (3.10).

Proof. By Lemma 3.3, the function $\|Ku_\alpha^\delta - g^\delta\|$ is monotonically increasing in α. Now the desired assertion follows directly from the continuity assumption. □

Remark 3.16.

(i) The uniqueness of the minimizer to the functional J_α guarantees the continuity of the functions $\bar{\phi}(\alpha)$ and $\bar{\psi}(\alpha)$ in α, cf. Theorem 3.2. Uniqueness in general is necessary for the continuity of the function $\bar{\phi}(\alpha)$; see Example 3.3 on the L^1-TV formulation.

(ii) Under the continuity assumption, the solutions to (3.10) can form a closed interval. Unique solvability can be deduced under conditions in Theorem 3.5, from which the strict monotonicity of the function $\bar{\phi}(\alpha)$ in α follows directly.

The next theorem shows the consistency of the discrepancy principle, i.e., the convergence of the solution $u_{\alpha(\delta)}^\delta$ to a ψ-minimizing solution as the noise level δ tends to zero.

Theorem 3.14. *Let Assumption 3.1 hold,* $\{g^{\delta_n}\}$ *be a sequence of noisy data with* $\delta_n = \|g^\dagger - g^{\delta_n}\| \to 0$ *as* $n \to \infty$, *and* $\{\alpha_n \equiv \alpha(\delta_n)\}$ *be determined by* (3.10). *Then the sequence* $\{u_{\alpha_n}\}$ *has a subsequence converging weakly to a* ψ*-minimizing solution, and if the* ψ*-minimizing solution* $u^\dagger$ *is unique, the whole sequence converges weakly. Further, if the functional* ψ *verifies the* H*-property, then the convergence is strong.*

Proof. Let $u^\dagger$ be a ψ-minimizing solution, and $\alpha_n = \alpha(\delta_n)$. The minimizing property of the inverse solution $u_{\alpha_n}^{\delta_n}$ implies

$$\begin{aligned}\tfrac{1}{2}\|Ku_{\alpha_n}^{\delta_n} - g^{\delta_n}\|^2 + \alpha_n\psi(u_{\alpha_n}^{\delta_n}) &\le \tfrac{1}{2}\|Ku^\dagger - g^{\delta_n}\|^2 + \alpha_n\psi(u^\dagger)\\ &\le \tfrac{1}{2}\delta_n^2 + \alpha_n\psi(u^\dagger),\end{aligned}$$

which together with (3.10) implies $\psi(u_{\alpha_n}^{\delta_n}) \le \psi(u^\dagger)$, i.e., the sequence $\{\psi(u_{\alpha_n}^{\delta_n})\}$ is uniformly bounded. Further,

$$\|Ku_{\alpha_n}^{\delta_n} - g^\dagger\| \le \|Ku_{\alpha_n}^{\delta_n} - g^{\delta_n}\| + \|g^{\delta_n} - g^\dagger\| \le (c_m + 1)\delta_n,$$

i.e., the sequence $\{\|Ku_{\alpha_n}^{\delta_n} - g^\dagger\|\}$ is also uniformly bounded. Now by Assumption 3.1, there exists a subsequence of $\{u_{\alpha_n}^{\delta_n}\}$, also denoted as $\{u_{\alpha_n}^{\delta_n}\}$, and some $u^* \in \mathcal{C}$, such that

$$u_{\alpha_n}^{\delta_n} \to u^* \text{ weakly in } X.$$

Now the weak lower-semicontinuity of ψ implies

$$\psi(u^*) \leq \liminf_{n\to\infty} \psi(u_{\alpha_n}^{\delta_n}) \leq \limsup_{n\to\infty} \psi(u_{\alpha_n}^{\delta_n}) \leq \psi(u^\dagger). \tag{3.11}$$

Moreover, there holds,

$$\begin{aligned}\|Ku^* - g^\dagger\| &= \liminf_{n\to\infty} \|Ku_{\alpha_n}^{\delta_n} - g^\dagger\| \\ &\leq \liminf_{n\to\infty}(\|Ku_{\alpha_n}^{\delta_n} - g^{\delta_n}\| + \|g^{\delta_n} - g^\dagger\|) \\ &\leq \liminf_{n\to\infty}(c_m + 1)\delta = 0,\end{aligned}$$

i.e., $\|Ku^* - g^\dagger\| = 0$, and thus $Ku^* = g^\dagger$. Therefore, u^* is a ψ-minimizing solution and $\psi(u^*) = \psi(u^\dagger)$. If the ψ-minimizing solution $u^\dagger$ is unique, then $u^* = u^\dagger$ and a standard subsequence argument yields that the whole sequence converges weakly to $u^\dagger$.

For every weakly convergent subsequence $\{u_{\alpha_n}^{\delta_n}\}$ with a limit $u^* \in \mathcal{C}$, we deduce from $\psi(u^*) = \psi(u^\dagger)$ and equation (3.11) that

$$\psi(u^\dagger) \leq \psi(u^*) \leq \liminf_{n\to\infty} \psi(u_{\alpha_n}^{\delta_n}) \leq \limsup_{n\to\infty} \psi(u_{\alpha_n}^{\delta_n}) \leq \psi(u^\dagger),$$

from which it follows that $\lim_{n\to\infty} \psi(u_{\alpha_n}^{\delta_n}) = \psi(u^*) = \psi(u^\dagger)$. This together with the H-property of ψ concludes the desired strong convergence. □

The next result shows the convergence rate under the classical source condition (3.6).

Theorem 3.15. *Let the source condition* (3.6) *hold, and α be determined by the discrepancy principle* (3.10). *Then there holds*

$$d_{\xi^\dagger,\mu^\dagger}(u_\alpha^\delta, u^\dagger) \leq (1 + c_m)\|w\|\delta.$$

Proof. The proof of Theorem 3.14 indicates that $\psi(u_\alpha^\delta) \leq \psi(u^\dagger)$ and the triangle inequality implies that

$$\|K(u_\alpha^\delta - u^\dagger)\| = \|Ku_\alpha^\delta - g^\dagger\| \leq (1 + c_m)\delta.$$

Now the source condition gives

$$\begin{aligned}d_{\xi^\dagger,\mu^\dagger}(u_\alpha^\delta, u^\dagger) &= \psi(u_\alpha^\delta) - \psi(u^\dagger) - \langle \xi^\dagger, u_\alpha^\delta - u^\dagger\rangle + \langle \mu^\dagger, u_\alpha^\delta - u^\dagger\rangle \\ &\leq \langle K^*w, u^\dagger - u_\alpha^\delta\rangle \leq \|w\|\|K(u^\dagger - u_\alpha^\delta)\| \leq (1 + c_m)\|w\|\delta.\end{aligned}$$

This concludes the proof of the theorem. □

Remark 3.17. It is known that in a Hilbert space setting, the best possible convergence rate for the discrepancy principle is $O(\delta^{1/2})$, and higher-order source conditions do not lead to a better rate. However, there are variants for which optimal convergence rates hold [269, 284]. It is unclear whether the same observation holds in a Banach space setting.

Remark 3.18. In practice, the estimate of the noise level δ can be imprecise. Hence it is important to know the sensitivity of the Morozov's choice $(u_\alpha^\delta, \alpha)$ with respect to δ. We illustrate this with the standard quadratic regularization, i.e.,

$$J_\alpha(u) = \tfrac{1}{2}\|Ku - g^\delta\|^2 + \tfrac{\alpha}{2}\|u\|^2.$$

The optimality condition together with the discrepancy principle reads

$$\begin{aligned} K^*Ku_\alpha^\delta + \alpha u_\alpha^\delta &= K^*g^\delta, \\ \|Ku_\alpha^\delta - g^\delta\|^2 &= \delta^2. \end{aligned}$$

Upon differentiating the system with respect to δ, we arrive at

$$\begin{aligned} K^*K\dot{u}_\alpha^\delta + \alpha\dot{u}_\alpha^\delta + \dot{\alpha}u_\alpha^\delta &= 0, \\ \langle K\dot{u}_\alpha^\delta, Ku_\alpha^\delta - g^\delta\rangle &= \delta, \end{aligned} \tag{3.12}$$

where

$$\dot{u}_\alpha^\delta = \frac{du_\alpha^\delta}{d\delta} \quad \text{and} \quad \dot{\alpha} = \frac{d\alpha}{d\delta}.$$

By taking inner product between the optimality condition and $\dot{u}_\alpha^\delta$, and comparing it with (3.12), we deduce

$$\alpha\langle\dot{u}_\alpha^\delta, u_\alpha^\delta\rangle = -\delta.$$

Equivalently, with the notation $\psi(u) = \frac{1}{2}\|u\|^2$, we have

$$\frac{d\bar{\psi}(\alpha)}{d\delta} = -\frac{\delta}{\alpha}. \tag{3.13}$$

Similarly, there holds

$$\frac{d\alpha}{d\delta} = \frac{1}{2\bar{\psi}(\alpha)}\left(\delta - \frac{1}{2}\frac{d}{d\delta}\|Ku_\alpha^\delta\|^2\right).$$

It follows from the monotonicity (increasing) of the function $\bar{\phi}(\alpha)$ in α that $\frac{d\alpha}{d\delta} \geq 0$, cf. Lemma 3.3. Hence, the above relation implies $\frac{d}{d\delta}\|Ku_\alpha^\delta\|^2 \leq 2\delta$. In practice, one observes that the term $\|Ku_\alpha^\delta\|$ is relatively insensitive to α due to the smoothing property of K, and thus we might deduce that

$$\frac{d\alpha}{d\delta} \approx \frac{\delta}{2\bar{\psi}(\alpha)}.$$

We note that the relation (3.13) can be derived in a very framework. By the value function calculus, cf. Theorem 3.7, we have $\frac{d\bar{\psi}}{d\bar{\phi}} = -\frac{1}{\alpha}$, and thus

$$\frac{d\bar{\psi}}{d\delta} = -\frac{1}{\alpha}\frac{d\bar{\phi}}{d\delta} = -\frac{\delta}{\alpha}.$$

Remark 3.19. Contemporaneously with Morozov, Arcangeli [10] published a discrepancy-like method. He proposed to choose the regularization parameter α by the condition

$$\|Ku_\alpha^\delta - g^\delta\| = \delta/\sqrt{\alpha}.$$

As was discussed earlier, provided that the functional J_α has a unique minimizer for every α, for fixed $\delta > 0$, the function

$$\Upsilon_{\mathrm{a}}(\alpha) = \sqrt{\alpha}\|Ku_\alpha^\delta - g^\delta\|$$

is continuous, increasing and

$$\lim_{\alpha\to 0} \Upsilon_{\mathrm{a}}(\alpha) = 0 \quad \text{and} \quad \lim_{\alpha\to\infty} \Upsilon_{\mathrm{a}}(\alpha) = \infty.$$

Therefore, there exists a value $\alpha = \alpha(\delta)$ satisfying the rule Υ_{a}. Next we claim that if the functional J_α with the exact data $g^\dagger$ satisfies that for every $\alpha > 0$, there holds $u_\alpha \neq u^\dagger$, then there holds

$$\lim_{\delta\to 0} \alpha(\delta) = 0 \quad \text{and} \quad \lim_{\delta\to 0} \frac{\delta^2}{\alpha(\delta)} = 0,$$

from which it follows directly that Arcangeli's rule is consistent, cf. Theorem 3.4. To show the first identity, we assume the contrary. Let $\{\delta_n\}$ be a sequence with $\delta_n \to 0$ and $\alpha(\delta_n) \to \alpha_0 > 0$. Then by the definition of the rule (possibly only on a subsequence)

$$\begin{aligned} 0 &= \lim_{n\to\infty} \sqrt{\alpha(\delta_n)}\|Ku_{\alpha(\delta_n)}^{\delta_n} - g^{\delta_n}\| \\ &= \sqrt{\alpha_0}\|Ku_{\alpha_0} - g^\dagger\|. \end{aligned}$$

Therefore, $\|Ku_{\alpha_0} - g^\dagger\| = 0$, which contradicts the assumption $u_{\alpha_0} \neq u^\dagger$. Meanwhile, it follows from the definition of the rule that

$$\begin{aligned} \tfrac{\delta^2}{\alpha(\delta)} &= \|Ku_{\alpha(\delta)}^\delta - g^\delta\|^2 \leq 2J_\alpha(u_{\alpha(\delta)}^\delta) \\ &\leq 2(\tfrac{1}{2}\|Ku^\dagger - g^\delta\| + \alpha(\delta)\psi(u^\dagger)) \\ &\leq \delta^2 + 2\alpha(\delta)\psi(u^\dagger). \end{aligned}$$

This together with the first identity shows the second identity $\delta_n^2/\alpha(\delta_n) \to 0$ as $n \to \infty$. We refer interested readers to [193] for related discussions on convergence rates.

Now we turn to efficient numerical algorithms, i.e., the model function approach and a quasi-Newton method, for solving equation (3.10) with $c_m = 1$.

The model function approach was first proposed by Ito and Kunisch in 1992 [160], and then improved in [206, 308, 173]. The basic idea is to express the discrepancy $\bar{\phi}(\alpha)$ in terms of the value function $F(\alpha)$, and then to approximate $F(\alpha)$ with rational functions, a.k.a. Padé approximations, which are called model functions.

By Lemma 3.5, if $\bar{\psi}(\alpha)$ is continuous in α, which we shall assume hereafter, equation (3.10) can be rewritten as

$$F(\alpha) - \alpha F'(\alpha) = \tfrac{\delta^2}{2}. \tag{3.14}$$

The value function $F(\alpha)$ is highly nonlinear. In the model function approach, we approximate it by the following model function $m_k(\alpha)$ at the k-th iteration,

$$m_k(\alpha) = b + \frac{c_k}{t_k + \alpha},$$

where b, c_k and t_k are constants to be determined. The constant b is determined by asymptotics, i.e., $m_k(0^+)$ or $m_k(+\infty)$. The latter choice merits better theoretical underpinnings and numerical performance [308], and thus we focus on this variant, i.e., fixing b at

$$b = \lim_{\alpha \to \infty} \bar{\phi}(\alpha).$$

By enforcing the Hermite interpolation conditions at α_k, i.e.,

$$m_k(\alpha_k) = F(\alpha_k) \quad \text{and} \quad m_k'(\alpha_k) = F'(\alpha_k),$$

we derive that

$$c_k = -\frac{(F(\alpha_k) - b)^2}{\bar{\psi}(\alpha_k)} \quad \text{and} \quad t_k = \frac{b - \bar{\phi}(\alpha_k)}{\bar{\psi}(\alpha_k)} - 2\alpha_k.$$

The sign of the parameter t_k is in general indefinite, and it is positive if and only if

$$b - \bar{\phi}(\alpha_k) - 2\alpha_k \bar{\psi}(\alpha_k) > 0. \tag{3.15}$$

Inequality (3.15) holds if the regularization parameter α_k is sufficiently small. If $t_k > 0$ holds, then the model function $m_k(\alpha)$ preserves the asymptotic behavior, monotonicity and concavity of $F(\alpha)$. With the approximation m_k in place of $F(\alpha)$ in (3.14), we arrive at an approximate discrepancy equation

$$m_k(\alpha) - \alpha m_k'(\alpha) = \tfrac{\delta^2}{2}. \tag{3.16}$$

Next we introduce

$$G(\alpha) = F(\alpha) - \alpha F'(\alpha) \quad \text{and} \quad G_k(\alpha) = m_k(\alpha) - \alpha m_k'(\alpha).$$

Observe that

$$G_k'(\alpha) = -\alpha m_k''(\alpha) = \frac{-2c_k\alpha}{(t_k+\alpha)^3} > 0, \tag{3.17}$$

i.e., the approximation $G_k(\alpha)$ preserves the monotonicity of $G(\alpha)$.

The next result shows the local solvability of (3.16).

Lemma 3.9. *Let t_k be positive, and F be continuously differentiable. If α_k is close to a solution of the discrepancy equation* (3.10)*, then there exists a unique positive solution to* (3.16).

Proof. The uniqueness follows from the strict monotonicity of $G_k'(\alpha)$, cf. (3.17). Equation (3.16) at the kth step reads

$$b + \frac{c_k}{t_k+\alpha} + \alpha\frac{c_k}{(t_k+\alpha)^2} = \frac{\delta^2}{2}$$

We discuss separately two cases, i.e. $\bar{\phi}(\alpha_k) > \frac{\delta^2}{2}$ and $\bar{\phi}(\alpha_k) < \frac{\delta^2}{2}$. In the former case, by the monotonicity, we need only to consider

$$\lim_{\alpha\to 0^+}\left(b + \frac{c_k}{t_k+\alpha} + \alpha\frac{c_k}{(t_k+\alpha)^2}\right) = b + \frac{c_k}{t_k}.$$

Therefore, a positive solution to (3.16) exists if and only if $b + \frac{c_k}{t_k} < \frac{\delta^2}{2}$, which by substituting the formulas for c_k and t_k is equivalent to

$$(b - \tfrac{\delta^2}{2})\left(\frac{b-\bar{\phi}(\alpha_k)}{F'(\alpha_k)} - 2\alpha_k\right) < \frac{(b-F(\alpha_k))^2}{F'(\alpha_k)}.$$

Now the identity $F' = \psi$ from Lemma 3.5 and rearranging the terms gives

$$(\bar{\phi}(\alpha_k) - \tfrac{\delta^2}{2})\left(b - \bar{\phi}(\alpha_k) - 2\alpha_k\bar{\psi}(\alpha_k)\right) < \alpha_k^2\bar{\psi}(\alpha_k)^2,$$

which together with the assumption that $\bar{\phi}(\alpha_k) \approx \frac{\delta^2}{2}$ yields the assertion. In the other case, we need only to consider

$$\lim_{\alpha\to+\infty}\left(b + \frac{c_k}{t_k+\alpha} + \alpha\frac{c_k}{(t_k+\alpha)^2}\right) = b > \tfrac{\delta^2}{2},$$

for which a positive solution is guaranteed. □

Remark 3.20. Lemma 3.9 remains true when the value of the constant b is set to any other constant larger than b, especially for large b, e.g., $b = 2\lim_{\alpha\to\infty}\bar{\phi}(\alpha)$, which automatically guarantees the positivity of t_k, by observing the monotonicity and asymptotes of $F(\alpha)$.

Lemma 3.9 indicates that equation (3.16) is only locally solvable, and thus the resulting algorithm is at best locally convergent. To circumvent the issue, we employ the idea of relaxation, i.e., replacing $G_k(\alpha)$ by

$$\widehat{G}_k(\alpha) = G_k(\alpha) + \widetilde{\alpha}_k(G_k(\alpha) - G_k(\alpha_k)).$$

Assume that $G_k(\alpha_k) > \delta^2/2$. The sign of the second term is determined by the sign of $\widetilde{\alpha}_k$ as $G_k(\alpha) \leq G_k(\alpha_k)$ for $\alpha \in [0, \alpha_k]$. The constant $\widetilde{\alpha}_k$ is chosen such that the equation

$$\widehat{G}_k(\alpha) = \tfrac{\delta^2}{2} \tag{3.18}$$

always has a unique solution. This can be achieved by prescribing $\widehat{G}_k(0) < \delta^2/2$, or specifically $\widehat{G}_k(0) = \gamma\delta^2$, $\forall\gamma \in [0, 1/2)$, which suggests

$$\widetilde{\alpha}_k = \tfrac{G_k(0)-\gamma\delta^2}{G_k(\alpha_k)-G_k(0)}.$$

From the monotonicity of $G_k(\alpha)$ we know $1+\widetilde{\alpha}_k > 0$ as long as $G(\alpha_k) > \delta^2$, which implies that $\widehat{G}'_k(\alpha) > 0$, i.e., it preserves the monotonicity. The complete algorithm is described in Algorithm 3.1. The stopping criterion at step 5 can be based on the relative change of the iterates $\{\alpha_k\}$.

Algorithm 3.1 Model function approach

1: Choose α_0 and K, and set $k = 0$.
2: **for** $k = 1, \ldots, K$ **do**
3: Solve (3.2) for $u_{\alpha_k}^\delta$, and update t_k and c_k;
4: Set the kth model function $m_k(\alpha)$, and solve (3.18) for α_{k+1}.
5: Check the stopping criterion.
6: **end for**
7: **output** approximations α_K and $u_{\alpha_K}^\delta$.

Lemma 3.10. *Let the function $\bar{\psi}(\alpha)$ be continuous, the condition* (3.15) *hold for all k, and the sequence α_k be convergent to $\bar{\alpha}$. Then*

$$\lim_{k\to\infty} G_{k+1}(\alpha_{k+1}) = \lim_{k\to\infty} G_k(\alpha_{k+1}).$$

Proof. By the definition of the model function $m_k(\alpha)$,

$$\begin{aligned} G_k(\alpha) &= b + \frac{c_k}{t_k+\alpha} + \alpha\frac{c_k}{(t_k+\alpha)^2} \\ &= b - (F(\alpha_k) - b)^2 \frac{b - \bar{\phi}(\alpha_k) - 2(\alpha_k - \alpha)\bar{\psi}(\alpha_k)}{(b - \bar{\phi}(\alpha_k) - (2\alpha_k - \alpha)\bar{\psi}(\alpha_k))^2}. \end{aligned}$$

Hence we have

$$
\begin{aligned}
&G_{k+1}(\alpha_{k+1}) - G_k(\alpha_{k+1}) \\
&= \bar{\phi}(\alpha_{k+1}) - b + (F(\alpha_k) - b)^2 \frac{b - \bar{\phi}(\alpha_k) - 2(\alpha_k - \alpha_{k+1})\bar{\psi}(\alpha_k)}{(b - \bar{\phi}(\alpha_k) - (2\alpha_k - \alpha_{k+1})\bar{\psi}(\alpha_k))^2}.
\end{aligned}
$$

Now by the continuity assumption on $\bar{\psi}(\alpha)$, $\bar{\phi}(\alpha)$ is also continuous, cf. Lemma 3.6. Hence, each term in $G_{k+1}(\alpha_{k+1}) - G_k(\alpha_{k+1})$ is continuous and has a limit. The assertion follows directly from the limit law. □

The next theorem shows the monotone convergence of Algorithm 3.1 in the case of 'large' initial guesses α_0, i.e., $G(\alpha_0) > \delta^2/2$.

Theorem 3.16. *Let $\bar{\phi}(\alpha)$ and $\bar{\psi}(\alpha)$ be continuous in α, and the solution $\alpha^\dagger$ to (3.10) be unique, and α_0 satisfy $G(\alpha_0) > \delta^2/2$. Then the sequence $\{\alpha_k\}_k$ generated by Algorithm 3.1 is well-defined. Moreover, the sequence is either finite, i.e., it terminates at some α_k satisfying $G(\alpha_k) \leq \delta^2/2$, or it is infinite and converges to $\alpha^\dagger$ strictly monotonically decreasingly.*

Proof. It suffices to show that if $\widehat{G}_k(\alpha_k) \leq \delta^2/2$ is never reached then α_k converges to $\alpha^\dagger$. So we may assume that $\widehat{G}_k(\alpha_k) > \delta^2/2$ for all k. This and the monotonicity of $\widehat{G}_k(\alpha)$ yield $\alpha_{k+1} < \alpha_k$. Secondly, the relation

$$
\widehat{G}_k(\alpha_k) = G_k(\alpha_k) = G(\alpha_k)
$$

together with the assumption implies $\alpha_k > \alpha^\dagger$ for all k. Therefore, the sequence $\{\alpha_k\}_k$ converges to some $\bar{\alpha} \geq \alpha^\dagger$ by the monotone convergence theorem. We next show that $\bar{\alpha} = \alpha^\dagger$. Taking the limit in α_k and noting the continuity assumption on $\bar{\psi}(\alpha)$ and $\bar{\phi}(\alpha)$, we observe that the sequences $\{c_k\}_k$ and $\{t_k\}_k$ both converge. Next, the relation $G(\alpha_{k+1}) = G_{k+1}(\alpha_{k+1})$ and the definition of $G_k(\alpha)$ yield

$$
G(\bar{\alpha}) = \lim_{k\to\infty} G(\alpha_{k+1}) = \lim_{k\to\infty} G_{k+1}(\alpha_{k+1}) = \lim_{k\to\infty} G_k(\alpha_{k+1}),
$$

where the last line follows from Lemma 3.10. Now in the equation $G_k(\alpha_{k+1}) + \widetilde{\alpha}_k(G_k(\alpha_{k+1}) - G_k(\alpha_k)) = \delta^2/2$, by the definitions of $G_k(\alpha)$ and $\widetilde{\alpha}_k$ and the convergence of α_k, we see

$$
\lim_{k\to\infty} \{G_k(\alpha_{k+1}) - G_k(\alpha_k)\} = 0,
$$

and that $\widetilde{\alpha}_k$ is convergent. This gives $G(\bar{\alpha}) = \delta^2/2$, and by the uniqueness assumption, we have $\bar{\alpha} = \alpha^\dagger$. □

Numerically, Algorithm 3.1 converges robustly and reasonably fast during the initial stage of the iteration process. In the implementation, we need an initial guess for the parameter α. The following result gives an upper bound [299] in the case of norm penalties.

Theorem 3.17. *Suppose that the functional*

$$J_\alpha(u) = \tfrac{1}{2}\|Ku - g^\delta\|^2 + \tfrac{\alpha}{p}\|u\|^p,$$

with $p \geq 1$, reaches its minimum at $u_\alpha^\delta \neq 0$ on X. Then

$$\alpha \leq \frac{\|K\|^p \|Ku_\alpha^\delta - g^\delta\|}{(\|g^\delta\| - \|Ku_\alpha^\delta - g^\delta\|)^{p-1}}.$$

Proof. If J_α reaches its minimum at u_α^δ on X, then it achieves its minimum on elements of the form $(1-s)u_\alpha^\delta$, $s \in \mathbb{R}$, at $s = 0$, i.e.,

$$J_\alpha((1-s)u_\alpha^\delta) - J_\alpha(u_\alpha^\delta) \geq 0 \quad \forall s \in \mathbb{R}.$$

Next by the triangle inequality $\|K(1-s)u_\alpha^\delta - g^\delta\| \leq \|Ku_\alpha^\delta - g^\delta\| + s\|Ku_\alpha^\delta\|$, we bound the first term $J_\alpha((1-s)u_\alpha^\delta)$ from above with $s \in [0,1]$ by

$$\begin{aligned} J_\alpha((1-s)u_\alpha^\delta) &= \tfrac{1}{2}\|Ku_\alpha^\delta - g^\delta - sKu_\alpha^\delta\|^2 + \tfrac{\alpha}{p}\|(1-s)u_\alpha^\delta\|^p \\ &\leq \tfrac{1}{2}\|Ku_\alpha^\delta - g^\delta\|^2 + s\|Ku_\alpha^\delta - g^\delta\|\|Ku_\alpha^\delta\| \\ &\quad + \tfrac{s^2}{2}\|Ku_\alpha^\delta\|^2 + \tfrac{\alpha}{p}(1-s)^p\|u_\alpha^\delta\|^p. \end{aligned}$$

Consequently,

$$\begin{aligned} 0 &\leq J_\alpha((1-s)u_\alpha^\delta) - J_\alpha(u_\alpha^\delta) \\ &\leq s\|Ku_\alpha^\delta - g^\delta\|\|Ku_\alpha^\delta\| + \tfrac{s^2}{2}\|Ku_\alpha^\delta\|^2 + \tfrac{\alpha}{p}((1-s)^p - 1)\|u_\alpha^\delta\|^p, \end{aligned}$$

which upon dividing both sides by s and letting $s \to 0^+$ yields

$$\alpha \leq \frac{\|Ku_\alpha^\delta - g^\delta\|\|Ku_\alpha^\delta\|}{\|u_\alpha^\delta\|^p}.$$

Further, with the inequality $\|Ku_\alpha^\delta\| \leq \|K\|\|u_\alpha^\delta\|$, we deduce

$$\alpha \leq \frac{\|Ku_\alpha^\delta - g^\delta\|\|K\|}{\|u_\alpha^\delta\|^{p-1}}.$$

Next we estimate $\|u_\alpha^\delta\|$ from below by

$$\|u_\alpha^\delta\| \geq \|Ku_\alpha^\delta\|/\|K\| \geq (\|g^\delta\| - \|Ku_\alpha^\delta - g^\delta\|)/\|K\|.$$

Substituting this lower bound for $\|u_\alpha^\delta\|$ yields the desired assertion. □

Remark 3.21. The space Y in Theorem 3.17 can be a general Banach space.

Corollary 3.8. *If the parameter α is selected on the basis of the discrepancy principle $\|Ku_\alpha^\delta - g^\delta\| = c_m\delta$, then*

$$\alpha \leq \frac{c_m\|K\|^p\delta}{(\|g^\delta\| - c_m\delta)^{p-1}}.$$

Remark 3.22. Theorem 3.17 indicates that the value determined by the principle is at most of order δ, i.e., $\mathcal{O}(\delta)$, which has been a quite popular a priori choice for inverse problems in Banach spaces. It shows also clearly the suboptimality of the principle for Tikhonov regularization in Hilbert spaces: It cannot achieve optimal convergence rates for the following source condition $u^\dagger = (K^*K)^\nu w$, when the source parameter ν lies within the range $]1/2, 1]$, since then the optimal convergence rate can only be realized for a choice larger than $\mathcal{O}(\delta)$ [88]. The parameter $\alpha(\delta)$ can be quite close to zero, thus a lower bound is often not available without further assumptions.

The local convergence of Algorithm 3.1 seems relatively slow, which necessitates developing faster algorithms. The secant method does not require derivative evaluation and merits local superlinear convergence for smooth equations. Hence it can be used for accelerating the algorithm, which lends itself to a hybrid algorithm, combining their virtues, i.e., the robustness of the model function approach and the fast local convergence of the secant method.

Consider the case of the L^2-L^2 model in a Hilbert spaces, i.e., $\phi(u, g^\delta) = \frac{1}{2}\|Ku - g^\delta\|^2$ and $\psi(u) = \frac{1}{2}\|Lu\|^2$ with L being a differential operator or identity operator. Then it is well known that the function $\bar{\phi}(\alpha)$ is convex in the α^{-1} coordinate; see the proposition below. This suggests implementing a quasi-Newton method in the α^{-1} coordinate since Newton type methods for convex functions are ensured to converge, and moreover, the convergence is monotone. For general nonsmooth models, the convexity may not hold, but numerical experiments with the secant method are very encouraging.

Proposition 3.3. *For $\phi(u, g^\delta) = \frac{1}{2}\|Ku - g^\delta\|^2$ and $\psi(u) = \frac{1}{2}\|Lu\|^2$, the function $\phi\left(u_{\alpha^{-1}}^\delta, g^\delta\right)$ is convex in α.*

Proof. It suffices to show that $\frac{d^2}{d\alpha^2}\bar{\phi}(\alpha^{-1}) \geq 0$. First, $u = u_{\alpha^{-1}}^\delta$ solves

$$(K^*K + \alpha^{-1}L^*L)u = K^*g^\delta.$$

Differentiating the equation with respect to α implies that $u' = \frac{du}{d\alpha}$ and $u'' = \frac{d^2u}{d\alpha^2}$ respectively satisfy

$$(K^*K + \alpha^{-1}L^*L)u' = \alpha^{-2}L^*Lu,$$
$$(K^*K + \alpha^{-1}L^*L)u'' = 2\alpha^{-2}L^*Lu' - 2\alpha^{-3}L^*Lu.$$

Upon denoting $Q_\alpha = (K^*K+\alpha^{-1}L^*L)^{-1}$ and the identity $I-\alpha^{-1}Q_\alpha L^*L = Q_\alpha K^*K$, we have

$$\begin{aligned} u' &= \alpha^{-2}Q_\alpha L^*Lu, \\ u'' &= 2\alpha^{-3}(\alpha^{-1}Q_\alpha L^*L - I)Q_\alpha L^*Lu \\ &= -2\alpha^{-3}Q_\alpha K^*KQ_\alpha L^*Lu. \end{aligned}$$

Now with the relation $K^*(Ku - g^\delta) = -\alpha^{-1}L^*Lu$, we arrive at

$$\begin{aligned} \frac{d^2}{d\alpha^2}\bar{\phi}(\alpha^{-1}) &= \langle K^*(Ku - g^\delta), u''\rangle + \langle Ku', Ku'\rangle \\ &= 3\alpha^{-4}\|KQ_\alpha L^*Lu\| \geq 0. \end{aligned}$$

This completes the proof of the proposition. □

3.4.2 *Hanke-Raus rule*

According to Theorem 3.8, the estimate for the total error $d_{\xi^\dagger,\mu^\dagger}(u_\alpha^\delta, u^\dagger)$ differs from that for the squared residual by a factor of $1/\alpha$:

$$\begin{aligned} \frac{\|Ku_\alpha^\delta - g^\delta\|^2}{\alpha} \leq \frac{(\delta + 2\alpha\|w\|)^2}{\alpha} &= \Big(\frac{\delta}{\sqrt{\alpha}} + 2\sqrt{\alpha}\|w\|\Big)^2 \\ &\approx \frac{1}{2}\Big(\frac{\delta}{\sqrt{\alpha}} + \sqrt{\alpha}\|w\|\Big)^2 \geq d_{\xi^\dagger,\mu^\dagger}(u_\alpha^\delta, u^\dagger). \end{aligned}$$

Since the value $\|Ku_\alpha^\delta - g^\delta\|^2/\alpha$ can be evaluated a posteriori without a knowledge of the noise level δ, one can use it as an estimate of the total error and to choose an appropriate regularization parameter α by minimizing the function

$$\Phi_{\mathrm{hr}}(\alpha) = \frac{\|Ku_\alpha^\delta - g^\delta\|^2}{\alpha}. \tag{3.19}$$

This resembles the parameter choice due to Hanke and Raus [125] for classical Tikhonov regularization. The developments here follow [169].

Before we proceed to the analysis of the rule Φ_{hr}, we recall that so-called Bakushinskiĭ's veto, i.e., theoretically speaking, all heuristic rules, which do not make use of the knowledge about the exact noise level δ, can suffer from nonconvergence in the framework of worst-case scenario analysis. Bakushinskiĭ's veto [18] reads:

Theorem 3.18. *Let G be a mapping from a subset D of one metric space (Y, ρ_Y) to another metric space (X, ρ_X). Assume that G is regularizable by a family R_δ of the regularized mapping from Y to X, i.e.,*

$$\sup_{\tilde{g}\in Y,\ \rho_Y(\tilde{g},g)\leq\delta} \rho_X(R_\delta(\tilde{g}), G(g)) \to 0 \text{ as } \delta \to 0^+ \tag{3.20}$$

for all $g \in D$. Then G is regularizable by a mapping R which does not depend on $\delta > 0$ if and only if G can be extended to all Y and the extension is continuous in D.

Proof. For the sufficiency part, we take $R = G$, the extension of G on Y, which is continuous on $D \subset Y$. For the necessity part, we first note that R is continuous on D since for $\tilde{g}_1, \tilde{g}_2 \in D$ with $\rho_Y(\tilde{g}_1, \tilde{g}_2) \leq \delta$

$$\rho_X(R(\tilde{g}_1), R(\tilde{g}_2)) \leq \rho_X(R(\tilde{g}_1), G(\tilde{g}_1)) + \rho_X(R(\tilde{g}_2), G(\tilde{g}_1)) \to 0 \text{ as } \delta \to 0^+$$

by the assumption. By taking $\tilde{g} = g$ in (3.20), the assumption implies that $G(g) = R(g)$, $g \in D$ and thus R is a desired extension. □

For example, let $K \in L(X,Y)$, $D = \text{range}(K)$ and $G(g) = K^{-1}g$, $g \in D$. If G is regularizable by any rule R on Y, especially $R(g^\delta) = (K^*K + \alpha(g^\delta)I)^{-1}K^*g^\delta$, then G must be continuous on D and thus range(K) is closed and K is boundedly invertible.

Remark 3.23. In some sense, Bakushinskiĭ's veto discourages the use of heuristic rules. However, this does not imply that heuristic rules are useless. In practice, one usually has additional knowledge for calibrating the model. It is known that weak assumptions on the exact data $g^\dagger$ as well as noisy data g^δ, hence leaving the worst-case scenario analysis, lead to provable error estimates [197, 244, 196]. Further, in a statistical context, Bakushinskiĭ's veto is not valid [24].

In view of Theorem 3.18, a priori error estimates for heuristic rules, including the Hanke-Raus rule, are usually not feasible, unless further conditions on the noise and exact data are imposed. First we derive a posteriori error estimates, by which we measure the distance between the approximate solution $u_{\alpha^*}^\delta$ and the exact one $u^\dagger$ in terms of the noise level $\delta = \|g^\delta - g^\dagger\|$, the residual $\delta_* = \|Ku_{\alpha^*}^\delta - g^\delta\|$ and other relevant quantities.

Theorem 3.19. *Let the source condition* (3.6) *hold, and α^* be defined as $\alpha^* \in \arg\min_{\alpha \in [0, \|K\|^2]} \Phi_{\text{hr}}(\alpha)$. If $\delta_* := \|Ku_{\alpha^*}^\delta - g^\delta\| \neq 0$, then there holds*

$$d_{\xi^\dagger, \mu^\dagger}(u_{\alpha^*}^\delta, u^\dagger) \leq C\left(1 + \left(\tfrac{\delta}{\delta_*}\right)^2\right) \max(\delta, \delta_*).$$

Proof. By Corollary 3.5, we have

$$d_{\xi^\dagger, \mu^\dagger}(u_\alpha^\delta, u^\dagger) \leq d_{\xi^\dagger, \mu^\dagger}(u_\alpha, u^\dagger) + d_{\xi_\alpha, \mu_\alpha}(u_\alpha^\delta, u_\alpha) + 6\|w\|\delta.$$

It suffices to estimate the Bregman distance terms. First we estimate the approximation error $d_{\xi^\dagger,\mu^\dagger}(u_{\alpha^*}, u^\dagger)$ for $\alpha = \alpha^*$. By Theorem 3.9, we obtain

$$\begin{aligned} d_{\xi^\dagger,\mu^\dagger}(u_{\alpha^*}, u^\dagger) &\le \|w\| \|Ku_{\alpha^*} - g^\dagger\| \\ &\le \|w\| \left(\|K(u_{\alpha^*} - u_{\alpha^*}^\delta)\| + \|Ku_{\alpha^*}^\delta - g^\delta\| + \delta \right) \\ &\le \|w\|(2\delta + \delta_* + \delta) \le 4\|w\| \max(\delta, \delta_*). \end{aligned}$$

Next we estimate the term $d_{\xi_{\alpha^*},\mu_{\alpha^*}}(u_{\alpha^*}^\delta, u_{\alpha^*})$. Using Theorem 3.9, we get

$$d_{\xi_{\alpha^*},\mu_{\alpha^*}}(u_{\alpha^*}^\delta, u_{\alpha^*}) \le \frac{\delta^2}{2\alpha^*} = \left(\frac{\delta}{\delta_*}\right)^2 \frac{\|Ku_{\alpha^*}^\delta - g^\delta\|^2}{2\alpha^*}. \tag{3.21}$$

By the definition of α^*, we only increase the right hand side if we replace α^* by any other $\bar{\alpha} \in [0, \|K\|^2]$. We use $\bar{\alpha} = \bar{c}\delta$ with $\bar{c} = \min(1, \delta^{-1})\|K\|^2$ and deduce from Theorem 3.8 that

$$\|Ku_{\bar{\alpha}}^\delta - g^\delta\| \le (1 + 2\bar{c}\|w\|)\delta.$$

Replacing α^* by $\bar{\alpha}$ in the right hand side of inequality (3.21), we have

$$d_{\xi_{\alpha^*}}(u_{\alpha^*}^\delta, u_{\alpha^*}) \le \left(\frac{\delta}{\delta_*}\right)^2 \frac{\|Ku_{\bar{\alpha}}^\delta - g^\delta\|^2}{2\bar{\alpha}} \le \left(\frac{\delta}{\delta_*}\right)^2 \frac{(1 + 2\bar{c}\|w\|)^2\delta}{2\bar{c}}.$$

By combining the above two estimates, we finally arrive at

$$\begin{aligned} d_{\xi^\dagger,\mu^\dagger}(u_{\alpha^*}^\delta, u^\dagger) &\le 4\|w\| \max(\delta, \delta_*) + \left(\frac{\delta}{\delta_*}\right)^2 \frac{(1 + 2\bar{c}\|w\|)^2}{2\bar{c}}\delta + 6\|w\|\delta \\ &\le C(1 + (\tfrac{\delta}{\delta_*})^2) \max(\delta, \delta_*) \end{aligned}$$

with $C = \max(10\|w\|, (1 + 2\bar{c}\|w\|)^2/(2\bar{c}))$ as desired. □

By stipulating additional conditions on the data g^δ, we can get rid of the prefactor $\frac{\delta}{\delta_*}$ in the estimates and even obtain convergence of the method. To show this, we denote by Q the orthogonal projection onto the orthogonal complement of the closure of range(K).

Corollary 3.9. *If for the noisy data g^δ, there exists some $\varepsilon > 0$ such that*

$$\|Q(g^\dagger - g^\delta)\| \ge \varepsilon \|g^\dagger - g^\delta\|,$$

then α^* according to (3.19) is positive. Moreover, under the conditions of Theorem 3.19, there holds

$$d_{\xi^\dagger,\mu^\dagger}(u_{\alpha^*}^\delta, u^\dagger) \le C\left(1 + \tfrac{1}{\varepsilon^2}\right) \max(\delta, \delta_*).$$

Proof. We observe

$$\begin{aligned} \|Ku_\alpha^\delta - g^\delta\| &\geq \|Q(Ku_\alpha^\delta - g^\delta)\| = \|Qg^\delta\| \\ &= \|Q(g^\delta - g^\dagger)\| \geq \varepsilon \|g^\delta - g^\dagger\|. \end{aligned} \tag{3.22}$$

This shows $\delta_* \geq \varepsilon\delta$ and especially that $\Phi_{\mathrm{hr}}(\alpha) \to +\infty$ as $\alpha \to 0$. Hence, there exists an $\alpha^* > 0$ minimizing $\Phi_{\mathrm{hr}}(\alpha)$ over $[0, \|K\|^2]$. The remaining assertion follows from (3.22) and the respective error estimate. □

The next theorem shows the convergence of the Hanke-Raus rule under the condition that $\|Q(g^\dagger - g^\delta)\| \geq \varepsilon \|g^\dagger - g^\delta\|$ holds uniformly for the data g^δ as δ tends to zero.

Theorem 3.20. *Let the functional J_α be coercive and have a unique minimizer. Further, assume that there exists an $\epsilon > 0$ such that for every $\delta > 0$:*

$$\|Q(g^\dagger - g^\delta)\| \geq \epsilon \|g^\dagger - g^\delta\|. \tag{3.23}$$

Then there holds

$$d_{\xi^\dagger,\mu^\dagger}(u^\delta_{\alpha^*(g^\delta)}, u^\dagger) \to 0 \quad \textit{for} \quad \delta \to 0.$$

Proof. By the definition of α^*, we observe that the sequence $\{\alpha^* \equiv \alpha^*(g^\delta)\}_{\delta\geq 0}$ is uniformly bounded and hence, there exists an accumulation point $\bar\alpha$. We distinguish the two cases $\bar\alpha = 0$ and $\bar\alpha > 0$.

We first consider the case $\bar\alpha = 0$. By Corollary 3.5, we split the error

$$d_{\xi^\dagger,\mu^\dagger}(u^\delta_{\alpha^*}, u^\dagger) \leq d_{\xi_{\alpha^*},\mu_{\alpha^*}}(u^\delta_{\alpha^*}, u_{\alpha^*}) + d_{\xi^\dagger,\mu^\dagger}(u_{\alpha^*}, u^\dagger) + 6\|w\|\delta, \tag{3.24}$$

and estimate the data and approximation errors separately.

For the error $d_{\xi_{\alpha^*},\mu_{\alpha^*}}(x^\delta_{\alpha^*}, u_{\alpha^*})$, we deduce from Theorem 3.9 and assumption (3.23) that

$$d_{\xi_{\alpha^*},\mu_{\alpha^*}}(u^\delta_{\alpha^*}, u_{\alpha^*}) \leq \frac{\delta^2}{2\alpha^*} \leq \frac{\|Ku^\delta_{\alpha^*} - g^\delta\|^2}{2\epsilon^2\alpha^*} = \frac{\Phi_{\mathrm{hr}}(\alpha^*)}{2\epsilon^2}.$$

Therefore, it suffices to show that $\Phi_{\mathrm{hr}}(\alpha^*)$ goes to zero as $\delta \to 0$. By Theorem 3.8, there holds for every $\alpha \in [0, \|K\|^2]$ that

$$\Phi_{\mathrm{hr}}(\alpha^*) \leq \Phi_{\mathrm{hr}}(\alpha) \leq \left(\frac{\delta}{\sqrt{\alpha}} + 2\|w\|\sqrt{\alpha}\right)^2.$$

Hence, we may choose $\alpha(\delta)$ such that $\alpha(\delta) \to 0$ and $\delta^2/\alpha(\delta) \to 0$ for $\delta \to 0$. This shows $\Phi_{\mathrm{hr}}(\alpha^*) \to 0$ for $\delta \to 0$.

For the approximation error $d_{\xi^\dagger,\mu^\dagger}(u_{\alpha^*}, u^\dagger)$, we deduce from the fact that $\bar\alpha = 0$ and Theorem 3.9 that

$$d_{\xi^\dagger,\mu^\dagger}(u_{\alpha^*}, u^\dagger) \leq \frac{\alpha^*\|w\|^2}{2} \to \frac{\bar\alpha\|w\|^2}{2} = 0 \quad \text{for} \quad \delta \to 0.$$

Hence, all three terms on the right hand side of inequality (3.24) tend to zero for $\delta \to 0$ as desired.

Next we consider the case $\bar{\alpha} > 0$. we use $\alpha^* \leq \|K\|^2$ to get

$$\Phi_{\rm hr}(\alpha^*) \geq \frac{\|Ku_{\alpha^*}^\delta - g^\delta\|^2}{\|K\|^2} \geq 0.$$

Since $\Phi_{\rm hr}(\alpha^*)$ goes to zero for $\delta \to 0$ we deduce that $\|Ku_{\alpha^*}^\delta - g^\delta\|$ tends to zero as well. Next by the minimizing property of $u_{\alpha^*}^\delta$, we have

$$\tfrac{1}{2}\|Ku_{\alpha^*}^\delta - g^\delta\|^2 + \alpha^*\psi(u_{\alpha^*}^\delta) \leq \tfrac{1}{2}\|Ku^\dagger - g^\delta\|^2 + \alpha^*\psi(u^\dagger).$$

Therefore, both sequences $\{\|Ku_{\alpha^*}^\delta - g^\delta\|\}$ and $\{\psi(u_{\alpha^*}^\delta)\}$ are uniformly bounded by the assumption $\bar{\alpha} > 0$. By the coercivity of the functional J_α, the sequence $\{u_{\alpha^*}^\delta\}$ is uniformly bounded, and thus there exists a subsequence, still denoted by $\{u_{\alpha^*}^\delta\}$, that converges weakly to some $\hat{u} \in \mathcal{C}$. By the weak lower semicontinuity of the norm and the functional ψ, we have

$$\|K\hat{u} - g^\dagger\| \leq \liminf_{\delta\to 0} \|Ku_{\alpha^*}^\delta - g^\delta\| = 0, \quad \psi(\hat{u}) \leq \liminf_{\delta\to 0} \psi(u_{\alpha^*}^\delta). \tag{3.25}$$

Consequently, for any u

$$\tfrac{1}{2}\|K\hat{u} - g^\dagger\|^2 + \bar{\alpha}\psi(\hat{u}) \leq \tfrac{1}{2}\|Ku - g^\dagger\|^2 + \bar{\alpha}\psi(u). \tag{3.26}$$

Hence $\hat{u}$ minimizes the functional $\frac{1}{2}\|Ku - g^\dagger\|^2 + \bar{\alpha}\psi(u)$, and by the uniqueness of the minimizer, $\hat{u} = u_{\bar{\alpha}}$, and the whole sequence converges weakly. Further, by taking $u = u_{\bar{\alpha}}$ in (3.26), we deduce

$$\tfrac{1}{2}\|Ku_{\bar{\alpha}} - g^\dagger\|^2 + \bar{\alpha}\psi(u_{\bar{\alpha}}) = \lim_{\delta\to 0}\{\tfrac{1}{2}\|Ku_{\alpha^*}^\delta - g^\delta\|^2 + \alpha^*\psi(u_{\alpha^*}^\delta)\}, \tag{3.27}$$

and the arguments in the proof of Theorem 3.2 yield

$$\psi(u_{\alpha^*}^\delta) \to \psi(u_{\bar{\alpha}}). \tag{3.28}$$

Next we show that $u_{\bar{\alpha}}$ is a ψ-minimizing solution to the equation $Ku = g^\dagger$. However, this follows directly from inequality (3.25) that $\|Ku_{\bar{\alpha}} - g^\dagger\| = 0$, and from inequality (3.26)

$$\psi(u_{\bar{\alpha}}) \leq \psi(u) \quad \forall u \quad \text{with } Ku = g^\dagger,$$

which in particular by choosing u in the set of ψ-minimizing solutions of $Ku = g^\dagger$ shows the claim. Now we deduce that

$$\lim_{\delta\to 0} d_{\xi^\dagger,\mu^\dagger}(u_{\alpha^*}^\delta, u^\dagger) = \lim_{\delta\to 0}\left(\psi(u_{\alpha^*}^\delta) - \psi(u^\dagger) - \langle \xi^\dagger - \mu^\dagger, u_{\alpha^*(g^\delta)}^\delta - u^\dagger\rangle\right) = 0,$$

by observing identity (3.28) and the weak convergence of the sequence $\{u_{\alpha^*(g^\delta)}^\delta\}$ to $u^\dagger$. This concludes the proof of the theorem. □

3.4.3 *Quasi-optimality criterion*

Now we turn to a third a posteriori choice rule, the quasi-optimality criterion. Throughout this part, we assume that there is no constraint $\mathcal{C}$. The motivation of the criterion is as follows. By Proposition 3.2 for any $q \in (0,1)$,

$$d_{\xi_\alpha^\delta}(u_{q\alpha}^\delta, u_\alpha^\delta) \leq \frac{(1-q)^2}{2q}\Phi_{\mathrm{hr}}(\alpha).$$

In particular, for a geometrically decreasing sequence of regularization parameters, the Bregman distance of two consecutive regularized solutions are bounded from above by a constant times the estimator Φ_{hr}. This suggests itself a parameter choice rule which resembles the classical quasi-optimality criterion [289, 291]. Specifically, for the given data g^δ and $q \in (0,1)$ we define a *quasi-optimality sequence* as

$$\mu_k^\delta = d_{\xi_{q^{k-1}}^\delta}(u_{q^k}^\delta, u_{q^{k-1}}^\delta).$$

The quasi-optimality criterion consists of choosing the regularization parameter $\alpha^{\mathrm{qo}} = q^k$ such that μ_k^δ is minimal over a given range $k \geq k_0$. The quasi-optimality sequence for the exact data $g^\dagger$ will be denoted by

$$\mu_k = d_{\xi_{q^{k-1}}}(u_{q^k}, u_{q^{k-1}}).$$

Remark 3.24. The classical quasi-optimality criterion [291] chooses α^{qo} such that the quantity $\|\alpha \frac{du_\alpha^\delta}{d\alpha}\|$ is minimal. For a general convex penalty ψ, this is not applicable since the mapping $\alpha \mapsto u_\alpha^\delta$ is in general not differentiable. For example, in the case of ℓ^1 regularization, the solution path, i.e., u_α^δ with respect to α, is piecewise linear [84]. Hence we opt for the discrete version which is also used in [290]. The literature on the quasi-optimality criterion in the Hilbert space setting is quite extensive; see [290, 291, 212, 211, 110, 21, 197, 196], where a priori convergence rates were established under certain conditions on the exact data and the noise (thereby leaving the worst-case scenario error estimate).

We start with some basic observations of the quasi-optimality sequence.

Lemma 3.11. *Let the source condition* (3.6) *hold. Then the quasi-optimality sequences* $\{\mu_k^\delta\}$ *and* $\{\mu_k\}$ *satisfy*

(1) $\lim_{k\to-\infty} \mu_k^\delta = 0$,

(2) $\lim_{k\to-\infty} \mu_k = 0$ *and* $\lim_{k\to\infty} \mu_k = 0$.

Proof. Appealing to Proposition 3.2, we have

$$\mu_k^\delta \le \frac{(1-q)^2}{2q} \frac{\|Ku_{q^{k-1}}^\delta - g^\delta\|^2}{q^{k-1}}.$$

Since the sequence $\{\|Ku_{q^{k-1}}^\delta - g^\delta\|\}_k$ stays bounded for $k \to -\infty$, the first claim follows directly. Setting $\delta = 0$ in assertion (1) shows the first statement of assertion (2). Now we use Theorem 3.9 and estimate

$$\mu_k \le \frac{(1-q)^2}{2} \frac{\|Ku_{q^{k-1}} - g^\dagger\|^2}{q^k} \le 2(1-q)^2\|w\|^2 q^{k-2}.$$

This shows the second statement of assertion (2). □

Now we show that the quasi-optimality sequences for exact and noisy data approximate each other for a vanishing noise level.

Lemma 3.12. *Let the source condition* (3.6) *hold. Then for any* $k_1 \in \mathbb{Z}$, *there holds*

$$\lim_{\delta\to 0} \sup_{k\le k_1} |\mu_k^\delta - \mu_k| = 0.$$

Proof. We will use the abbreviations $u_k^\delta = u_{q^k}^\delta$, $u_k = u_{q^k}$, $\xi_k^\delta = -K^*(Ku_k^\delta - g^\delta)$ and $\xi_k = -K^*(Ku_k - g)$. By the definition of μ_k^δ and μ_k, we have

$$\begin{aligned}
|\mu_k^\delta - \mu_k| &= |d_{\xi_{k-1}^\delta}(u_k^\delta, u_{k-1}^\delta) - d_{\xi_k}(u_k, u_{k-1})| \\
&= \big|\psi(u_k^\delta) - \psi(u_{k-1}^\delta) - \langle \xi_{k-1}^\delta, u_k^\delta - u_{k-1}^\delta\rangle \\
&\qquad - \psi(u_k) + \psi(u_{k-1}) + \langle \xi_{k-1}, u_k - u_{k-1}\rangle\big| \\
&= |d_{\xi_k}(u_k^\delta, u_k) - d_{\xi_{k-1}}(u_{k-1}^\delta, u_{k-1}) \\
&\qquad + \langle \xi_{k-1} - \xi_{k-1}^\delta, u_k^\delta - u_{k-1}^\delta\rangle + \langle \xi_k - \xi_{k-1}, u_k^\delta - u_k\rangle|.
\end{aligned}$$

Now we estimate all four terms separately. Using Theorem 3.9 we can bound the first two terms by

$$d_{\xi_k}(u_k^\delta, u_k) \le \frac{\delta^2}{2q^k}, \qquad d_{\xi_{k-1}}(u_{k-1}^\delta, u_{k-1}) \le \frac{\delta^2}{2q^{k-1}}.$$

For the third term, we get from Theorem 3.9 and Proposition 3.2

$$\begin{aligned}
&\langle \xi_{k-1} - \xi_{k-1}^\delta, u_k^\delta - u_{k-1}^\delta\rangle \\
&= \langle K(u_{k-1} - u_{k-1}^\delta) + (g^\delta - g^\dagger), K(u_k^\delta - u_{k-1}^\delta)\rangle / q^{k-1} \\
&\le (\|K(u_{k-1} - u_{k-1}^\delta)\| + \delta)\|K(u_k^\delta - u_{k-1}^\delta)\|/q^{k-1} \\
&\le 6\delta(1-q)(\delta + 2q^k\|w\|)/q^{k-1}.
\end{aligned}$$

Similarly, we can estimate the last term by

$$\langle \xi_k - \xi_{k-1}, u_k^\delta - u_k \rangle \leq 8\delta(1-q)\|w\|/q.$$

Hence, all four terms are bounded for $k \leq k_1$ and decrease to zero as $\delta \to 0$. This proves the claim. $\square$

In general, the quasi-optimality sequences $\{\mu_k^\delta\}$ and $\{\mu_k\}$ can vanish for finite indices k. Fortunately, their positivity can be guaranteed for a class of p-convex functionals ψ, cf. Remark 3.11.

Lemma 3.13. *Let the functional ψ be p-convex, $\psi(u) = 0$ only for $u = 0$ and satisfy that for any u the value $\langle \xi, u \rangle$ is independent of the choice of $\xi \in \partial\psi(u)$. If the data $g^\dagger$ (respectively g^δ) admits nonzero $\tilde{\alpha}$ for which $u_{\tilde{\alpha}} \neq 0$, then $\mu_k > 0$ (respectively $\mu_k^\delta > 0$) for all $k \geq [\ln \tilde{\alpha}/\ln q]$.*

Proof. It suffices to prove the assertion on μ_k. By the optimality condition for u_α, we have

$$-K^*(Ku_\alpha - g^\dagger) \in \alpha\partial\psi(u_\alpha).$$

By assumption, the value $\langle \xi_\alpha, u_\alpha \rangle$ is independent of the choice of $\xi_\alpha \in \partial\psi(u_\alpha)$ and hence, taking duality pairing with u_α gives for any $\xi_\alpha \in \partial\psi(u_\alpha)$

$$\langle Ku_\alpha, Ku_\alpha - g^\dagger \rangle + \alpha\langle \xi_\alpha, u_\alpha \rangle = 0.$$

For non-zero u_α we have that $\langle \xi_\alpha, u_\alpha \rangle$ is non-zero and hence, we get

$$\alpha = \frac{\langle Ku_\alpha, Ku_\alpha - g^\dagger \rangle}{\langle \xi_\alpha, u_\alpha \rangle}. \tag{3.29}$$

Next by the assumption that the data $g^\dagger$ admits nonzero $\tilde{\alpha}$ for which $u_{\tilde{\alpha}} \neq 0$, then for any $\alpha < \tilde{\alpha}$, 0 cannot be a minimizer of the Tikhonov functional. To see this, we assume that 0 is a minimizer, i.e.,

$$\begin{aligned} \tfrac{1}{2}\|g^\dagger\|^2 &= \tfrac{1}{2}\|K0 - g^\dagger\|^2 + \alpha\psi(0) \leq \tfrac{1}{2}\|Ku_{\tilde{\alpha}} - g^\dagger\|^2 + \alpha\psi(u_{\tilde{\alpha}}) \\ &< \tfrac{1}{2}\|Ku_{\tilde{\alpha}} - g^\dagger\|^2 + \tilde{\alpha}\psi(u_{\tilde{\alpha}}), \end{aligned}$$

by the strict positivity of ψ for nonzero u. This contradicts the optimality of $u_{\tilde{\alpha}}$. Now let $\alpha_1, \alpha_2 < \tilde{\alpha}$ be distinct. Then both sets $S(\alpha_1)$ and $S(\alpha_2)$ contain no zero element. Next we show that the two sets are disjointed. Assume that $S(\alpha_1)$ and $S(\alpha_2)$ intersects nontrivially, i.e., there exists some nonzero $\tilde{u}$ such that $\tilde{u} \in S(\alpha_1) \cap S(\alpha_2)$. Then by equation (3.29) and choosing any $\tilde{\xi} \in \partial\psi(\tilde{u})$, we have

$$\alpha_1 = \frac{\langle K\tilde{u}, K\tilde{u} - g^\dagger \rangle}{\langle \tilde{\xi}, \tilde{u} \rangle} = \alpha_2,$$

which is in contradiction with the distinctness of α_1 and α_2. Therefore, for distinct $\alpha_1, \alpha_2 < \tilde{\alpha}$, $S(\alpha_1) \cap S(\alpha_2) = \emptyset$. Consequently, we have

$$\|u_{\alpha_1} - u_{\alpha_2}\| > 0.$$

Now by the p-convexity of ψ, we deduce for $q^{k-1} \leq \tilde{\alpha}$ that

$$\mu_k = d_{\xi_{q^{k-1}}}(u_{q^k}, u_{q^{k-1}}) \geq C\|u_{q^k} - u_{q^{k-1}}\|^p > 0,$$

which shows the assertion for μ_k. □

Remark 3.25. The assumptions on ψ in Lemma 3.13 are satisfied for many commonly used regularization functionals, e.g., $\|u\|_{\ell^p}^p$, $\|u\|_{L^p(\Omega)}^p$ with $p > 1$ and elastic-net [170]. However, the special case of $\|u\|_{\ell^1}$ is not covered. Indeed, the ℓ^1 minimization can retrieve the support of the exact solution for sufficiently small noise level δ and α [219]. Hence, both μ_k^δ and μ_k vanish for sufficiently large k, due to the lack of p-convexity. The bound $\tilde{\alpha}$ depends on $g(g^\delta)$, and for nonvanishing $g(g^\delta)$ can be either positive or $+\infty$. The choice of k_0 should be related to $\tilde{\alpha}$ such that $\mu_{k_0}^\delta(\mu_{k_0})$ is nonzero.

Corollary 3.10. *Let the source condition* (3.6) *hold. Then under the conditions of Lemma 3.13 and* $k_0 \geq [\ln \tilde{\alpha}]/\ln q$, *the parameter* α^{qo} *chosen by the quasi-optimality criterion satisfies that for any sequence* $\delta_n \to 0$ *there holds* $\alpha^{\text{qo}} \to 0$.

Proof. By definition, $\alpha^{\text{qo}} = q^{k^*}$ where k^* is such that μ_k^δ is minimal. By the triangle inequality, $\mu_k^\delta \leq \mu_k + |\mu_k^\delta - \mu_k|$. Now, let $\epsilon > 0$. By Lemma 3.11, $\mu_k \to 0$ for $k \to \infty$ and hence, there exists an integer $\underline{k}$ such that $\mu_{\underline{k}} \leq \epsilon/2$. Moreover, by Lemma 3.12, for any k_1 there is $\bar{\delta} > 0$ such that $|\mu_k^\delta - \mu_k| \leq \epsilon/2$ for all $k \leq k_1$, in particular with $\underline{k}$. Hence $\mu_k^\delta \leq \mu_k + |\mu_k - \mu_k^\delta| < \varepsilon$ for the same value of $\underline{k}$.

By Lemma 3.13, for any finite integer k_1, the set $\{\mu_k\}_{k=k_0}^{k_1}$ is finite and positive, and thus there exists a constant $\sigma > 0$ such that $\mu_k > \sigma$ for $k = k_0, \ldots, k_1$. Lemma 3.12 indicates that μ_k^δ is larger than $\sigma/2$ for $k = k_0, \ldots, k_1$ and sufficiently small δ. Thus the sequence $\{\alpha^{\text{qo}}\}_{\delta_n}$ can contain terms on $\{q^k\}_{k=k_0}^{k_1}$ only if δ is not too small, since μ_k^δ goes to zero as δ tends to zero. Since the choice k_1 is arbitrarily, this implies the desired assertion. □

As was remarked earlier, it is in general impossible to show the convergence of u_α^δ to $u^\dagger$ for a heuristic parameter choice in the context of worst-case scenario analysis. For the quasi-optimality criterion, Glasko et al [110] introduced the notion of auto-regularizable set as a condition on the exact as well as noisy data. In the case of the continuous quasi-optimality criterion this is the set of g^δ such that

$$\frac{\|\alpha\frac{du_\alpha^\delta}{d\alpha} - \alpha\frac{du_\alpha}{d\alpha}\|}{\|u_\alpha^\delta - u_\alpha\|} \geq q > 0$$

holds uniformly in α and δ. This abstract condition on the exact data has been replaced by a condition on the noise in [197, 21].

In our setting, the following sets are helpful for proving convergence.

Definition 3.4. For $r > 0$, $q \in (0,1)$, $K : X \to Y$ and $g^\dagger \in \text{range}(K)$ we define the sets

$$\mathcal{D}_r = \Big\{g^\delta \in Y : |d_{\xi^\delta_{q^{k-1}}}(u^\delta_{q^k}, u^\delta_{q^{k-1}}) - d_{\xi_{q^{k-1}}}(u_{q^k}, u_{q^{k-1}})| \\ \geq r d_{\xi_{q^k}}(u^\delta_{q^k}, u_{q^k})\ \forall k\Big\}.$$

The condition $g^\delta \in \mathcal{D}_r$ can be regarded as a discrete analogue of the above-mentioned auto-regularizable condition. With the set $\mathcal{D}_r$ at hand, we can now show another result on the asymptotic behavior of the quasi-optimality sequence. The condition is that the noisy data belongs to some set $\mathcal{D}_r$.

Lemma 3.14. *Let the source condition* (3.6) *hold*, $k_0 \geq [\ln\tilde{\alpha}/\ln q]$, $g^\delta \in \mathcal{D}_r$ *for some* $r > 0$ *and* $\psi(u_\alpha^\delta) \to \infty$ *for* $\alpha \to 0$. *Then* $\mu_k^\delta \to \infty$ *for* $k \to \infty$.

Proof. We observe that

$$\begin{aligned} r d_{\xi_{q^k}}(u^\delta_{q^k}, u_{q^k}) &\leq |d_{\xi^\delta_{q^{k-1}}}(u^\delta_{q^k}, u^\delta_{q^{k-1}}) - d_{\xi_{q^{k-1}}}(u_{q^k}, u_{q^{k-1}})| \\ &= |\mu_k^\delta - \mu_k|. \end{aligned} \tag{3.30}$$

By the definition of the Bregman distance and Theorem 3.9 we have for $\xi_\alpha = -K^*(Ku_\alpha - g^\dagger)/\alpha$ that

$$\begin{aligned} d_{\xi_\alpha}(u_\alpha^\delta, u_\alpha) &= \psi(u_\alpha^\delta) - \psi(u_\alpha) - \langle \xi_\alpha, u_\alpha^\delta - u_\alpha\rangle \\ &= \psi(u_\alpha^\delta) - \psi(u_\alpha) + \tfrac{1}{\alpha}\langle Ku_\alpha - g^\dagger, K(u_\alpha^\delta - u_\alpha)\rangle \\ &\geq \psi(u_\alpha^\delta) - \psi(u_\alpha) - 4\|w\|\delta. \end{aligned}$$

Since $\psi(u_\alpha)$ is bounded for $\alpha \to 0$, $d_{\xi_\alpha}(u_\alpha^\delta, u_\alpha) \to \infty$ for $\alpha \to 0$. This means that for $k \to \infty$ there holds that $d_{\xi_{q^k}}(u^\delta_{q^k}, u_{q^k}) \to \infty$ and since $\mu_k \to 0$, the claim follows from (3.30). □

Now we state the convergence for the quasi-optimality criterion.

Theorem 3.21. *Let the source condition* (3.6) *hold*, $k_0 \geq [\ln\tilde{\alpha}/\ln q]$, *and* $\{\delta_n\}_n$, $\delta_n > 0$, *be a sequence converging to zero such that* $g^{\delta_n} \to g^\dagger \in$

range(K) and $g^{\delta_n} \in \mathcal{D}_r$ for some $r > 0$. Let $\{\alpha_n^{\rm qo} = \alpha_n^{\rm qo}(g^{\delta_n})\}_n$ be chosen by the quasi-optimality criterion. Then

$$\lim_{n\to\infty} d_{\xi^\dagger}(u_{\alpha_n^{\rm qo}}^{\delta_n}, u^\dagger) = 0.$$

Proof. We denote $\alpha_n^{\rm qo}$ by q^{k_n}. Then by using Corollary 3.5, we derive

$$\begin{aligned} d_{\xi^\dagger}(u_{q^{k_n}}^{\delta_n}, u^\dagger) &\leq d_{\xi_{q^{k_n}}}(u_{q^{k_n}}^{\delta_n}, u_{q^{k_n}}) + d_{\xi^\dagger}(u_{q^{k_n}}, u^\dagger) + 6\|w\|\delta_n \\ &\leq \frac{1}{r}|d_{\xi^\delta_{q^{k_n-1}}}(u^\delta_{q^{k_n}}, u^\delta_{q^{k_n-1}}) - d_{\xi_{q^{k_n-1}}}(u_{q^{k_n}}, u_{q^{k_n-1}})| \\ &\quad + d_{\xi^\dagger}(u_{q^{k_n}}, u^\dagger) + 6\|w\|\delta_n \\ &= \frac{1}{r}|\mu^\delta_{k_n} - \mu_{k_n}| + d_{\xi^\dagger}(u_{q^{k_n}}, u^\dagger) + 6\|w\|\delta_n. \end{aligned}$$

Now all three terms on the right hand side tend to zero for $n \to \infty$ (the first due to Lemma 3.12 and the second due to $q^{k_n} = \alpha_n^{\rm qo} \to 0$ by Corollary 3.10). □

Remark 3.26.

- The important question on how the sets $\mathcal{D}_r$ look like, and especially, under what circumstance they are non-empty, remains open. In [110], Glasko and Kriksin use spectral theory to investigate this issue – a tool which is unfortunately unavailable in Banach space setting.
- In the presence of a convex constraint $\mathcal{C}$, one needs an extension of Lemma 3.13 to the constrained case.

Remark 3.27. The efficient implementation of the Hanke-Raus rule and the quasi-optimality criterion is still unclear. It seems that both objective functions are plagued with local minima, and hence their minimization is not straightforward, and has not received due attention.

3.5 Augmented Tikhonov regularization

In this part, we describe the augmented Tikhonov regularization, a recent framework derived from hierarchical Bayesian modeling, and discuss its built-in mechanism for parameter choice: balancing principle. The approach was first derived in [174], and further analyzed in [156, 154]. Finally, we present analytical results for the balancing principle, especially variational characterization, a posteriori error estimates and efficient implementation.

3.5.1 *Augmented Tikhonov regularization*

We begin with finite-dimensional linear inverse problems

$$\mathbf{K}\mathbf{u} = \mathbf{g}^\delta,$$

where $\mathbf{K} \in \mathbb{R}^{n\times m}$, $\mathbf{u} \in \mathbb{R}^m$ and $\mathbf{g}^\delta \in \mathbb{R}^n$ represent discrete form of continuous operator K, unknown u and data g^δ. Using a hierarchical Bayesian formulation (see Section 7.1 for detailed derivations), one arrives at the following functional $J(\mathbf{u}, \lambda, \tau)$ for the maximum a posteriori estimator:

$$J(\mathbf{u}, \lambda, \tau) = \frac{\tau}{2}\|\mathbf{K}\mathbf{u} - \mathbf{g}^\delta\|^2 + \frac{\lambda}{2}\|\mathbf{L}\mathbf{u}\|^2 + b_0\tau - a_0' \ln \tau + b_1\lambda - a_1' \ln \tau,$$

where $a_0' = \frac{n}{2} - 1 + a_0$ and $a_1' = \frac{m'}{2} - 1 + a_1$, and $\mathbf{L} \in \mathbb{R}^{m'\times m}$ is the regularizing matrix and it is of full column rank. Each term in the functional $J(\mathbf{u}, \lambda, \tau)$ admits the following statistical interpretation

(1) The first term follows from an independent identically distributed Gaussian assumption on the noise in the data $\mathbf{g}^\delta$, and τ is the inverse variance (a.k.a. precision) of the Gaussian noise.
(2) The second term assumes a Markov random field on the unknown, with the interaction structure encoded by the matrix $\mathbf{L}$, and the scalar λ determines the strength of interactions between neighboring sites.
(3) The third and fourth terms assume a Gamma distribution on the precision τ and scale λ, with parameter pair being (a_0, b_0) and (a_1, b_1), respectively, i.e., $\tau \sim G(\tau; a_0, b_0)$ and $\lambda \sim G(\lambda; a_1, b_1)$.

The functional $J(\mathbf{u}, \lambda, \tau)$ is called augmented Tikhonov regularization in that the first two terms of the functional reproduce the classical Tikhonov regularization (with a regularization parameter $\alpha = \lambda\tau^{-1}$), while the remaining terms provide the mechanism to automatically determine the noise precision τ and the parameter λ. One distinct feature of the augmented approach is that it determines the solution $\mathbf{u}$ and the regularization parameter simultaneously from the data $\mathbf{g}^\delta$.

By taking the limit of the discrete functional $J(\mathbf{u}, \lambda, \tau)$, we arrive at a general augmented Tikhonov functional

$$J(u, \lambda, \tau) = \tau\phi(u, g^\delta) + \lambda\psi(u) + b_0\lambda - a_0 \ln \lambda + b_1\tau - a_1 \ln \tau, \tag{3.31}$$

which extends the Tikhonov functional J_α defined in (3.2) in the hope of automatically choosing also a regularization parameter. The parameter pairs (a_0, b_0) and (a_1, b_1) in (3.31) should be the limit of the discrete values.

Since the functional $J(u, \lambda, \tau)$ might be nonsmooth and nonconvex, we resort to optimality in a generalized sense.

Definition 3.5. A tuple $(u^*, \lambda^*, \tau^*) \in X \times \mathbb{R}^+ \times \mathbb{R}^+$ is called a critical point of the functional $J(u, \lambda, \tau)$ if it satisfies

$$\begin{aligned} u^* &= \arg\min_{u \in X} \left\{ \phi(u, g^\delta) + \lambda^* (\tau^*)^{-1} \psi(u) \right\}, \\ &\psi(u^*) + b_0 - a_0 \tfrac{1}{\lambda^*} = 0, \\ &\phi(u^*, g^\delta) + b_1 - a_1 \tfrac{1}{\tau^*} = 0. \end{aligned} \tag{3.32}$$

It follows from the optimality system (3.32) that the automatically determined regularization parameter $\alpha^* = \lambda^*(\tau^*)^{-1}$ satisfies

$$\alpha^* := \lambda^* \cdot (\tau^*)^{-1} = \gamma \frac{\bar{\phi}(\alpha^*) + b_1}{\bar{\psi}(\alpha^*) + b_0}, \tag{3.33}$$

where the constant $\gamma = a_0/a_1$, which is determined by the statistical a priori knowledge (shape parameters in the Gamma distributions).

We now establish an alternative variational characterization. To this end, we introduce a functional $\mathcal{G}$ by

$$\mathcal{G}(u) = \ln \phi((u, g^\delta) + b_1) + \gamma \ln(\psi(u) + b_0).$$

Clearly, under the assumption $b_1 > 0$ or $\lim_{\alpha \to 0^+} \phi(u_\alpha^\delta, g^\delta) > 0$, the existence of a minimizer to the functional $\mathcal{G}$ follows directly since it is bounded from below, weakly lower semi-continuous and tends to $+\infty$ as $\|u\| \to +\infty$.

Theorem 3.22. *If the functional J_α is convex and ϕ and ψ are continuously differentiable, then $u_{\alpha^*}^\delta$ is a Tikhonov solution with α^* computed by the rule* (3.31) *if and only if it is a critical point of the functional $\mathcal{G}$.*

Proof. Under the premise that the functionals $\phi(u, g^\delta)$ and $\psi(u)$ are differentiable, a critical point u^* of the functional $\mathcal{G}$ satisfies

$$(\phi(u^*, g^\delta) + b_1)^{-1} \phi'(u^*, g^\delta) + \tfrac{\gamma}{\psi(u^*) + b_0} \psi'(u^*) = 0.$$

Setting $\alpha^* = \gamma(\phi(u^*, g^\delta) + b_1)(\psi(u^*) + b_0)^{-1}$ gives

$$\phi'(u^*, g^\delta) + \alpha^* \psi'(u^*) = 0,$$

i.e., u^* is a critical point of the functional J_{α^*}. If the functional J_{α^*} is convex, then u^* is also a minimizer of the functional J_{α^*} and $u^* = u_{\alpha^*}^\delta$. □

Balancing principle Now we discuss briefly the connections of the augmented approach to existing choice rules for the classical L^2-L^2 model.

One distinct feature of the augmented approach is that it utilizes both the fidelity $\bar{\phi}(\alpha)$ and penalty $\bar{\psi}(\alpha)$ to determine the regularization parameter α. In the literature, there are several rules based on the simultaneous

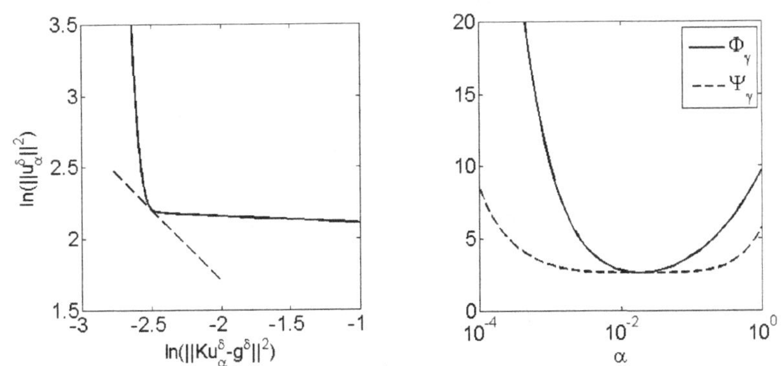

Fig. 3.1 Parameter choice for the L^2-L^2 model. Left: the L-curve ; Right: criteria Φ_γ and Ψ_γ, with $\gamma = 1$. In the left panel, the solid and dashed lines denote the L-curve and the tangent line at the "corner", respectively.

use of the quantities $\bar{\phi}(\alpha)$ and $\bar{\psi}(\alpha)$. The most prominent one is the L-curve criterion [127] (see Fig. 3.1 for an illustration), which plots these two quantities in a doubly logarithmic scale, and selects the regularization parameter corresponding to the "corner" of the curve, often indicated by the point with the maximum curvature. Alternatively, Regińska [257] proposed the local-minimum criterion, which minimizes the function $\bar{\phi}(\alpha)^\gamma\bar{\psi}(\alpha)$. Its first-order optimality condition reads

$$-\bar{\phi}(\alpha) + \alpha\gamma\bar{\psi}(\alpha) = 0. \tag{3.34}$$

It can be deemed as an algebraic counterpart of the L-curve criterion: it selects the point on the curve whose tangent has a slope $-\gamma$. We refer to Fig. 3.1 for an illustration. Obviously, (3.34) is a special case of the augmented Tikhonov functional. We shall call equation (3.34) by balancing principle since it attempts to balance the fidelity $\bar{\phi}(\alpha)$ and the penalty $\bar{\psi}(\alpha)$ with γ being the relative weight. It was independently derived several times in the literature [257, 181, 206]. In [181], the case of $\gamma = 1$ was called the zero-crossing method. It is very popular in the biomedical engineering community, e.g., electrocardiography [181], and some analysis of the method was provided in [182]. Some a posteriori error estimates were derived for the local minimum criterion in [220]. We note that the terminology balancing principle occasionally refers to a principle due to Lepskii [213, 230, 208].

Example 3.5. In this example, we provide a derivation of the balancing principle via the model function approach, cf. Section 3.4.1, for the L^2-L^2

model, following [206]. Here one uses a model function

$$m_k(\alpha) = C_k \frac{\alpha}{\alpha + T_k},$$

where the parameters C_k and T_k are determined from the interpolation conditions of the value function $F(\alpha)$ and its derivative at α_k, i.e.,

$$C_k = \frac{F^2(\alpha_k)}{\bar{\phi}(\alpha_k)} \quad \text{and} \quad T_k = \frac{\alpha_k^2 \bar{\psi}(\alpha_k)}{\bar{\phi}(\alpha_k)}.$$

One heuristic procedure for estimating the noise level δ is as follows: first one computes the y-intercept y_0 of the tangent to the curve $y = m_k(\alpha)$ at α_k, and then finds α_{k+1} from $m_k(\alpha_{k+1}) = \sigma y_0$, with the parameter $\sigma > 1$ (instead of $\sigma \leq 1$ in [206]). Clearly, the intercept $y_0 = \bar{\phi}(\alpha_k)$, and thus

$$\frac{F^2(\alpha_k)}{\bar{\phi}(\alpha_k)} \frac{\alpha_{k+1}}{\frac{\alpha_k^2 \bar{\psi}(\alpha_k)}{\bar{\phi}(\alpha_k)} + \alpha_{k+1}} = \sigma \bar{\phi}(\alpha_k),$$

Upon convergence, the limit α^* satisfies

$$\alpha^* \bar{\psi}(\alpha^*) = (\sigma - 1)\bar{\phi}(\alpha^*).$$

This is identical with equation (3.34). The preceding derivation provides a geometrical interpretation of the balancing principle.

Note that with noninformative priors on λ and τ, we deduce from (3.32) that augmented Tikhonov regularization satisfies also (3.34). Hence, we arrive at the following interesting observation:

Observation 3.1. Hierarchical Bayesian inference with noninformative hyperpriors is balancing.

Remark 3.28. Generally, the choice of "hyper-parameter" γ in the balancing rule, should be model-dependent, due to the very distinct features of different models. In practice, it might be determined by a simulation-based approach: we generate artificial examples using the prescribed operator but with fictitious (expected) exact and noisy data and then perform the inversion for calibrating the hyperparameter γ.

3.5.2 *Variational characterization*

In this part, we analyze the balancing principle, which finds $\alpha > 0$ such that

$$\bar{\phi}(\alpha) = \gamma \alpha \bar{\psi}(\alpha). \tag{3.35}$$

An equivalent characterization (for the L^2-L^2 model) is the local minimum criterion introduced in [257], which determines an optimal $\alpha > 0$ by minimizing

$$\Psi_\gamma(\alpha) = \bar{\phi}(\alpha)^\gamma \bar{\psi}(\alpha). \tag{3.36}$$

This criterion is not very convenient for analysis, as we shall see below. Hence we introduce an alternative variational characterization of the principle. Throughout, we assume $F(\alpha)$ is positive for all $\alpha > 0$, which holds for all commonly used models. The rule finds an $\alpha > 0$ by minimizing

$$\Phi_\gamma(\alpha) = \frac{F^{1+\gamma}(\alpha)}{\alpha}. \tag{3.37}$$

The rule Φ_γ follows from (3.35): If the function F is differentiable, then Lemma 3.5 implies $\bar{\phi}(\alpha) - \gamma\alpha\bar{\psi}(\alpha) = F(\alpha) - (1+\gamma)\alpha F'(\alpha) = 0$. Hence, the identity

$$(1+\gamma)\frac{dF}{F} = \frac{d\alpha}{\alpha}$$

holds, and upon integration, we obtain the rule (3.37).

Remark 3.29. The criterion Ψ_γ can also be derived from (3.35). By Theorem 3.7, (3.35) can be rewritten as $\gamma\frac{d\bar{\phi}}{\bar{\phi}} = -\frac{d\bar{\psi}}{\bar{\psi}}$, whose primitive is $\ln \bar{\phi}^\gamma \bar{\psi}$.

As to the existence of a minimizer of the rule Φ_γ, we have the following remark. If there exists an element $\tilde{u} \in X$ such that $\psi(\tilde{u}) = 0$, then the optimality of u_α^δ implies directly the uniform boundedness of $F(\alpha)$. Then, obviously, $\alpha = +\infty$ is a global minimizer of Φ_γ. The existence of a finite minimizer to Φ_γ is not always guaranteed. However, this can be remedied by a simple modification

$$\tilde{\Phi}_\gamma(\alpha) = \frac{(F(\alpha) + b_0\alpha)^{1+\gamma}}{\alpha}, \tag{3.38}$$

where $b_0 > 0$ is a small number, and its optimality condition reads

$$\alpha = \frac{1}{\gamma}\frac{\bar{\phi}(\alpha)}{\bar{\psi}(\alpha) + b_0},$$

i.e., (3.33) with $b_1 = 0$. Obviously, $\lim_{\alpha\to+\infty} \tilde{\Phi}_\gamma(\alpha) = +\infty$. Under the condition $\lim_{\alpha\to 0^+} \bar{\phi}(\alpha) > 0$, it follows that $\lim_{\alpha\to 0^+} \tilde{\Phi}_\gamma(\alpha) = +\infty$. Thus, a finite minimizer $0 < \alpha < \infty$ is guaranteed.

It is worth noting that uniqueness of the minimizer to J_α is not needed for properly defining Φ_γ since it solely depends on the continuous function F. Hence the problem of minimizing Φ_γ over any bounded closed intervals is meaningful. However, the uniqueness of the minimizer u_α^δ to J_α is generally necessary in order to properly define (3.36) and (3.35). They are ill-defined if the functions $\bar{\phi}(\alpha)$ and $\bar{\psi}(\alpha)$ are multi-valued/discontinuous. We shall show below that Φ_γ indeed extends Ψ_γ to nonsmooth models in that they share the set of local minimizers when F'' exists and is negative. However, the rule Φ_γ admits (1) easier theoretical justifications (a posteriori error estimates), and (2) an efficient numerical algorithm with practically very desirable monotone convergence. There seem no known results on either a posteriori error estimates for the criterion Ψ_γ or its efficient numerical implementation.

A first result characterizes the minimizer and sheds insight into the mechanism of the rule Φ_γ: the necessary optimality condition coincides with (3.35), and hence it does implement the balancing idea.

Theorem 3.23. *If $\alpha > 0$ is a local minimizer of $\Phi_\gamma(\alpha)$, then F is differentiable at α and* (3.35) *holds for all minimizers $u_\alpha^\delta \in S(\alpha)$.*

Proof. If α is a local minimizer of Φ_γ, then

$$D^+\Phi_\gamma(\alpha) \geq 0 \quad \text{and} \quad D^-\Phi_\gamma(\alpha) \leq 0$$

where

$$D^\pm\Phi_\gamma(\alpha) = ((1+\gamma)\alpha D^\pm F(\alpha) - F(\alpha))\frac{F^\gamma}{\alpha^2}.$$

However, by Theorem 3.4, $D^+F(\alpha) \leq D^-F(\alpha)$, which implies $D^+F(\alpha) = D^-F(\alpha)$. Thus F is differentiable at α and

$$\Phi_\gamma'(\alpha) = (\gamma\alpha\,\bar{\psi}(\alpha) - \bar{\phi}(\alpha))\frac{F^\gamma(\alpha)}{\alpha^2} = 0. \tag{3.39}$$

Simplifying the formula gives (3.35). □

Remark 3.30. Theorem 3.23 holds also for nonlinear and/or constrained problems.

In the next three results, we show that the rule Φ_γ is an extension of the criterion Ψ_γ. To this end, we scale Φ_γ by the factor $c_\gamma = \frac{\gamma^\gamma}{(1+\gamma)^{1+\gamma}}$ and denote $c_\gamma\Phi_\gamma$ by Φ_γ. A first result compares Φ_γ with Ψ_γ.

Theorem 3.24. *For any $\gamma > 0$, there holds*

$$\Psi_\gamma(\alpha) \leq \Phi_\gamma(\alpha) \quad \textit{for all } \alpha.$$

The equality is achieved if and only if α solves (3.35).

Proof. Recall that for any $a, b \geq 0$ and $p, q > 1$ with $\frac{1}{p} + \frac{1}{q} = 1$, there holds $ab \leq \frac{a^p}{p} + \frac{b^q}{q}$, and the equality holds if and only if $a^p = b^q$. Let $p = \frac{1+\gamma}{\gamma}$ and $q = 1 + \gamma$. Applying the inequality with $a = \bar{\phi}^{\frac{\gamma}{1+\gamma}} \alpha^{-\frac{\gamma}{2(1+\gamma)}}$ and $b = (\gamma\bar{\psi})^{\frac{1}{1+\gamma}} \alpha^{\frac{1}{2(1+\gamma)}}$ gives

$$\bar{\phi}^{\frac{\gamma}{1+\gamma}} (\gamma\bar{\psi})^{\frac{1}{1+\gamma}} \alpha^{\frac{1-\gamma}{2(1+\gamma)}} \leq \frac{\gamma}{1+\gamma} \frac{\bar{\phi} + \alpha\bar{\psi}}{\alpha^{\frac{1}{2}}} = \frac{\gamma}{1+\gamma} \frac{F(\alpha)}{\alpha^{\frac{1}{2}}}.$$

Collecting terms in the inequality yields

$$\bar{\phi}^{\frac{\gamma}{1+\gamma}} \bar{\psi}^{\frac{1}{1+\gamma}} \leq \gamma^{-\frac{1}{1+\gamma}} \frac{\gamma}{1+\gamma} \frac{F(\alpha)}{\alpha^{\frac{1}{1+\gamma}}}.$$

Hence, we have

$$\Psi_\gamma(\alpha) \leq \frac{\gamma^\gamma}{(1+\gamma)^{1+\gamma}} \frac{F^{1+\gamma}(\alpha)}{\alpha} = \Phi_\gamma(\alpha).$$

The equality holds if and only if

$$[\bar{\phi}^{\frac{\gamma}{1+\gamma}} \alpha^{-\frac{\gamma}{2(1+\gamma)}}]^{\frac{1+\gamma}{\gamma}} = [(\gamma\bar{\psi})^{\frac{1}{1+\gamma}} \alpha^{\frac{1}{2(1+\gamma)}}]^{1+\gamma},$$

which yields precisely the relation (3.35). □

Next we show that the rule Φ_γ is indeed an extension of Ψ_γ: The set of strict local minimizers of Ψ_γ coincides with that of Φ_γ if $D^\pm F' < 0$.

Theorem 3.25. *Let F' exist and be continuous, and $D^\pm F' < 0$. Then $\alpha > 0$ is a strict local minimizer of Ψ_γ if and only if it is a strict local minimizer of Φ_γ.*

Proof. Let $\alpha > 0$ be a strict local minimizer of Ψ_γ, i.e., there exists an open neighborhood $\mathcal{U}$ of α such that $\Psi_\gamma(\alpha) < \Psi_\gamma(\zeta)$ for all $\zeta \in \mathcal{U} \setminus \{\alpha\}$.

We first claim that $\bar{\phi}(\alpha) \equiv F - \alpha F' \neq 0$. If this does not hold, then (3.3) and the nonnegativity of $\bar{\phi}(\zeta)$ indicate that $\bar{\phi}(\zeta) = 0$ for all $0 < \zeta < \alpha$. Hence $\frac{F'(\zeta)}{F(\zeta)} = \frac{1}{\zeta}$ and $F(\zeta) = C\zeta$, contradicting the assumption $D^\pm F'(\zeta) < 0$. This assertion and the necessary optimality conditions $D^+\Psi_\gamma(\alpha) \geq 0$ and $D^-\Psi_\gamma(\alpha) \leq 0$, where

$$D^\pm \Psi_\gamma(\alpha) = -(F - \alpha F')^{\gamma-1} D^\pm F'[(1+\gamma)\alpha F' - F], \tag{3.40}$$

imply

$$(1+\gamma)\alpha F' - F = \gamma\alpha\bar{\psi}(\alpha) - \bar{\phi}(\alpha) = 0.$$

Now Theorem 3.24 yields

$$\Phi_\gamma(\alpha) = \Psi_\gamma(\alpha) < \Psi_\gamma(\zeta) \leq \Phi_\gamma(\zeta) \quad \forall \zeta \in \mathcal{U} \setminus \{\alpha\},$$

i.e., α is a strict local minimizer of Φ_γ.

Next let α be a strict local minimizer of Φ_γ. Then by Theorem 3.24, $\Psi'_\gamma(\alpha) = 0$, and there exits a small $h > 0$ such that

$$\Phi'_\gamma(\zeta) < 0, \quad \alpha - h < \zeta < \alpha \quad \text{and} \quad \Phi'_\gamma(\zeta) > 0, \quad \alpha < \zeta < \alpha + h. \quad (3.41)$$

In view of (3.39), (3.40) and the condition $D^\pm F' < 0$, we deduce

$$\Phi'_\gamma(\zeta) D^+\Psi_\gamma(\zeta) > 0 \quad \text{and} \quad \Phi'_\gamma(\zeta) D^-\Psi_\gamma(\zeta) > 0.$$

Hence we have $D^\pm\Psi_\gamma(\zeta) < 0$ if $\alpha - h < \zeta < \alpha$ and $D^\pm\Psi_\gamma(\zeta) > 0$ if $\alpha < \zeta < \alpha + h$, i.e., α is a strict local minimizer of Ψ_γ. □

Numerically, we observe that Ψ_γ is flatter than Φ_γ in the neighborhood of a local minimizer; see Fig. 3.1. The following result justifies the observation.

Theorem 3.26. *Let F'' exist and be continuous. If the inequality $\Phi''_\gamma > 0$ holds at a local minimizer α^*, then $\frac{\Psi'_\gamma}{\Phi'_\gamma} < 1$ in its neighborhood.*

Proof. Direct computation gives

$$\Phi''_\gamma = \frac{\gamma^\gamma}{(1+\gamma)^{1+\gamma}} \frac{F^{\gamma-1}}{\alpha^3} \left[\gamma(1+\gamma)(\alpha F')^2 + (1+\gamma)\alpha^2 F'' F \right.$$
$$\left. -2(1+\gamma)\alpha F' F + 2F^2\right].$$

Consequently,

$$\begin{aligned} \frac{\alpha^2 \Phi''_\gamma}{\Phi_\gamma} &= \gamma(1+\gamma)\frac{(\alpha F')^2}{F^2} + (1+\gamma)\frac{\alpha^2 F''}{F} - 2(1+\gamma)\frac{\alpha F'}{F} + 2 \\ &= \gamma(1+\gamma)\theta^2 + (1+\gamma)\frac{\alpha^2 F''}{F} - 2(1+\gamma)\theta + 2, \end{aligned}$$

where $\theta = \frac{\alpha F'}{F}$. Upon rearranging the terms, we get

$$-(1+\gamma)\frac{\alpha^2 F''}{F} = \gamma(1+\gamma)\theta^2 - 2(1+\gamma)\theta + 2 - \frac{\alpha^2\Phi''_\gamma}{\Phi_\gamma}.$$

Assisted with this identity, we deduce

$$\begin{aligned} \frac{\Psi'_\gamma}{\Phi'_\gamma} &= [-F'']\bar{\phi}^{\gamma-1}\alpha^2(\gamma+1)^{\gamma+1}\gamma^{-\gamma}F^{-\gamma} \\ &= \left(1+\gamma^{-1}\right)^\gamma \cdot (1+\gamma)\frac{-\alpha^2 F''}{F} \cdot \left(\frac{\bar{\phi}}{F}\right)^{\gamma-1} \\ &= \left(1+\gamma^{-1}\right)^\gamma \cdot \left[\gamma(1+\gamma)\theta^2 - 2(1+\gamma)\theta + 2 - \frac{\alpha^2\Phi''_\gamma}{\Phi_\gamma}\right] \cdot (1-\theta)^{\gamma-1}. \end{aligned}$$

At a local minimizer α^*, we have $\theta = \frac{1}{1+\gamma}$, and thus

$$\frac{\Psi'_\gamma}{\Phi'_\gamma} = 1 - \left(1+\gamma^{-1}\right)\frac{\alpha^2\Phi''_\gamma}{\Phi_\gamma} < 1,$$

by noting the assumption $\Phi''_\gamma(\alpha) > 0$ at α^*. The remaining assertion follows from the continuity condition of F''. □

As was mentioned in Remark 3.28, the hyperparameter γ in principle should be calibrated from synthetic data, or estimated from a priori statistical knowledge of the problem. Here we provide a sensitivity analysis of the selected regularization parameter α with respect to γ.

Remark 3.31. We assume that the functions $\bar{\phi}$ and $\bar{\psi}$ are differentiable. It follows from the balancing equation (3.35) and the chain rule that

$$\frac{d\alpha}{d\gamma}(\bar{\phi}' - \alpha\gamma\bar{\psi}' - \gamma\bar{\psi}) - \alpha\bar{\psi} = 0,$$

and meanwhile the second-order derivative of $\Phi_\gamma(\alpha)$ at a local minimizer α^* is given by

$$\frac{d^2}{d\alpha^2}\Phi_\gamma(\alpha)|_{\alpha=\alpha^*} = \frac{\gamma\bar{\psi} + \gamma\alpha^*\bar{\psi}' - \bar{\phi}'}{(\alpha^*)^2}F^\gamma(\alpha^*).$$

At a strict local minimizer, we have $\frac{d^2\Phi_\gamma}{d\alpha^2}(\alpha^*) > 0$, and thus

$$\frac{d\alpha}{d\gamma} < 0,$$

i.e., the selected parameter α^* is a decreasing function of γ. Next we derive a semi-analytic and semi-heuristic formula. Note that by Theorem 3.7, there holds

$$\bar{\phi}' - \alpha\gamma\bar{\psi}' - \gamma\bar{\psi} = -(1+\gamma)\alpha\bar{\psi}' - \gamma\bar{\psi},$$

and that $\bar{\psi}' \leq 0$ and $\bar{\psi} \geq 0$. Therefore, overall its definitiveness is unclear. However, in practice, it may happen that

$$-\beta\alpha\bar{\psi}' \leq \bar{\psi}.$$

Then we expect an estimate of the form $\frac{d\bar{\psi}}{\bar{\psi}} \leq -\beta\frac{d\alpha}{\alpha}$ i.e., $\bar{\psi} \leq \alpha^{-\beta}$. Hence, under the condition that $-(1+\gamma)\alpha\bar{\psi}' - \gamma\bar{\psi} \approx -\tilde{\beta}\bar{\psi}$, we have

$$-\frac{d\alpha}{d\gamma}\tilde{\beta} - \alpha\bar{\psi} \approx 0,$$

i.e., $\alpha \approx \alpha(\bar{\gamma})e^{-\tilde{\beta}(\gamma-\bar{\gamma})}$. In other words, the regularization decays (locally) exponentially with the increase of the parameter γ.

Now we provide justifications of the balancing principle by deriving a posteriori error estimates for the case of a Hilbertian Y and $\phi(u, g^\delta) = \frac{1}{2}\|Ku - g^\delta\|^2$. First, let X be a Hilbert space, $\psi(u) = \frac{1}{2}\|u\|^2$ and there is no constraint.

Theorem 3.27. *Assume that the exact solution $u^\dagger \in X_{\mu,\rho} = \{u \in X : u = (K^*K)^\mu w,\ \|w\| \leq \rho\}$ for some $0 < \mu \leq 1$. Let α^* be determined by the rule Φ_γ, and $\delta_* := \|g^\delta - Ku_{\alpha^*}^\delta\|$. Then the following estimate holds*

$$\|u^\dagger - u_{\alpha^*}^\delta\| \leq C\left(\rho^{\frac{1}{1+2\mu}} + \frac{F(\delta^{\frac{2}{1+2\mu}})^{\frac{1+\gamma}{2}}}{F(\alpha^*)^{\frac{1+\gamma}{2}}}\right)\max\{\delta, \delta_*\}^{\frac{2\mu}{1+2\mu}}. \tag{3.42}$$

Proof. We decompose the error $u^\dagger - u_\alpha^\delta$ into

$$u^\dagger - u_\alpha^\delta = r_\alpha(K^*K)u^\dagger + g_\alpha(K^*K)K^*(g^\dagger - g^\delta),$$

where $g_\alpha(t) = \frac{1}{\alpha+t}$ and $r_\alpha(t) = 1 - tg_\alpha(t) = \frac{\alpha}{\alpha+t}$. In view of the source condition and the interpolation inequality, we derive

$$\begin{aligned}
\|r_\alpha(K^*K)u^\dagger\| &= \|r_\alpha(K^*K)(K^*K)^\mu w\| \\
&\leq \|(K^*K)^{\frac{1}{2}+\mu} r_\alpha(K^*K)w\|^{\frac{2\mu}{2\mu+1}} \|r_\alpha(K^*K)w\|^{\frac{1}{2\mu+1}} \\
&= \|r_\alpha(KK^*)Ku^\dagger\|^{\frac{2\mu}{2\mu+1}} \|r_\alpha(K^*K)w\|^{\frac{1}{2\mu+1}} \\
&\leq c\left(\|r_\alpha(KK^*)g^\delta\| + \|r_\alpha(KK^*)(g^\delta - g^\dagger)\|\right)^{\frac{2\mu}{2\mu+1}} \|w\|^{\frac{1}{2\mu+1}},
\end{aligned}$$

where the constant c depends only on the maximum of r_α over $[0, \|K\|^2]$. This together with the relation $r_{\alpha^*}(KK^*)g^\delta = g^\delta - Ku_{\alpha^*}^\delta$ yields

$$\|r_{\alpha^*}(K^*K)u^\dagger\| \leq c(\delta_* + c\delta)^{\frac{2\mu}{2\mu+1}} \rho^{\frac{1}{2\mu+1}} \leq c_1 \max\{\delta, \delta_*\}^{\frac{2\mu}{2\mu+1}} \rho^{\frac{1}{2\mu+1}}.$$

It remains to estimate the term $\|g_{\alpha^*}(K^*K)K^*(g^\delta - g^\dagger)\|$. The standard estimate [88] yields

$$\|g_{\alpha^*}(K^*K)K^*(g^\delta - g^\dagger)\| \leq c\frac{\delta}{\sqrt{\alpha^*}}.$$

However, by the minimizing property of α^*, we have

$$\frac{F(\alpha^*)^{1+\gamma}}{\alpha^*} \leq \frac{F(\hat{\alpha})^{1+\gamma}}{\hat{\alpha}}.$$

We may take $\hat{\alpha} = \delta^{\frac{2}{1+2\mu}}$, and then we have

$$\frac{1}{\alpha^*} \leq \frac{F(\delta^{\frac{2}{1+2\mu}})^{1+\gamma}}{F(\alpha^*)^{1+\gamma}} \frac{1}{\delta^{\frac{2}{1+2\mu}}}.$$

Combining the preceding estimates, we arrive at

$$\|u^\dagger - u_{\alpha^*}^\delta\| \le c_1 \max\{\delta,\delta_*\}^{\frac{2\mu}{2\mu+1}} \rho^{\frac{1}{2\mu+1}} + c\frac{F(\delta^{\frac{2}{1+2\mu}})^{\frac{1+\gamma}{2}}}{F(\alpha^*)^{\frac{1+\gamma}{2}}}\delta^{\frac{2\mu}{1+2\mu}}$$

$$\le C\left(\rho^{\frac{1}{1+2\mu}} + \frac{F(\delta^{\frac{2}{1+2\mu}})^{\frac{1+\gamma}{2}}}{F(\alpha^*)^{\frac{1+\gamma}{2}}}\right)\max\{\delta,\delta_*\}^{\frac{2\mu}{1+2\mu}}.$$

This shows the desired a posteriori error estimate. □

Next we a second a posteriori error estimate for a general convex penalty.

Theorem 3.28. *Let condition* (3.6) *hold, and* $\delta_* = \|Ku_{\alpha^*}^\delta - g^\delta\|$. *Then for each* α^* *given by the rule* Φ_γ, *the following estimate holds*

$$d_{\xi^\dagger,\mu^\dagger}(u_{\alpha^*}^\delta, u^\dagger) \le C\left(1 + \frac{F(\delta)^{1+\gamma}}{F(\alpha^*)^{1+\gamma}}\right)\max\{\delta,\delta_*\}.$$

Proof. By Corollary 3.5, we have for any α

$$d_{\xi^\dagger,\mu^\dagger}(u_\alpha^\delta, u^\dagger) \le d_{\xi_\alpha,\mu_\alpha}(u_\alpha^\delta, u_\alpha) + d_{\xi^\dagger,\mu^\dagger}(u_\alpha, u^\dagger) + 6\|w\|\delta.$$

By the proof of Theorem 3.19, we have

$$d_{\xi^\dagger,\mu^\dagger}(u_{\alpha^*}, u^\dagger) \le 4\|w\| \max(\delta,\delta_*).$$

It remains to estimate the term $d_{\xi_\alpha,\mu_\alpha}(u_\alpha^\delta, u_\alpha)$. By Theorem 3.9, we have $d_{\xi_{\alpha^*},\mu_{\alpha^*}}(u_{\alpha^*}^\delta, u_{\alpha^*}) \le \frac{\delta^2}{2\alpha^*}$. The minimizing property of α^* implies that for $\hat\alpha$

$$\frac{F(\alpha^*)^{1+\gamma}}{\alpha^*} \le \frac{F(\hat\alpha)^{1+\gamma}}{\hat\alpha}, \quad \text{i.e.} \quad \frac{1}{\alpha^*} \le \frac{F(\hat\alpha)^{1+\gamma}}{F(\alpha^*)^{1+\gamma}}\frac{1}{\hat\alpha}.$$

We may take $\hat\alpha = \delta$ and combine the above two inequalities to arrive at

$$d_{\xi_{\alpha^*},\mu_{\alpha^*}}(u_{\alpha^*}^\delta, u_{\alpha^*}) \le \frac{F(\delta)^{1+\gamma}}{F(\alpha^*)^{1+\gamma}}\frac{\delta}{2}.$$

Now summarizing these three estimates gives

$$d_{\xi^\dagger,\mu^\dagger}(u_{\alpha^*}^\delta, u^\dagger) \le C\left(1 + \frac{F(\delta)^{1+\gamma}}{F(\alpha^*)^{1+\gamma}}\right)\max\{\delta,\delta_*\},$$

with $C = \max\{10\|w\|, \frac{1}{2}\}$. □

Remark 3.32. If the discrepancy δ_* is of order δ, Theorems 3.27 and 3.28 imply that the approximation $u_{\alpha^*}^\delta$ with α^* chosen by the balancing principle converges to the exact solution $u^\dagger$ at the same rate as a priori parameter choice rules under identical source conditions. If δ_* does not decrease as quickly as δ, then the convergence would be suboptimal. More dangerous is the case that δ_* decreases more quickly. Then the prefactor may blow up, and the approximation may diverge. Hence the value of δ_* should always be monitored as an a posteriori check: The computed approximation $u_{\alpha^*}^\delta$ should be discarded if δ_* is deemed too small. This applies to other heuristic rules, e.g., the Hanke-Raus rule.

3.5.3 *Fixed point algorithm*

Now we develop a fixed point algorithm, cf. Algorithm 3.2, to determine an optimal α by the rule Φ_γ. We provide a local convergence analysis of the algorithm for the following scenario.

Assumption 3.2. Let the interval $[\zeta_0, \zeta_1]$ satisfy: *(a)* $\bar{\psi}(\zeta_1) > 0$; and *(b)* There exists an optimal parameter α_b in $[\zeta_0, \zeta_1]$ such that $D^\pm\Phi_\gamma(\alpha) < 0$ for $\zeta_0 \leq \alpha < \alpha_b$ and $D^\pm\Phi_\gamma(\alpha) > 0$ for $\alpha_b < \alpha \leq \zeta_1$.

In view of Lemma 3.3, $\bar{\psi}(\zeta_1) > 0$ implies $\bar{\psi}(\alpha) > 0$ for any $\alpha \in [0, \zeta_1]$. Assumption 3.2(a) guarantees the well-definedness of the algorithm, and it holds for a broad class of nonsmooth models, e.g., the L^2-ℓ^1 and L^2-TV model. Assumption 3.2(b) ensures that there exists only one local minimizer α_b to Φ_γ in the interval $[\zeta_0, \zeta_1]$. In Algorithm 3.2, the stopping criterion can be based on monitoring the relative change of consecutive iterates (either u_α^δ or α). It will terminate automatically when two consecutive iterates coincide, i.e., $\alpha_{k_0+1} = \alpha_{k_0}$ for some k_0 (then the sequence $\{\alpha_k\}$ is finite).

Algorithm 3.2 Fixed point algorithm

1: Set $k = 0$ and choose α_0;
2: **for** $k = 0, 1, \ldots, K$ **do**
3: Solve for $u_{\alpha_k}^\delta$ by (3.2) with $\alpha = \alpha_k$, i.e.,

$$u_{\alpha_k}^\delta \in \arg\min_{u \in C} \{\phi(u, g^\delta) + \alpha_k \psi(u)\}.$$

4: Update the regularization parameter α_{k+1} by

$$\alpha_{k+1} = \frac{1}{\gamma} \frac{\phi(u_{\alpha_k}^\delta, g^\delta)}{\psi(u_{\alpha_k}^\delta)}.$$

5: Check the stopping criterion.
6: **end for**
7: Return approximation $(u_{\alpha_K}^\delta, \alpha_K)$.

The following lemma provides a monotonicity of the regularization parameter computed by Algorithm 3.2, which is key to analyzing its convergence. We introduce

$$r(\alpha) = \bar{\phi}(\alpha) - \gamma\alpha\bar{\psi}(\alpha)$$

for $u_\alpha^\delta \in S(\alpha)$ used in the representation $\bar{\phi}$ and $\bar{\psi}$. The identity $r(\alpha) = 0$ implies that (3.35) holds.

Lemma 3.15. *Let Assumption 3.2 hold, and* $\alpha_0 \in [\zeta_0, \zeta_1]$. *Then the sequence* $\{\alpha_k\}_k$ *generated by Algorithm 3.2 satisfies*

(a) *It is either finite or infinite and strictly monotone, and increasing (decreasing) if* $r(\alpha_0) > 0$ $(r(\alpha_0) < 0)$;

(b) *It is contained in* $[\zeta_0, \alpha_b]$ $([\alpha_b, \zeta_1])$ *if* $r(\alpha_0) > 0$ $(r(\alpha_0) < 0)$.

Proof. We consider only the case $r(\alpha_0) > 0$ since the other case follows analogously. The proof proceeds by induction on k. By definition, $\alpha_1 = (\gamma\bar{\psi}(\alpha_0))^{-1}\bar{\phi}(\alpha_0) > \alpha_0$. Now let us assume $\alpha_k > \alpha_{k-1}$. By Lemma 3.3, we have

$$\begin{aligned}\alpha_{k+1} - \alpha_k &= \gamma^{-1}\frac{\bar{\phi}(\alpha_k)}{\bar{\psi}(\alpha_k)} - \gamma^{-1}\frac{\bar{\phi}(\alpha_{k-1})}{\bar{\psi}(\alpha_{k-1})} \\ &= \frac{[\bar{\phi}(\alpha_k) - \phi(\bar{\alpha}_{k-1})]\bar{\psi}(\alpha_{k-1})}{\gamma\bar{\psi}(\alpha_{k-1})\bar{\psi}(\alpha_k)} \\ &\quad + \frac{\bar{\phi}(\alpha_{k-1})[\bar{\psi}(\alpha_{k-1}) - \bar{\psi}(\alpha_k)]}{\gamma\bar{\psi}(\alpha_{k-1})\bar{\psi}(\alpha_k)} \geq 0.\end{aligned}$$

Hence the sequence $\{\alpha_k\}$ is monotonically increasing. Here are two possible situations: (1) $\alpha_{k+1} = \alpha_k$, the algorithm terminates and the sequence is finite; (2) $\alpha_{k+1} > \alpha_k$, and the induction continues. This shows the assertion in part (a).

By virtue of the preceding proof, $\{\alpha_k\}$ is strictly monotone for all $k \leq k_0$ (k_0 is possibly $+\infty$). Thus the identity

$$\alpha_{k+1} - \alpha_k = (\gamma\bar{\psi}(\alpha_k))^{-1}r(\alpha_k)$$

implies $r(\alpha_k) > 0$ for all $k \leq k_0 - 1$. Obviously, it suffices to show (b) for $\{\alpha_k\}_0^{k_0}$. Theorem 3.4 and Lemma 3.3 indicate

$$\begin{aligned}D^+\Phi_\gamma(\alpha) &\leq \alpha^{-2}F^\gamma(\alpha)(\gamma\alpha\bar{\psi}(\alpha) - \bar{\phi}(\alpha)) \\ &= \alpha^{-2}F^\gamma(\alpha)(-r(\alpha)) \leq D^-\Phi_\gamma(\alpha).\end{aligned} \tag{3.43}$$

Consequently, $D^+\Phi_\gamma(\alpha_0) < 0$, which together with Assumption 3.2 implies $\zeta_0 \leq \alpha_0 < \alpha_b$. Now let us assume $\zeta_0 \leq \alpha_k < \alpha_b$. We claim that

$$r(\alpha) > 0 \quad \text{for all} \quad \alpha \in (\alpha_k, \alpha_{k+1}). \tag{3.44}$$

If this claim is not true, then there exists an $\hat{\alpha} \in (\alpha_k, \alpha_{k+1})$ such that $r(\hat{\alpha}) \leq 0$, i.e., $(\gamma\bar{\psi}(\hat{\alpha}))^{-1}\bar{\phi}(\hat{\alpha}) \leq \hat{\alpha}$. However, Lemma 3.3 implies

$$\alpha_{k+1} = (\gamma\bar{\psi}(\alpha_k))^{-1}\bar{\phi}(\alpha_k) \leq (\gamma\bar{\psi}(\hat{\alpha}))^{-1}\bar{\phi}(\hat{\alpha}) \leq \hat{\alpha}.$$

This is a contradiction. Finally, we assert $\alpha_{k+1} \leq \alpha_b$, which would conclude the proof. Suppose the contrary, i.e., $\alpha_b < \alpha_{k+1}$, then (3.44) implies $r(\alpha_b) > 0$. This contradicts the assumption $r(\alpha_b) = 0$, and thus $\zeta_0 \leq \alpha_{k+1} \leq \alpha_b$. □

The monotonicity result holds for any minimizer u_α^δ of J_α and the uniqueness of the solution is not required. If $r(\alpha_0) < 0$, then the sequence $\{\alpha_k\}$ is decreasing and is bounded from below by zero, and thus, $\alpha_k \downarrow \alpha^*$. Meanwhile, it is monotonically increasing for $r(\alpha_0) > 0$. Then $\alpha_k \uparrow \alpha^* < +\infty$ if it is also bounded. It may tend to $+\infty$, reminiscent of the fact that $+\infty$ is a global minimizer. Algorithm 3.2 adapts straightforwardly to (3.38) by changing Step 4 to

$$\alpha_{k+1} = \frac{1}{\gamma} \frac{\bar{\phi}(\alpha_k)}{\bar{\psi}(\alpha_k) + \beta_1}.$$

Theorem 3.29 below remains valid for this modification.

Now we prove the descent property of the fixed point algorithm.

Theorem 3.29. *Let Assumption 3.2 hold, and $\alpha_0 \in [\zeta_0, \zeta_1]$. Then the following statements hold for the sequence $\{\alpha_k\}$ generated by Algorithm 3.2*

(a) The sequence $\{\Phi_\gamma(\alpha_k)\}$ is monotonically decreasing.
(b) The sequence $\{\alpha_k\}$ converges to the local minimizer α_b.

Proof. We focus on $r(\alpha_0) > 0$. By Lemma 3.15, it suffices to consider $\alpha_k < \alpha_{k+1}$. The concavity of F implies its local Lipschitz continuity (see pp. 236 of [91]), and hence Φ_γ is locally Lipschitz continuous and absolutely continuous on any bounded closed interval. Moreover, Φ'_γ exists almost everywhere with $\Phi'_\gamma(\alpha) = \alpha^{-2}F^\gamma(\alpha)(-r(\alpha))$, and it is locally integrable. Therefore, the following identity holds

$$\Phi_\gamma(\alpha_{k+1}) = \Phi_\gamma(\alpha_k) + \int_{\alpha_k}^{\alpha_{k+1}} \Phi'_\gamma(\alpha)\, d\alpha,$$

which together with (3.44) yields $\Phi_\gamma(\alpha_{k+1}) < \Phi_\gamma(\alpha_k)$. This shows part (a).

By Lemma 3.15(b), there exists a limit $\alpha^* \in [\zeta_0, \zeta_1]$ to the sequence. We shall show $D^-\Phi_\gamma(\alpha^*) = 0$. We treat two cases separately: A finite sequence $\{\alpha_k\}_{k=1}^{k_0}$ and $\alpha_k < \alpha_{k+1}$ for all k. For the former, the relations (3.43) and $\gamma\alpha\bar{\psi}(\alpha_{k_0}) - \bar{\phi}(\alpha_{k_0}) = 0$ indicate $D^\pm\Phi_\gamma(\alpha_{k_0}) = 0$. By Assumption 3.2, α_{k_0} is the local minimizer α_b. For the latter, Theorem 3.4 indicates

$$\begin{aligned} \frac{1}{\gamma} \frac{F(\alpha_k) - \alpha_k D^- F(\alpha_k)}{D^- F(\alpha_k)} \leq \alpha_{k+1} &= \frac{1}{\gamma} \frac{\bar{\phi}(\alpha_k)}{\bar{\psi}(\alpha_k)} \\ &\leq \frac{1}{\gamma} \frac{F(\alpha_k) - \alpha_k D^+ F(\alpha_k)}{D^+ F(\alpha_k)}. \end{aligned}$$

By virtue of the left continuity of D^-F and the inequalities $D^-F(\zeta) \leq D^+F(\alpha) \leq D^-F(\alpha)$ for all $\alpha < \zeta$, we deduce $\lim_{k\to\infty} D^\pm F(\alpha_k) =$

$D^- F(\alpha^*)$, and

$$\alpha^* = \frac{1}{\gamma} \frac{F(\alpha^*) - \alpha^* D^- F(\alpha^*)}{D^- F(\alpha^*)},$$

i.e., $D^-\Phi_\gamma(\alpha^*) = 0$. Hence, the limit α^* is the local minimizer $\alpha_{\rm b}$. □

3.6 Multi-parameter Tikhonov regularization

In this part, we turn to multi-parameter Tikhonov regularization

$$J_{\boldsymbol{\alpha}}(u) = \phi(u, g^\delta) + \boldsymbol{\alpha} \cdot \boldsymbol{\psi}(u) \quad \text{over } u \in \mathcal{C}, \tag{3.45}$$

where $\mathcal{C} \subset X$ is a closed convex (feasible solution) set, and takes its minimizer, denoted by $u_{\boldsymbol{\alpha}}^\delta$ ($u_{\boldsymbol{\alpha}}$ in case of the exact data $g^\dagger$), as a solution. The notation $\boldsymbol{\alpha} \cdot \boldsymbol{\psi}(u)$ denotes the dot product between the regularization parameters $\boldsymbol{\alpha} = (\alpha_1, \ldots, \alpha_n)^{\rm t} \in \mathbb{R}_+^n$ and the vector-valued regularization functionals $\boldsymbol{\psi}(u) = (\psi_1(u), \ldots, \psi_n(u))^{\rm t}$.

In comparison with Tikhonov regularization discussed earlier, the model (3.45) involves multiple regularization. This is motivated by empirical observations. For example, images typically exhibit multiple distinct features/structures. However, single regularization generally favors one feature over others, and thus inappropriate for simultaneously preserving multiple features. A reliable and simultaneous recovery of several distinct features calls for multiple regularization. This idea has been pursued earlier in the literature. For instance, in [162] the authors proposed a model to preserve both flat and gray regions in natural images by combining total variation regularization with Sobolev smooth regularization. One classical example of multi-parameter Tikhonov regularization is elastic net [313, 170].

There are a number of theoretical developments on multi-parameter regularization in a Hilbert space setting. In [26, 41], the L-hypersurface and generalized cross validation were suggested for determining the regularization parameters for finite-dimensional linear systems, respectively. In [62], a multi-resolution analysis for ill-posed linear operator equations was given, and some convergence rates results were shown; see [221] also for error estimates for the discrepancy principle. In this part, we extend the augmented approach in Section 3.5 to multi-parameter regularization, following [153, 155]. We derive the balancing principle and a novel hybrid principle, i.e., balanced discrepancy principle, analyze their theoretical properties, and discuss efficient numerical implementations.

3.6.1 *Balancing principle*

First we extend the augmented Tikhonov regularization to the context of multiple penalties. For the multi-parameter model (3.45), it can be derived analogously from hierarchical Bayesian inference, and the resulting augmented Tikhonov functional $J(u, \boldsymbol{\lambda}, \tau)$ is given by

$$J(u, \boldsymbol{\lambda}, \tau) = \tau\phi(u, g^\delta) + \boldsymbol{\lambda} \cdot \boldsymbol{\psi}(u) + \sum_{i=1}^{n}(b_i\lambda_i - a_i \ln \lambda_i) + b_0\tau - a_0 \ln \tau,$$

where (a_i, b_i) are hyperparameter pairs for $\boldsymbol{\lambda}$ and τ, reflecting a priori statistical knowledge about these parameters. Let $\alpha_i = \frac{\lambda_i}{\tau}$. Then the necessary optimality condition of any minimizer $(u^\delta_{\boldsymbol{\alpha}}, \boldsymbol{\lambda}, \tau)$ to the functional $J(u, \boldsymbol{\lambda}, \tau)$ is given by

$$\begin{cases} u^\delta_{\boldsymbol{\alpha}} = \arg\min_{u \in \mathcal{C}} \left\{ \phi(u, g^\delta) + \boldsymbol{\alpha} \cdot \boldsymbol{\psi}(u) \right\}, \\ \lambda_i = \dfrac{a_i}{\psi_i(u^\delta_{\boldsymbol{\alpha}}) + b_i}, \quad i = 1, \ldots, n, \\ \tau = \dfrac{a_0}{\phi(u^\delta_{\boldsymbol{\alpha}}, g^\delta) + b_0}. \end{cases} \tag{3.46}$$

Upon rewriting the system, we arrive at the following system for $(u^\delta_{\boldsymbol{\alpha}}, \boldsymbol{\alpha})$

$$\begin{cases} u^\delta_{\boldsymbol{\alpha}} = \arg\min_{u \in \mathcal{C}} \left\{ \phi(u, g^\delta) + \boldsymbol{\alpha} \cdot \boldsymbol{\psi}(u) \right\}, \\ \alpha_i = \dfrac{1}{\gamma_i} \dfrac{\phi(u^\delta_{\boldsymbol{\alpha}}, g^\delta) + b_0}{\psi_i(u^\delta_{\boldsymbol{\alpha}}) + b_i}, \quad i = 1, \ldots, n, \end{cases} \tag{3.47}$$

where the scalars γ_i are defined by $\gamma_i = \frac{a_0}{a_i}$. The system (3.47) represents the balancing principle for selecting an optimal $\boldsymbol{\alpha}$.

Now we introduce the value function $F(\boldsymbol{\alpha})$ for $J_{\boldsymbol{\alpha}}$:

$$F(\boldsymbol{\alpha}) = \inf_{u \in \mathcal{C}} J_{\boldsymbol{\alpha}}(u).$$

The following lemma provides the essential calculus for multi-parameter Tikhonov regularization, where the subscript α_i in F_{α_i} denotes the derivative of $F(\boldsymbol{\alpha})$ with respect to α_i. The proof is analogous to the single-parameter counterpart, and hence omitted.

Lemma 3.16. *The function $F(\boldsymbol{\alpha})$ is monotone and concave, and hence almost everywhere differentiable. Further, if it is differentiable, then*

$$F_{\alpha_i}(\boldsymbol{\alpha}) = \psi_i(u^\delta_{\boldsymbol{\alpha}}).$$

Now we can provide an alternative characterization. For simplicity, we focus on the case $n = 2$. We consider the function $\overline{\Phi}$ defined by

$$\overline{\Phi}(\boldsymbol{\alpha}) = \frac{\overline{F}(\boldsymbol{\alpha})^{1+\frac{1}{\gamma_1}+\frac{1}{\gamma_2}}}{\alpha_1^{\frac{1}{\gamma_1}} \alpha_2^{\frac{1}{\gamma_2}}},$$

where $\overline{F}(\boldsymbol{\alpha}) = \overline{\phi}(\boldsymbol{\alpha}) + \boldsymbol{\alpha} \cdot \overline{\boldsymbol{\psi}}(\boldsymbol{\alpha})$ with $\overline{\phi}(\boldsymbol{\alpha}) = \phi(u_{\boldsymbol{\alpha}}^\delta, g^\delta) + b_0$ and $\overline{\psi}_i(\boldsymbol{\alpha}) = \psi_i(u_{\boldsymbol{\alpha}}^\delta) + b_i$. The necessary optimality condition for $\overline{\Phi}(\boldsymbol{\alpha})$, if $F(\boldsymbol{\alpha})$ is everywhere differentiable, is given by

$$\begin{cases} \dfrac{\partial \overline{\Phi}}{\partial \alpha_1} = \dfrac{\overline{F}(\boldsymbol{\alpha})^{\frac{1}{\gamma_1}+\frac{1}{\gamma_2}}}{\alpha_1^{\frac{1}{\gamma_1}} \alpha_2^{\frac{1}{\gamma_2}}} \dfrac{(-\gamma_2 \overline{F}(\boldsymbol{\alpha}) + \alpha_1(\gamma_1 + \gamma_2 + \gamma_1\gamma_2)\overline{F}_{\alpha_1}(\boldsymbol{\alpha}))}{\gamma_1\gamma_2\alpha_1} = 0, \\ \dfrac{\partial \overline{\Phi}}{\partial \alpha_2} = \dfrac{\overline{F}(\boldsymbol{\alpha})^{\frac{1}{\gamma_1}+\frac{1}{\gamma_2}}}{\alpha_1^{\frac{1}{\gamma_1}} \alpha_2^{\frac{1}{\gamma_2}}} \dfrac{(-\gamma_1 \overline{F}(\boldsymbol{\alpha}) + \alpha_2(\gamma_1 + \gamma_2 + \gamma_1\gamma_2)\overline{F}_{\alpha_2}(\boldsymbol{\alpha}))}{\gamma_1\gamma_2\alpha_2} = 0, \end{cases}$$

which, in view of Lemma 3.16, is equivalent to

$$\begin{cases} -\gamma_2\overline{\phi} + \alpha_1\gamma_1(1+\gamma_2)\overline{\psi}_1 - \alpha_2\gamma_2\overline{\psi}_2 = 0, \\ -\gamma_1\overline{\phi} - \alpha_1\gamma_1\overline{\psi}_1 + \alpha_2(1+\gamma_1)\gamma_2\overline{\psi}_2 = 0. \end{cases}$$

Solving the system with respect to $\overline{\psi_i}$, one obtains that for $i = 1, 2$

$$\alpha_i = \frac{1}{\gamma_i}\frac{\overline{\phi}}{\overline{\psi_i}} = \frac{1}{\gamma_i}\frac{\phi(u_{\boldsymbol{\alpha}}^\delta, g^\delta) + b_0}{\psi_i(u_{\boldsymbol{\alpha}}^\delta) + b_i}.$$

Hence, the optimality system of $\overline{\Phi}$ coincides with that of the augmented functional. In summary, we have the following equivalence.

Proposition 3.4. *Let the value function $F(\boldsymbol{\alpha})$ be differentiable everywhere. Then the critical points of the function $\overline{\Phi}$ are solutions to the optimality system* (3.47) *of the augmented Tikhonov functional $J(u, \boldsymbol{\lambda}, \tau)$.*

The function $\overline{\Phi}$ is rather flexible and provides a wide variety of possibilities: the vector $\boldsymbol{\gamma}$ may be fine tuned to achieve specific desirable properties. In the case of one single $\gamma_i = \gamma$ in $\overline{\Phi}$ for some fixed positive constant γ and also $b_i = b$, the function $\overline{\Phi}$ can be compactly written as

$$\Phi_\gamma(\boldsymbol{\alpha}) = \frac{(F(\boldsymbol{\alpha}) + b_0 + b\|\boldsymbol{\alpha}\|_1)^{n+\gamma}}{\prod_{i=1}^n \alpha_i}. \tag{3.48}$$

In view of the concavity in Lemma 3.16, the problem of minimizing Φ_γ over any bounded and closed region in $\mathbb{R}_+^n$ is well defined.

The next result characterizes a strict minimizer to Φ_γ. It also explains the term balancing principle: the selected parameter $\boldsymbol{\alpha}^*$ balances the regularization $\boldsymbol{\psi}$ and the fidelity ϕ with a weight γ.

Proposition 3.5. *If the value function $F(\boldsymbol{\alpha})$ is differentiable at a minimizer of Φ_γ, then the following optimality system holds*

$$\gamma\alpha_i\bar{\psi}_i(\boldsymbol{\alpha}) = \bar{\phi}(\boldsymbol{\alpha}). \tag{3.49}$$

Conversely, if $\boldsymbol{\alpha}$ is a strict minimizer of Φ_γ, then F is differentiable at $\boldsymbol{\alpha}$.

In practice, it is important to incorporate all prior information about the sought-for solution u whenever it is available, which can be achieved by imposing further constraints. For example, for the noise level information $\phi(u, g^\delta) \leq c$, the constraint $\mathcal{C}$ reads $\mathcal{C} = \{\phi(u, g^\delta) \leq c\}$, and by the Lagrangian approach, the augmented Tikhonov functional is given by

$$\begin{aligned} J(u, \boldsymbol{\lambda}, \tau, \mu) = \tau\phi(u, g^\delta) &+ \boldsymbol{\lambda} \cdot \boldsymbol{\psi}(u) + \sum_{i=1}^{n}(b_i\lambda_i - a_i \ln \lambda_i) \\ &+ b_0\tau - a_0 \ln \tau + \tau\langle\phi(u, g^\delta) - c, \mu\rangle, \end{aligned}$$

where the unknown scalar $\mu \geq 0$ is the Lagrange multiplier for the inequality constraint $\phi(u, g^\delta) \leq c$. The respective optimality system reads

$$\begin{cases} u_{\boldsymbol{\alpha}}^\delta = \arg\min\limits_u \left\{\phi(u, g^\delta) + \boldsymbol{\alpha} \cdot \boldsymbol{\psi}(u) + \langle\phi(u, g^\delta) - c, \mu\rangle\right\}, \\ \lambda_i = \dfrac{a_i}{\psi_i(u_{\boldsymbol{\alpha}}^\delta) + b_i}, \\ \tau = \dfrac{a_0}{(1 + \mu)\phi(u_{\boldsymbol{\alpha}}^\delta, g^\delta) + b_0}, \\ c \geq \phi(u_{\boldsymbol{\alpha}}^\delta, g^\delta). \end{cases}$$

Then the inequality constraint $\phi(x, g^\delta) \leq c$ and the balancing principle are simultaneously satisfied: for $i = 1, 2, \ldots, n$

$$\gamma_i\alpha_i(\psi_i(u_{\boldsymbol{\alpha}}^\delta) + b_i) = (1 + \mu)\phi(u_{\boldsymbol{\alpha}}^\delta, g^\delta) + b_0. \tag{3.50}$$

In case of one single regularization, the identity $\gamma_1\alpha_1(\psi_1(u_{\boldsymbol{\alpha}}^\delta) + b_1) = (1 + \mu)\phi(u_{\boldsymbol{\alpha}}^\delta, g^\delta) + b_0$ does not provide any additional constraint since the Lagrange multiplier $\mu \geq 0$ is unknown. If the constraint $\phi(u_{\boldsymbol{\alpha}}^\delta, g^\delta) \leq c$ is active, i.e., $\phi(u_{\boldsymbol{\alpha}}^\delta, g^\delta) = c$, then it directly leads to the discrepancy principle. Note that, in case of multiple regularization, the identity $\phi(u_{\boldsymbol{\alpha}}^\delta, g^\delta) = c$ alone does not uniquely determine the vector $\boldsymbol{\alpha}$, and hence an auxiliary rule must be supplied in order to uniquely select an appropriate regularization parameter. The derivations suggest incorporating (3.50), which might help resolve the nonuniqueness issue inherent to the discrepancy principle. In particular, this gives rise to a new principle

$$\begin{cases} \phi(u_{\boldsymbol{\alpha}}^\delta, g^\delta) = c, \\ \gamma_i\alpha_i(\psi_i(u_{\boldsymbol{\alpha}}^\delta) + b_i) = \text{const}, \quad i = 1, \ldots, n. \end{cases}$$

We shall name the new principle by the balanced discrepancy principle since it augments the discrepancy principle with a balancing principle.

3.6.2 *Error estimates*

In this part, we shall derive error estimates in case that the space Y is a Hilbert space and the fidelity $\phi(u, g^\delta)$ is given by $\phi(u, g^\delta) = \frac{1}{2}\|Ku - g^\delta\|^2$, with a focus on the case of two penalty terms. Then the balanced discrepancy principle reads: for some $c_m \geq 1$ (with $b_i = 0$)

$$\begin{cases} \|Ku_{\boldsymbol{\alpha}}^\delta - g^\delta\| = c_m \delta, \\ \gamma_i \alpha_i \psi_i(u_{\boldsymbol{\alpha}}^\delta) = \text{const}, \quad i = 1, 2. \end{cases} \tag{3.51}$$

First, we discussed the consistency and an a priori error estimate for the hybrid principle (3.51). The consistency is ensured under a priori positive upper and lower bounds on the ratio of the selected parameter $\boldsymbol{\alpha}$.

Assumption 3.3. For any $\boldsymbol{\alpha} > 0$, the functional $J_{\boldsymbol{\alpha}}(u)$ is coercive and its level set $\{u \in \mathcal{C} : J_{\boldsymbol{\alpha}}(u) \leq c\}$ for any $c > 0$ is weakly compact, and the functionals ϕ and ψ_i are weakly lower semi-continuous.

Theorem 3.30. *Let Assumption 3.3 hold, and* $t(\delta) = \frac{\alpha_1(\delta)}{\alpha_1(\delta)+\alpha_2(\delta)}$. *Let the sequence* $\{\boldsymbol{\alpha}(\delta)\}_\delta$ *be determined by* (3.51). *If a subsequence* $\{\boldsymbol{\alpha}(\delta)\}_\delta$ *converges and* $\tilde{t} \equiv \lim_{\delta\to 0} t(\delta) \in (0,1)$, *then the subsequence* $\{u_{\boldsymbol{\alpha}(\delta)}^\delta\}_\delta$ *contains a subsequence weakly convergent to a* $[\tilde{t}, 1-\tilde{t}] \cdot \boldsymbol{\psi}$*-minimizing solution and*

$$\lim_{\delta\to 0} [t(\delta), 1-t(\delta)] \cdot \boldsymbol{\psi}(u_{\boldsymbol{\alpha}}^\delta) = [\tilde{t}, 1-\tilde{t}] \cdot \boldsymbol{\psi}(u^\dagger).$$

Proof. The minimizing property of the minimizer $u_{\boldsymbol{\alpha}}^\delta$ indicates

$$\begin{aligned} \tfrac{1}{2}\|Ku_{\boldsymbol{\alpha}}^\delta - g^\delta\|^2 + \boldsymbol{\alpha}\cdot\boldsymbol{\psi}(u_{\boldsymbol{\alpha}}^\delta) &\leq \tfrac{1}{2}\|Ku^\dagger - g^\delta\|^2 + \boldsymbol{\alpha}\cdot\boldsymbol{\psi}(u^\dagger) \\ &\leq \tfrac{1}{2}\delta^2 + \boldsymbol{\alpha}\cdot\boldsymbol{\psi}(u^\dagger). \end{aligned}$$

By virtue of the constraint $\|Ku_{\boldsymbol{\alpha}}^\delta - g^\delta\| = c_m\delta$, we deduce

$$\boldsymbol{\alpha}\cdot\boldsymbol{\psi}(u_{\boldsymbol{\alpha}}^\delta) \leq \boldsymbol{\alpha}\cdot\boldsymbol{\psi}(u^\dagger).$$

Now the assumption $\tilde{t} \in (0,1)$ implies that the sequence $\{\psi_i(u_{\boldsymbol{\alpha}}^\delta)\}_\delta$ is uniformly bounded. Hence the weak compactness of the functional $J_{\boldsymbol{\alpha}}$ in Assumption 3.1 indicates that there exists a subsequence, also denoted by $\{u_{\boldsymbol{\alpha}}^\delta\}_\delta$, and some u^*, such that $u_{\boldsymbol{\alpha}}^\delta \to u^*$ weakly. The weak lower semicontinuity of the functional ϕ and the triangle inequality imply

$$\begin{aligned} 0 \leq \|Ku^* - g^\dagger\| &\leq \liminf_{\delta\to 0}(\|Ku^\dagger - g^\delta\| + \|Ku_{\boldsymbol{\alpha}}^\delta - g^\delta\|) \\ &\leq \liminf_{\delta\to 0}(1+c_m)\delta = 0, \end{aligned}$$

i.e., $Kx^* = g^\dagger$. Again, by the weak lower semicontinuity of ψ_i, we deduce

$$\begin{aligned}[\tilde{t}, 1-\tilde{t}] \cdot \boldsymbol{\psi}(u^*) &\leq \liminf_{\delta\to 0}\, [t(\delta), 1-t(\delta)] \cdot \boldsymbol{\psi}(u_{\boldsymbol{\alpha}}^\delta) \\ &\leq \limsup_{\delta\to 0}\, [t(\delta), 1-t(\delta)] \cdot \boldsymbol{\psi}(u_{\boldsymbol{\alpha}}^\delta) \\ &\leq \lim_{\delta\to 0} [t(\delta), 1-t(\delta)] \cdot \boldsymbol{\psi}(u^\dagger) = [\tilde{t}, 1-\tilde{t}] \cdot \boldsymbol{\psi}(u^\dagger),\end{aligned}$$

which implies that u^* is a $[\tilde{t}, 1-\tilde{t}] \cdot \boldsymbol{\psi}$-minimizing solution. The desired identity follows from the above inequalities with u^* in place of $u^\dagger$. □

Remark 3.33. By definition, $\gamma_1\alpha_1(\delta)\psi_1(u_{\boldsymbol{\alpha}}^\delta) = \gamma_2\alpha_2(\delta)\psi_2(u_{\boldsymbol{\alpha}}^\delta)$, the condition in Theorem 3.30 amounts to the uniform boundedness of $\psi_i(u_{\boldsymbol{\alpha}}^\delta)$, which seems plausible in practice.

Now we state a convergence rates result.

Theorem 3.31. *If the exact solution $u^\dagger$ satisfies the following source condition: for any $t \in [0,1]$, there exists a $w_t \in Y$ such that*

$$K^* w_t = \xi_t \in \partial\left([t, 1-t] \cdot \boldsymbol{\psi}(u^\dagger)\right).$$

Then for any $\boldsymbol{\alpha}^*$ determined by the principle (3.51), there holds

$$d_{\xi_{t^*}}(u_{\boldsymbol{\alpha}^*}^\delta, u^\dagger) \leq (1+c_m)\|w_{t^*}\|\delta,$$

where the weight $t^* \equiv t^*(\delta) \in [0,1]$ is given by $\frac{\alpha_1^*(\delta)}{\alpha_1^*(\delta)+\alpha_2^*(\delta)}$.

Proof. In view of the minimizing property of the approximation $u_{\boldsymbol{\alpha}^*}^\delta$ and the constraint $\|Ku_{\boldsymbol{\alpha}^*}^\delta - g^\delta\| = c_m\delta$, we have

$$[t^*, 1-t^*] \cdot \boldsymbol{\psi}(u_{\boldsymbol{\alpha}^*}^\delta) \leq [t^*, 1-t^*] \cdot \boldsymbol{\psi}(u^\dagger).$$

The source condition implies that there exists a $\xi_{t^*} \in \partial\left([t^*, 1-t^*] \cdot \boldsymbol{\psi}(u^\dagger)\right)$ and $w_{t^*} \in Y$ such that $\xi_{t^*} = K^* w_{t^*}$. This and the Cauchy-Schwarz inequality imply

$$\begin{aligned}d_{\xi_{t^*}}(u_{\boldsymbol{\alpha}^*}^\delta, u^\dagger) &= [t^*, 1-t^*] \cdot \boldsymbol{\psi}(u_{\boldsymbol{\alpha}^*}^\delta) - [t^*, 1-t^*] \cdot \boldsymbol{\psi}(u^\dagger) - \langle \xi_{t^*}, u_{\boldsymbol{\alpha}^*}^\delta - u^\dagger\rangle \\ &\leq -\langle \xi_{t^*}, u_{\boldsymbol{\alpha}^*}^\delta - u^\dagger\rangle = -\langle K^* w_{t^*}, u_{\boldsymbol{\alpha}^*}^\delta - u^\dagger\rangle \\ &= -\langle w_{t^*}, K(u_{\boldsymbol{\alpha}^*}^\delta - u^\dagger)\rangle \leq \|w_{t^*}\|\|K(u_{\boldsymbol{\alpha}^*}^\delta - u^\dagger)\| \\ &\leq \|w_{t^*}\|\left(\|Ku_{\boldsymbol{\alpha}^*}^\delta - g^\delta\| + \|g^\delta - Ku^\dagger\|\right) \leq (1+c_m)\|w_{t^*}\|\delta.\end{aligned}$$

This shows the desired estimate. □

Remark 3.34. For the error estimate in Theorem 3.31, the order of convergence relies on the constraint (discrepancy principle), whereas the weight t^* is determined by the reduced balancing system (3.50). Hence the reduced system (3.50) helps resolve the vast nonuniqueness issue.

Next we give a posteriori error estimates for the balancing principle. We shall focus on the specific choice $b = b_0 = 0$ for the principle, i.e.,

$$\Phi_\gamma(\boldsymbol{\alpha}) = \frac{F^{2+\gamma}(\boldsymbol{\alpha})}{\alpha_1\alpha_2}. \tag{3.52}$$

We treat two scenarios separately: quadratic regularization and general convex regularization. In the former case, we consider the choice $\psi_1(u) = \frac{1}{2}\|L_1 u\|^2$ and $\psi_2(u) = \frac{1}{2}\|L_2 u\|^2$ with linear operators L_i fulfilling $\ker(L_i) \cap \ker(K) = \{0\}$, $i = 1, 2$. A typical choice of this type is that ψ_1 and ψ_2 imposes the L^2-norm and higher-order Sobolev smoothness, respectively, e.g., $\psi_1(u) = \frac{1}{2}\|u\|^2_{L^2(\Omega)}$ and $\psi_2(u) = \frac{1}{2}\|u\|^2_{H^1(\Omega)}$. We shall use a weighted norm $\|\cdot\|_t$ defined by

$$\|u\|_t^2 = t\|L_1 u\|^2 + (1-t)\|L_2 u\|^2,$$

where the weight $t \equiv t(\boldsymbol{\alpha}) \in [0, 1]$ is defined as before. Also we denote by

$$Q_t = tL_1^* L_1 + (1-t)L_2^* L_2, \quad L_t = Q_t^{1/2}, \quad \text{and} \quad \tilde{K}_t = KL_t^{-1}.$$

Obviously, $\|u\|_t = \|L_t u\|$.

Theorem 3.32. *Let $\mu \in [0, 1]$ be fixed. If the exact solution $u^\dagger$ satisfies the following source condition: for any $t \in [0, 1]$, there exists a $w_t \in Y$ such that $L_t u^\dagger = (\tilde{K}_t \tilde{K}_t)^\mu w_t$, then for any $\boldsymbol{\alpha}^*$ selected by the rule Φ_γ, the following estimate holds*

$$\|u^\delta_{\boldsymbol{\alpha}^*} - u^\dagger\|_{t^*} \leq C\left(\|w_{t^*}\|^{\frac{1}{2\mu+1}} + \frac{F^{\frac{2+\gamma}{4}}(\delta^{\frac{2}{2\mu+1}}\mathbf{e})}{F^{\frac{2+\gamma}{4}}(\boldsymbol{\alpha}^*)}\right)\max\{\delta_*, \delta\}^{\frac{2\mu}{2\mu+1}},$$

where $\mathbf{e} = (1, 1)^{\mathrm{t}}$ and $\delta_* = \|Ku^\delta_{\boldsymbol{\alpha}^*} - g^\delta\|$.

Proof. First, we decompose the error $u^\delta_{\boldsymbol{\alpha}} - u^\dagger$ into

$$u^\delta_{\boldsymbol{\alpha}} - u^\dagger = (u^\delta_{\boldsymbol{\alpha}} - u_{\boldsymbol{\alpha}}) + (u_{\boldsymbol{\alpha}} - u^\dagger).$$

It suffices to bound the two terms separately. First we estimate the error $u^\delta_{\boldsymbol{\alpha}} - u_{\boldsymbol{\alpha}}$. It follows from the optimality conditions for $u_{\boldsymbol{\alpha}}$ and $u^\delta_{\boldsymbol{\alpha}}$ that

$$(K^*K + \alpha_1 L_1^* L_1 + \alpha_2 L_2^* L_2)(u_{\boldsymbol{\alpha}} - u^\delta_{\boldsymbol{\alpha}}) = K^*(g^\dagger - g^\delta).$$

Multiplying the identity with $u_{\boldsymbol{\alpha}} - u^\delta_{\boldsymbol{\alpha}}$ and using the Cauchy-Schwarz and Young's inequalities give

$$\begin{aligned}\|K(u^\delta_{\boldsymbol{\alpha}} - u_{\boldsymbol{\alpha}})\|^2 + \alpha_1\|L_1(u^\delta_{\boldsymbol{\alpha}} - u_{\boldsymbol{\alpha}})\|^2 + \alpha_2\|L_2(u^\delta_{\boldsymbol{\alpha}} - u_{\boldsymbol{\alpha}})\|^2 \\ = \langle K(u^\delta_{\boldsymbol{\alpha}} - u_{\boldsymbol{\alpha}}), g^\dagger - g^\delta\rangle \\ \leq \|K(u^\delta_{\boldsymbol{\alpha}} - u_{\boldsymbol{\alpha}})\|^2 + \tfrac{1}{4}\|g^\dagger - g^\delta\|^2.\end{aligned}$$

From this we derive

$$s(t\|L_1(u_{\boldsymbol{\alpha}}^\delta - u_{\boldsymbol{\alpha}})\|^2 + (1-t)\|L_2(u_{\boldsymbol{\alpha}} - u_{\boldsymbol{\alpha}}^\delta)\|^2) \leq \tfrac{1}{4}\|g^\dagger - g^\delta\|^2,$$

where $s = \alpha_1 + \alpha_2$, i.e.,

$$\|u_{\boldsymbol{\alpha}}^\delta - u_{\boldsymbol{\alpha}}\|_t \leq \frac{\|g^\delta - g^\dagger\|}{2\sqrt{s}} \leq \frac{\delta}{2\sqrt{s}} \leq \frac{\delta}{2\sqrt{\max_i \alpha_i}}.$$

Meanwhile, the minimizing property of $\boldsymbol{\alpha}^*$ to Φ_γ implies that for any $\hat{\boldsymbol{\alpha}}$

$$\frac{F^{2+\gamma}(\boldsymbol{\alpha}^*)}{\max(\alpha_i^*)^2} \leq \frac{F^{2+\gamma}(\boldsymbol{\alpha}^*)}{\alpha_1^*\alpha_2^*} \leq \frac{F^{2+\gamma}(\hat{\boldsymbol{\alpha}})}{\hat{\alpha}_1\hat{\alpha}_2}.$$

In particular, we may take $\hat{\boldsymbol{\alpha}} = \delta^{\frac{2}{2\mu+1}}\boldsymbol{e}$ to get

$$\frac{1}{\sqrt{\max_i \alpha_i^*}} \leq \frac{F^{\frac{2+\gamma}{4}}(\delta^{\frac{2}{2\mu+1}}\boldsymbol{e})}{F^{\frac{2+\gamma}{4}}(\boldsymbol{\alpha}^*)}\delta^{-\frac{1}{2\mu+1}}.$$

Consequently, we have

$$\|u_{\boldsymbol{\alpha}^*}^\delta - u_{\boldsymbol{\alpha}}\|_{t^*} \leq \frac{F^{\frac{2+\gamma}{4}}(\delta^{\frac{2}{2\mu+1}}\boldsymbol{e})}{F^{\frac{2+\gamma}{4}}(\boldsymbol{\alpha}^*)}\delta^{\frac{2\mu}{2\mu+1}}.$$

Next we estimate the approximation error $u_{\boldsymbol{\alpha}} - u^\dagger$. To this end, we observe

$$\begin{aligned} u_{\boldsymbol{\alpha}} - u^\dagger &= (K^*K + \alpha_1 L_1^*L_1 + \alpha_2 L_2^*L_2)^{-1}(\alpha_1 L_1^*L_1 + \alpha_2 L_2^*L_2)u^\dagger \\ &= s(K^*K + sQ_t)^{-1}Q_t u^\dagger \\ &= sL_t^{-1}(L_t^{-1}K^*KL_t^{-1} + sI)^{-1}L_t u^\dagger. \end{aligned}$$

Hence, there holds

$$L_t(u_{\boldsymbol{\alpha}} - u^\dagger) = s(\tilde{K}_t^*\tilde{K}_t + sI)L_t u^\dagger.$$

Consequently, we deduce from the source condition and the moment inequality

$$\begin{aligned} \|u_{\boldsymbol{\alpha}} - u^\dagger\|_t &= \|L_t(u_{\boldsymbol{\alpha}} - u^\dagger)\| = \|s(\tilde{K}_t^*\tilde{K}_t + sI)^{-1}L_t u^\dagger\| \\ &= \|s(\tilde{K}_t^*\tilde{K}_t + sI)^{-1}(\tilde{K}_t^*\tilde{K}_t)^\mu w_t\| \\ &\leq \|s(\tilde{K}_t^*\tilde{K}_t + sI)^{-1}(\tilde{K}_t^*\tilde{K}_t)^{\frac{1}{2}+\mu} w_t\|^{\frac{2\mu}{2\mu+1}}\|s(\tilde{K}_t^*\tilde{K}_t + sI)^{-1}w_t\|^{\frac{1}{2\mu+1}} \\ &= \|s(\tilde{K}_t^*\tilde{K}_t + sI)^{-1}\tilde{K}_t L_t u^\dagger\|^{\frac{2\mu}{2\mu+1}}\|s(\tilde{K}_t^*\tilde{K}_t + sI)^{-1}w_t\| \\ &\leq c(\|s(\tilde{K}_t\tilde{K}_t^* + sI)^{-1}g^\delta\| + \|s(\tilde{K}_t\tilde{K}_t^* + sI)^{-1}(g^\delta - g^\dagger)\|)^{\frac{2\mu}{2\mu+1}} \\ &\qquad \times \|w_t\|^{\frac{1}{2\mu+1}}, \end{aligned}$$

where the constant $c(\leq 1)$ depends only on the maximum of $r_s(t) = \frac{s}{s+t}$ over $[0, \|\tilde{K}_t\|^2]$. Further, we note the relation

$$\begin{aligned} s(\tilde{K}_t\tilde{K}_t^* + sI)^{-1}g^\delta &= g^\delta - (\tilde{K}_tK_t^* + sI)^{-1}\tilde{K}_t\tilde{K}_t^*g^\delta \\ &= g^\delta - \tilde{K}(\tilde{K}_t^*\tilde{K}_t + sI)^{-1}\tilde{K}_t^*g^\delta \\ &= g^\delta - K(K^*K + sQ_t)^{-1}K^*g^\delta = g^\delta - Ku_{\boldsymbol{\alpha}}^\delta. \end{aligned}$$

Hence, we deduce

$$\|u_{\boldsymbol{\alpha}^*} - u^\dagger\|_{t^*} \leq c(\delta_* + c\delta)^{\frac{2\mu}{2\mu+1}}\|w_t\|^{\frac{1}{2\mu+1}} \leq c_1 \max\{\delta_*, \delta\}^{\frac{2\mu}{2\mu+1}}.$$

Combining the above two estimates yields the desired estimate. □

Similarly, we can state an estimate for general convex penalties.

Theorem 3.33. *Let $u^\dagger$ satisfy the following source condition: for any $t \in [0,1]$ there exists a $w_t \in Y$ such that $K^*w_t = \xi_t \in \partial\left([t, 1-t] \cdot \psi(u^\dagger)\right)$. Then for every $\boldsymbol{\alpha}^*$ determined by the rule Φ_γ, the following estimate holds*

$$d_{\xi_{t^*}}(u_{\boldsymbol{\alpha}^*}^\delta, u^\dagger) \leq C\left(\|w_{t^*}\| + \frac{F^{1+\frac{\gamma}{2}}(\delta\mathbf{e})}{F^{1+\frac{\gamma}{2}}(\boldsymbol{\alpha}^*)}\right)\max(\delta, \delta_*),$$

with $\mathbf{e} = (1,1)^{\mathrm{t}}$, $\delta_* = \|Ku_{\boldsymbol{\alpha}^*}^\delta - g^\delta\|$ and $t^* = \alpha_1^*/(\alpha_1^* + \alpha_2^*)$.

Remark 3.35. For general multiple regularization, there holds

$$d_{\xi_{\boldsymbol{\beta}^*}}(u_{\boldsymbol{\alpha}^*}^\delta, u^\dagger) \leq C\left(\|w_{t^*}\| + \frac{F^{1+\frac{\gamma}{n}}(\delta\mathbf{e})}{F^{1+\frac{\gamma}{n}}(\boldsymbol{\alpha}^*)}\right)\max(\delta, \delta_*),$$

under the following source condition: there exists a $w_{\boldsymbol{\beta}^*} \in Y$ such that

$$K^*w_{\boldsymbol{\beta}^*} = \xi_{\boldsymbol{\beta}^*} \in \partial(\boldsymbol{\beta}^* \cdot \psi(u^\dagger)),$$

where the parameter $\boldsymbol{\alpha}^*$ is determined by the rule Φ_γ, and $\boldsymbol{\beta}^* = \boldsymbol{\alpha}^*/\|\boldsymbol{\alpha}^*\|_1$.

3.6.3 *Numerical algorithms*

In this section, we describe algorithms for computing the regularization parameter by the hybrid principle, and the balancing principle, in the case of two regularization and with the choice $b = 0$.

In practice, the application of the hybrid principle invokes solving the nonlinear system (3.51), which is nontrivial due to its potential nonsmoothness and high degree of nonlinearity. We propose the following Broyden's method [44] for its efficient solution; see Algorithm 3.3 for a complete description.

Algorithm 3.3 Broyden's method for system (3.51).

1: Set $k = 0$ and choose $\boldsymbol{\alpha}^0$.
2: Compute the Jacobian $\mathbf{J}_0 = \nabla \mathbf{T}(\boldsymbol{\alpha}^0)$ and equation residual $\mathbf{T}(\boldsymbol{\alpha}^0)$.
3: **for** $k = 1, \ldots, K$ **do**
4: Calculate the quasi-Newton update $\Delta\boldsymbol{\alpha} = -\mathbf{J}_{k-1}^{-1}\mathbf{T}(\boldsymbol{\alpha}^{k-1})$
5: Update the regularization parameter $\boldsymbol{\alpha}$ by $\boldsymbol{\alpha}^k = \boldsymbol{\alpha}^{k-1} + \Delta\boldsymbol{\alpha}$
6: Evaluate the equation residual $\mathbf{T}(\boldsymbol{\alpha}^k)$ and set $\Delta\mathbf{T} = \mathbf{T}(\boldsymbol{\alpha}^k) - \mathbf{T}(\boldsymbol{\alpha}^{k-1})$
7: Compute Jacobian update

$$\mathbf{J}_k = \mathbf{J}_{k-1} + \frac{1}{\|\Delta\boldsymbol{\alpha}\|^2}[\Delta\mathbf{T} - \mathbf{J}_k\Delta\boldsymbol{\alpha}] \cdot \Delta\boldsymbol{\alpha}^{\mathrm{t}}.$$

8: Check the stopping criterion.
9: **end for**
10: Output the solution

Numerically, we reformulate system (3.51) equivalently as

$$\mathbf{T}(\boldsymbol{\alpha}) \equiv \begin{pmatrix} \bar{\phi}(\boldsymbol{\alpha}) - \delta^2 + \alpha_2\bar{\psi}_2(\boldsymbol{\alpha}) - \alpha_1\bar{\psi}_1(\boldsymbol{\alpha}) \\ \bar{\phi}(\boldsymbol{\alpha}) - \delta^2 + \alpha_1\bar{\psi}_1(\boldsymbol{\alpha}) - \alpha_2\bar{\psi}_2(\boldsymbol{\alpha}) \end{pmatrix} = 0. \tag{3.53}$$

In Algorithm 3.3, the Jacobian $\mathbf{J}_0$ can be approximated by finite difference. Step 7 is known as Broyden update. The stopping criterion can be based on monitoring the residual norm $\|\mathbf{T}(\boldsymbol{\alpha})\|$. Our experiences indicate that the algorithm converges fast and steadily, however, a convergence analysis is still missing.

Now we describe a fixed point algorithm for computing a minimizer of the rule Φ_γ. One basic version is listed in Algorithm 3.4, which extends Algorithm 3.2 to the multi-parameter model. An equally interesting variant of Algorithm 3.4 is to update the parameter $\boldsymbol{\alpha}$ by

$$\alpha_1^{k+1} = \frac{1}{\gamma}\frac{\phi(u_{\boldsymbol{\alpha}^k}^\delta, g^\delta)}{\psi_1(u_{\boldsymbol{\alpha}^k}^\delta)} \quad \text{and} \quad \alpha_2^{k+1} = \frac{1}{\gamma}\frac{\phi(u_{\boldsymbol{\alpha}^k}, g^\delta)}{\psi_2(u_{\boldsymbol{\alpha}^k})}. \tag{3.54}$$

Our experiences indicate that this variant converges fairly steadily and fast, but its theoretical analysis remains unclear.

To analyze Algorithm 3.4, we first introduce an operator T by (with $\bar{\phi}(\boldsymbol{\alpha}) := \phi(u_{\boldsymbol{\alpha}}^\delta, g^\delta)$ and $\bar{\psi}_i(\boldsymbol{\alpha}) = \psi_i(u_{\boldsymbol{\alpha}}^\delta)$):

$$T(\boldsymbol{\alpha}) = (1+\gamma)^{-1} \begin{pmatrix} \dfrac{\bar{\phi}(\boldsymbol{\alpha}) + \alpha_2\bar{\psi}_2(\boldsymbol{\alpha})}{\bar{\psi}_1(\boldsymbol{\alpha})} \\ \dfrac{\bar{\phi}(\boldsymbol{\alpha}) + \alpha_1\bar{\psi}_1(\boldsymbol{\alpha})}{\bar{\psi}_2(\boldsymbol{\alpha})} \end{pmatrix}.$$

Algorithm 3.4 Fixed point algorithm for minimizing (3.48).

1: Set $k = 0$ and choose $\boldsymbol{\alpha}^0$.

2: Solve for $u_{\boldsymbol{\alpha}^k}^\delta$ by the Tikhonov regularization

$$u_{\boldsymbol{\alpha}^k}^\delta = \arg\min_u \left\{ \phi(u, g^\delta) + \boldsymbol{\alpha}^k \cdot \boldsymbol{\psi}(u) \right\}.$$

3: Update the regularization parameter $\boldsymbol{\alpha}^{k+1}$ by

$$\alpha_1^{k+1} = \frac{1}{1+\gamma} \frac{\phi(u_{\boldsymbol{\alpha}^k}^\delta, g^\delta) + \alpha_2^k \psi_2(u_{\boldsymbol{\alpha}^k}^\delta)}{\psi_1(u_{\boldsymbol{\alpha}^k}^\delta)},$$
$$\alpha_2^{k+1} = \frac{1}{1+\gamma} \frac{\phi(u_{\boldsymbol{\alpha}^k}^\delta, g^\delta) + \alpha_1^k \psi_1(u_{\boldsymbol{\alpha}^k}^\delta)}{\psi_2(u_{\boldsymbol{\alpha}^k}^\delta)}.$$

4: Check the stopping criterion.

By Lemma 3.3, the following monotonicity result holds. Here the subscript $-i$ refers to the index different from i.

Lemma 3.17. *The function $\bar{\psi}_i(\boldsymbol{\alpha})$ is monotonically decreasing in α_i, and*

$$\frac{\partial}{\partial \alpha_i}(\bar{\phi}(\boldsymbol{\alpha}) + \alpha_{-i}\bar{\psi}_{-i}(\boldsymbol{\alpha})) + \alpha_i \frac{\partial}{\partial \alpha_i}\bar{\psi}_i(\boldsymbol{\alpha}) = 0, \quad i = 1, 2.$$

We have the following monotone result for the fixed point operator T.

Proposition 3.6. *Let the function $F(\boldsymbol{\alpha})$ be twice differentiable for any $\boldsymbol{\alpha}$. Then the map $T(\boldsymbol{\alpha})$ is monotone if $F^2 F_{\alpha_1\alpha_1} F_{\alpha_2\alpha_2} > (F_{\alpha_1} F_{\alpha_2} - F F_{\alpha_1\alpha_2})^2$.*

Proof. Let $A(\boldsymbol{\alpha}) = \bar{\phi} + \alpha_2 \bar{\psi}_2$ and $B(\boldsymbol{\alpha}) = \bar{\phi} + \alpha_1 \bar{\psi}_1$. By Lemma 3.17, there hold the following relations

$$\frac{\partial A}{\partial \alpha_1} + \alpha_1 \frac{\partial \bar{\psi}_1}{\partial \alpha_1} = 0 \quad \text{and} \quad \frac{\partial B}{\partial \alpha_2} + \alpha_2 \frac{\partial \bar{\psi}_2}{\partial \alpha_2} = 0. \tag{3.55}$$

With the help of these two relations, we deduce

$$\begin{aligned} \frac{\partial}{\partial \alpha_1} \frac{A}{\bar{\psi}_1} &= \frac{\bar{\psi}_1 \frac{\partial A}{\partial \alpha_1} - A \frac{\partial \bar{\psi}_1}{\partial \alpha_1}}{\bar{\psi}_1^2} = \frac{\bar{\psi}_1(-\alpha_1 \frac{\partial \bar{\psi}_1}{\partial \alpha_1}) - A \frac{\partial \bar{\psi}_1}{\partial \alpha_1}}{\bar{\psi}_1^2} \\ &= -\frac{1}{\bar{\psi}_1^2} \frac{\partial \bar{\psi}_1}{\partial \alpha_1}(\alpha_1 \bar{\psi}_1 + A) = -\frac{F}{\bar{\psi}_1^2} \frac{\partial \bar{\psi}_1}{\partial \alpha_1}, \end{aligned}$$

and

$$\begin{aligned}\frac{\partial}{\partial\alpha_2}\frac{A}{\bar{\psi}_1} &= \frac{\bar{\psi}_1\frac{\partial A}{\partial\alpha_2} - A\frac{\partial\bar{\psi}_1}{\partial\alpha_2}}{\bar{\psi}_1^2} \\ &= \frac{\bar{\psi}_1\frac{\partial}{\partial\eta_2}(F-\alpha_1\bar{\psi}_1) - (F-\alpha_1\bar{\psi}_1)\frac{\partial\bar{\psi}_1}{\partial\alpha_2}}{\bar{\psi}_1^2} \\ &= \frac{1}{\bar{\psi}_1^2}\left[\bar{\psi}_1\bar{\psi}_2 - F\frac{\partial\bar{\psi}_1}{\partial\alpha_2}\right],\end{aligned}$$

where we have used Lemma 3.16. Similarly, we have

$$\frac{\partial}{\partial\alpha_2}\frac{B}{\bar{\psi}_2} = -\frac{F}{\bar{\psi}_2^2}\frac{\partial\bar{\psi}_2}{\partial\alpha_2} \quad \text{and} \quad \frac{\partial}{\partial\alpha_1}\frac{B}{\bar{\psi}_2} = \frac{1}{\bar{\psi}_2^2}\left[\bar{\psi}_1\bar{\psi}_2 - F\frac{\partial\bar{\psi}_2}{\partial\alpha_1}\right].$$

Therefore, the Jacobian ∇T of the operator $T(\boldsymbol{\alpha})$ is given by

$$\nabla T = (1+\gamma)^{-1}\begin{pmatrix} -\frac{F}{\bar{\psi}_1^2}\frac{\partial\bar{\psi}_1}{\partial\alpha_1} & \frac{1}{\bar{\psi}_1^2}\left[\bar{\psi}_1\bar{\psi}_2 - F\frac{\partial\bar{\psi}_1}{\partial\alpha_2}\right] \\ \frac{1}{\bar{\psi}_2^2}\left[\bar{\psi}_1\bar{\psi}_2 - F\frac{\partial\bar{\psi}_2}{\partial\alpha_1}\right] & -\frac{F}{\bar{\psi}_2^2}\frac{\partial\bar{\psi}_2}{\partial\alpha_2}\end{pmatrix}.$$

The monotonicity of the function $\bar{\psi}_i(\boldsymbol{\alpha})$ with α_i from Lemma 3.17 implies that $-\frac{\partial\bar{\psi}_i}{\partial\alpha_i} \geq 0$. Therefore, it suffices to show that the determinant $|\nabla T|$ of the Jacobian ∇T is positive. However, it follows from Lemma 3.16 that the identity

$$\frac{\partial\bar{\psi}_1}{\partial\alpha_2} = F_{\alpha_2\alpha_2} = \frac{\partial\bar{\psi}_2}{\partial\alpha_1}$$

holds, and hence the determinant $|\nabla T|$ is given by

$$|\nabla T| = (1+\gamma)^{-1}\frac{1}{\bar{\psi}_1^2\bar{\psi}_2^2}\left[F^2\frac{\partial\bar{\psi}_1}{\partial\alpha_1}\frac{\partial\bar{\psi}_2}{\partial\alpha_2} - \left(\bar{\psi}_1\bar{\psi}_2 - F\frac{\partial\bar{\psi}_2}{\partial\alpha_1}\right)^2\right].$$

Hence, the nonnegativity of $|\nabla T|$ follows from the assumption $F^2F_{\alpha_1\alpha_1}F_{\alpha_2\alpha_2} - (F_{\alpha_1}F_{\alpha_2} - FF_{\alpha_1\alpha_2})^2 > 0$. This concludes the proof. □

Bibliographical notes

In our presentation we have largely focused on quadratic fidelity in Hilbert spaces. A direct generalization is powered norms, for which convergence analysis is also relatively complete. The case of L^1 fitting is very popular, especially in image processing [70], due to its robustness with respect to impulsive type noises/data outliers [147]; see [246] for some interesting discussions and [144] for an improved convergence analysis. Apart from the

L^1 fidelity, the other popular nonquadratic fidelity is the Kullback-Leibler divergence for Poisson data. It has been analyzed in [259] for linear inverse problems. A general convergence theory for convergence rate analysis was discussed in [92, 93]. Further, we would like to point out that we have not touched the important topic of stochastic models, such as Hilbert processes. We refer to [29, 306, 98] for relevant error estimates.

Very recently in [241, 96], Naumova et al analyzed a slightly different multi-penalty formulation

$$\|K(u_1 + u_2) - g^\delta\|^2 + \alpha\|u_1\|^2 + \beta\|Lu_2\|^2,$$

where L is a densely defined operator. This formulation enforces constraints separately on the components, which is different from the multi-parameter model discussed in Section 3.6. In particular, they showed the optimality of convergence rates for the classical discrepancy principle for the formulation. Similar models have been extensively used in the imaging community [55].

We would like to note that the iterative refinement of Tikhonov regularization, i.e., iterated Tikhonov regularization, can improve the reconstruction accuracy under certain circumstances. In a Hilbert space setting, it was studied in [87, 268, 124]. Recently, Jin [178, 180] presented some first study of the method for nonsmooth penalties. This is closely related to iterative regularization methods, e.g., the Landweber method and Gauss-Newton method. Iterative methods achieves regularization by early stopping. These methods have been also intensively studied in the context of nonsmooth regularization, and we refer to the monograph [272] for a comprehensive treatment, [271, 189, 134, 135, 177, 179, 227] for the Landweber method and [189, 188, 145, 190] for Gauss-Newton method.

One especially popular nonsmooth regularization model is sparsity constraints, which originate from lasso in statistics [285]. It can take several different forms: ℓ^1-penalty [77], ℓ^q-penalty ($0 < q < 1$) (also known as bridge penalty in statistics), and measures (including Radon measures [71, 39], total variation [265, 55] and total generalized variation [37] etc.). In particular, the analysis of the ℓ^1- and ℓ^q-regularization has received intensive studies, and various refined estimates have been developed [218, 117, 38, 114, 256, 115, 118, 194, 297, 49, 8] for a very incomplete list, and the survey [172] for its application to parameter identifications.

Chapter 4

Tikhonov Theory for Nonlinear Inverse Problems

In this chapter, we consider ill-posed nonlinear operator equations

$$K(u) = g^\dagger, \tag{4.1}$$

where the nonlinear operator $K : X \to Y$ is Fréchet (Gâteaux) differentiable, and the spaces X and Y are Banach spaces, with the norm both denoted by $\|\cdot\|$. In practice, one has only access to the noisy data g^δ whose accuracy relative to the exact data

$$g^\dagger = K(u^\dagger),$$

where $u^\dagger \in X$ is the exact solution, is measured by the noise level

$$\delta = \|g^\dagger - g^\delta\|.$$

In practice, the unknown u may be subjected to pointwise constraint, e.g., $u \geq c$ almost everywhere. This is especially true for distributed coefficient estimation in differential equations to ensure the well-definedness of the operator K; see, e.g., [19] for relevant examples. We denote the constraint set by $\mathcal{C} \subset X$, and assume that it is closed and convex and the sought-for solution $u^\dagger \in \mathcal{C}$.

The now classical approach for obtaining an accurate yet stable approximation is Tikhonov regularization, which consists of minimizing the following Tikhonov functional

$$J_\alpha(u) = \|K(u) - g^\delta\|^p + \alpha\psi(u), \tag{4.2}$$

where the two terms are the fidelity incorporating the information in the data g^δ and a regularization for stabilizing the problem, respectively. A minimizer of J_α over the set $\mathcal{C}$ will be denoted by u_α^δ, and by u_α in case of exact data. As was discussed in Chapter 3, a correct choice of the fidelity should faithfully reflect the statistical characteristics of noise corrupting the data. Meanwhile, the penalty ψ is chosen to reflect a priori knowledge (such as smoothness) and constraint on the expected solutions, and nonsmooth penalties are now very popular and commonly employed.

In this chapter, we shall analyze the Tikhonov model J_α, e.g., existence, consistency, stability and convergence rates, and demonstrate the application of the approach to distributed parameter identification problems, which can serve as prototypical examples of nonlinear inverse problems. The outline of the chapter is as follows. We begin with the well-posedness of the Tikhonov model J_α in a general setting, following the classical works [89, 274] and the recent works [141]. The rest of the chapter focuses on convergence rate analysis. We first recall the Hilbert space setting from [89] and subsequent developments. Then we present a new approach from optimization viewpoint [152] in Section 4.3. The optimization approach will be illustrated on a class of nonlinear parameter identification problems in Section 4.4. Extensions of the classical approach to the Banach space setting, especially via variational inequalities, are discussed in Section 4.5. Finally, in Section 4.6, we discuss an alternative approach via conditional stability [65].

4.1 Well-posedness

In this section, we address the well-posedness, i.e., the existence, stability and consistency of minimizers to the Tikhonov functional J_α. Throughout, we make the following assumption.

Assumption 4.1. The operator $K : X \to Y$, X being reflexive, and the nonnegative functional $\psi : Y \to \mathbb{R}^+$ satisfy

(i) The functional $J_\alpha(u)$ is coercive, i.e., $J_\alpha(u_n) \to \infty$ as $\|u_n\|_X \to \infty$.
(ii) The operator $K : X \to Y$ is sequentially weakly closed, i.e., $u_n \to u^*$ weakly in X implies that $F(u_n) \to F(u^*)$ weakly in Y.
(iii) The functional ψ is proper, convex and weakly lower semicontinuous.

Remark 4.1. The coercivity assumption on J_α can be replaced by a compactness assumption on the set $\mathcal{C}$. In case that the space X is nonreflexive, the weak convergence should be replaced with weak $*$ convergence.

Remark 4.2. The assumption that F is weakly sequentially closed is nontrivial to verify for most nonlinear inverse problems. This is due to the low regularity of the solution to the differential equations with rough coefficients; see Example 4.1 below for an illustration. However, there exist a variety of stronger conditions, which imply sequential weak closeness. In particular, all linear operators and all weak-weak continuous operators are weakly sequentially closed.

Theorem 4.1. *Let Assumption 4.1 hold. Then for every $\alpha > 0$, there exists a minimizer to J_α.*

Proof. Since $\mathcal{C} \neq \emptyset$ and $g^\delta \in Y$, there exists at least one $\tilde{u} \in \mathcal{C}$ such that $J_\alpha(\tilde{u}) < \infty$. Hence, there is a sequence $\{u_n\} \subset \mathcal{C}$ such that

$$\lim_{n\to\infty} J_\alpha(u_n) = \inf_{u\in\mathcal{C}} J_\alpha(u).$$

By Assumption 4.1(i), the sequence $\{u_n\}$ is uniformly bounded in X and contains a subsequence, still denoted by $\{u_n\}$, and some $u^* \in \mathcal{C}$ (by the convexity and closedness of $\mathcal{C}$) such that $u_n \to u^*$ weakly. Since the functional ψ is weakly lower semicontinuous, cf. Assumption 4.1(iii), we have

$$\psi(u^*) \leq \liminf_{n\to\infty} \psi(u_n).$$

By Assumption 4.1(ii), K is weakly closed, i.e., $K(u_n) - g^\delta \to K(u^*) - g^\delta$ weakly in Y. Now by the weak lower semicontinuity of norms

$$\|K(u^*) - g^\delta\|^p \leq \liminf_{n\to\infty} \|K(u_n) - g^\delta\|^p.$$

Combining the preceding two estimates implies that u^* is a minimizer. □

Now we give an example on verifying the sequential weak closedness of a nonlinear operator K.

Example 4.1. This example considers the recovery of a Robin boundary condition from boundary observation. Let $\Omega \subset \mathbb{R}^2$ be an open bounded domain with a Lipschitz boundary Γ consisting of two disjoint parts Γ_i and Γ_c. We consider the equation

$$\begin{cases} -\Delta y = 0 & \text{in } \Omega, \\ \dfrac{\partial y}{\partial n} = f & \text{on } \Gamma_c, \\ \dfrac{\partial y}{\partial n} + uy = 0 & \text{on } \Gamma_i. \end{cases} \tag{4.3}$$

The inverse problem consists in recovering the Robin coefficient u defined on Γ_i from noisy observational data g^δ on the boundary Γ_c, i.e., K maps $u \in X = L^2(\Gamma_i)$ to $\gamma_{\Gamma_c} y(u) \in Y = L^2(\Gamma_c)$, where γ_{Γ_c} denotes the Dirichlet trace operator, and similarly γ_{Γ_i}, and $y(u)$ is the solution to (4.3). We shall seek u in the admissible set $\mathcal{C} = \{u \in L^\infty(\Gamma_i) : u \geq c\} \subset X$ for some fixed $c > 0$, by means of Tikhonov regularization

$$\int_{\Gamma_c} (K(u) - g^\delta)^2 ds + \alpha \int_{\Gamma_i} u^2 ds.$$

We shall denote the mapping of $u \in \mathcal{C}$ to the solution $y \in H^1(\Omega)$ of (4.3) by $y(u)$. By Lax-Milgram theorem, for any $u \in \mathcal{C}$, problem (4.3) has a unique solution $y \in H^1(\Omega)$ and

$$\|y\|_{H^1(\Omega)} \leq C\|f\|_{H^{-\frac{1}{2}}(\Gamma_c)}. \tag{4.4}$$

Next we verify the weak sequentially closedness of K. Let the sequence $\{u_n\} \subset \mathcal{C}$ converging weakly in $L^2(\Gamma_i)$ to $u^* \in \mathcal{C}$. We need to show

$$K(u_n) \to K(u^*) \quad \text{in } L^2(\Gamma_c).$$

For $u_n \in U$, set $y_n = y(u_n) \in H^1(\Omega)$. By the a priori estimate from (4.4), the sequence $\{y_n\}$ is uniformly bounded in $H^1(\Omega)$ and has a convergent subsequence, also denoted by $\{y_n\}$, such that there exists $y^* \in H^1(\Omega)$ with

$$y_n \rightharpoonup y^* \text{ in } H^1(\Omega).$$

The trace theorem and the Sobolev embedding theorem [1] imply

$$y_n \to y^* \text{ in } L^p(\Gamma_c)$$

for any $p < +\infty$. In particular, we will take $p = 4$. Then we have

$$\left| \int_{\Gamma_i} u_n(y_n - y^*) v ds \right| \leq \|u_n\|_{L^2(\Gamma_i)} \|y_n - y^*\|_{L^4(\Gamma_i)} \|v\|_{L^4(\Gamma_i)} \to 0$$

by the weak convergence of $\{u_n\}$ in $L^2(\Gamma_i)$ and the strong convergence of $\{y_n\}$ in $L^4(\Gamma_i)$. Therefore, we have

$$\begin{aligned} \lim_{n\to\infty} \int_{\Gamma_i} u_n y_n v ds &= \lim_{n\to\infty} \left(\int_{\Gamma_i} u_n(y_n - y^*) v ds + \int_{\Gamma_i} u_n y^* v ds \right) \\ &= \int_{\Gamma_i} u^* y^* v ds. \end{aligned}$$

Now passing to the limit in the weak formulation indicates that y^* satisfies

$$\int_\Omega \nabla y^* \cdot \nabla v dx + \int_{\Gamma_i} u^* y^* v ds = \int_{\Gamma_c} f v ds \quad \text{for all } v \in H^1(\Omega),$$

i.e., $y^* = y(u^*)$. Since every subsequence has itself a subsequence converging weakly in $H^1(\Omega)$ to $y(u^*)$, the whole sequence converges weakly. The sequential weak closedness of $K : u \mapsto y(u)|_{\Gamma_c}$ then follows from the trace theorem and Sobolev compact embedding theorem.

Now we turn to the stable dependence of the Tikhonov minimizer u_α^δ with respect to perturbations in the data g^δ.

Theorem 4.2. *Let Assumption 4.1 hold. Let $\{g_n\}$ be a sequence converging to g^δ in Y, and $\{u_n\}$ be the sequence of minimizers to J_α, with g_n in place of g^δ. Then the sequence $\{u_n\}$ contains a weakly convergent subsequence, and the limit is a minimizer to the functional J_α. Further, if the minimizer is unique, then the whole sequence converges weakly, and if the functional ψ satisfies the H-property, then the convergence is strong.*

Proof. Let $u_\alpha^\delta \in \mathcal{C}$ be a minimizer to the functional J_α. By the minimizing property of u_n, we have

$$\|K(u_n) - g_n\|^p + \alpha\psi(u_n) \leq \|K(u_\alpha^\delta) - g_n\|^p + \alpha\psi(u_\alpha^\delta).$$

This together with the inequality $(a+b)^p \leq 2^{p-1}(a^p + b^p)$ for $a, b \geq 0$ and $p \geq 1$ yields

$$\begin{aligned}
&\|K(u_n) - g^\delta\|^p + \alpha\psi(u_n) \\
&\leq 2^{p-1}(\|K(u_n) - g_n\|^p + \alpha\psi(u_n) + \|g_n - g^\delta\|^p) \\
&\leq 2^{p-1}(\|K(u_\alpha^\delta) - g_n\|^p + \alpha\psi(u_\alpha^\delta) + \|g_n - g^\delta\|^p).
\end{aligned}$$

Now by the coercivity of J_α from Assumption 4.1(i), the sequence $\{u_n\} \subset \mathcal{C}$ contains a subsequence, also denoted by $\{u_n\}$, and some $u^* \in \mathcal{C}$ such that $u_n \to u^*$ weakly. By the sequential weak closedness of the operator K, cf. Assumption 4.1(ii), $K(u_n) \to K(u^*)$ weakly in Y. Now by the norm convergence of g_n to g^δ, $K(u_n) - g_n \to K(u^*) - g^\delta$ weakly. This together with the weak lower semicontinuity of norms yield

$$\|K(u^*) - g^\delta\|^p \leq \liminf_{n\to\infty} \|K(u_n) - g_n\|^p. \tag{4.5}$$

Now it follows from the weak lower semicontinuity of the functional ψ, cf. Assumption 4.1(iii) that

$$\begin{aligned}
\|K(u^*) - g^\delta\|^p + \alpha\psi(u^*) &\leq \liminf_{n\to\infty} \|K(u_n) - g_n\|^p + \alpha \liminf_{n\to\infty} \psi(u_n) \\
&\leq \liminf_{n\to\infty} (\|K(u_n) - g_n\|^p + \alpha\psi(u_n)) \\
&\leq \lim_{n\to\infty} (\|K(u_\alpha^\delta) - g_n\|^p + \alpha\psi(u_\alpha^\delta)) \\
&\leq \|K(u_\alpha^\delta) - g^\delta\|^p + \alpha\psi(u_\alpha^\delta).
\end{aligned}$$

By the minimizing property of u_α^δ to J_α, u^* is a minimizer to J_α. Further by taking $u_\alpha^\delta = u^*$ in the inequalities yield

$$\lim_{n\to\infty} \|K(u_n) - g_n\|^p + \alpha\psi(u_n) = \|K(u^*) - g^\delta\|^p + \alpha\psi(u^*). \tag{4.6}$$

If the minimizer u_α^δ is unique, then by a standard subsequence argument, the whole sequence converges to u_α^δ. Next we show that $\psi(u_n)$ converges to $\psi(u^*)$ by contradiction. Assume that the contrary holds. Then by the lower semicontinuity of ψ, we have

$$c := \limsup_{n\to\infty} \psi(u_n) > \psi(u^*).$$

We take a subsequence $\{u_n\}$ such that $\psi(u_n) \to c$. Then from (4.6), on this subsequence we have

$$\begin{aligned}\lim_{n\to\infty} \|K(u_n) - g_n\|^p &= \lim_{n\to\infty} (\|K(u_n) - g_n\|^p + \alpha\psi(u_n)) - \alpha \lim_{n\to\infty} \psi(u_n) \\ &= \|K(u^*) - g^\delta\|^p + \alpha(\psi(u^*) - c) < \|K(u^*) - g^\delta\|^p,\end{aligned}$$

which contradicts (4.5), and thus $\psi(u_n) \to \psi(u^*)$. This together with the H-property of ψ yields the desired assertion. □

The Tikhonov minimizer u_α^δ is also stable with respect to other model perturbations, e.g., the regularization parameter α.

Theorem 4.3. *Let Assumption 4.1 hold. Let $\{\alpha_n\} \subset \mathbb{R}^+$ be a sequence converging to $\alpha > 0$ in Y, and $\{u_{\alpha_n}^\delta\}$ be the sequence of minimizers to J_{α_n}. Then the sequence $\{u_{\alpha_n}^\delta\}$ contains a weakly convergent subsequence, and the limit is a minimizer to the functional J_α. Further, if the minimizer is unique, then the whole sequence converges weakly, and if the functional ψ satisfies the H-property, then the convergence is strong.*

Proof. Let $u_\alpha^\delta \in \mathcal{C}$ be a minimizer to the functional J_α. By the minimizing property of $u_n \equiv u_{\alpha_n}^\delta$, we have

$$\|K(u_n) - g^\delta\|^p + \alpha_n\psi(u_n) \le \|K(u_\alpha^\delta) - g^\delta\|^p + \alpha_n\psi(u_\alpha^\delta).$$

Since the sequence $\alpha_n \to \alpha > 0$, we may assume that $\alpha_{\min} := \inf_n \alpha_n$ and $\alpha_{\max} := \sup \alpha_n$ satisfy $0 < \alpha_{\min}, \alpha_{\max} < \infty$. Then we have

$$\|K(u_n) - g^\delta\|^p + \alpha_{\min}\psi(u_n) \le \|K(u_\alpha^\delta) - g^\delta\|^p + \alpha_{\max}\psi(u_\alpha^\delta),$$

i.e., the sequence $\{J_{\alpha_{\min}}(u^n)\}$ is uniformly bounded. The rest of the proof is essentially identical with that for Theorem 4.2, and hence omitted. □

Like in the linear case, the proper generalized solution concept is the ψ-minimizing solution:

Definition 4.1. An element $u^\dagger \in \mathcal{C}$ is called an ψ-minimizing solution if

$$\psi(u^\dagger) \le \psi(u), \quad \forall u \in \{u \in \mathcal{C} : K(u) = g^\dagger\}.$$

The next result shows the existence of a ψ-minimizing solution.

Theorem 4.4. *Let Assumption 4.1 hold, and there exist a solution to* (4.1). *Then there exists at least one ψ-minimizing solution.*

Proof. Suppose that there does not exist a ψ-minimizing solution in $\mathcal{C}$. Then there exists a sequence $\{u_n\} \subset \mathcal{C}$ of solutions to (4.1) such that $\psi(u_n) \to c$ and

$$c < \psi(u), \quad \forall u \in \{u \in \mathcal{C} : K(u) = g^\dagger\}.$$

Therefore, the functional $J_\alpha(u_n)$ with $g^\dagger$ in place of g^δ is uniformly bounded, and by the coercivity of J_α, cf. Assumption 4.1(i), the sequence $\{u_n\}$ contains a subsequence, also denoted by $\{u_n\}$, and some $u^* \in \mathcal{C}$ such that $u_n \to u^*$ weakly. By the weak lower semicontinuity of the functional ψ in Assumption 4.1(iii), there holds $\psi(u^*) = \lim_{n\to\infty} \psi(u_n) = c$. Now, by the weak sequential closedness of K from Assumption 4.1(ii), we have $K(u^*) = g^\dagger$. This contradicts the assumption, and completes the proof. □

Finally we give a consistency result, i.e., the convergence of minimizers u_α^δ to a ψ-minimizing solution as the noise level δ tends to zero. This is the central property for a regularization scheme. Assumption 4.1 in connection with an appropriate rule for choosing α is sufficient.

Theorem 4.5. *Let Assumption 4.1 hold. Let the sequence $\{\delta_n\}$ be convergent to zero, and g^{δ_n} satisfy $\|g^\dagger - g^{\delta_n}\| = \delta_n$. Further, the parameter $\alpha(\delta)$ is chosen such that*

$$\lim_{\delta\to 0} \alpha(\delta) = 0 \quad \textit{and} \quad \lim_{\delta\to 0} \frac{\delta^p}{\alpha(\delta)} = 0.$$

Let $\{u_{\alpha(\delta_n)}^{\delta_n}\}$ be a sequence of minimizers to $J_{\alpha(\delta_n)}$ with g^{δ_n} in place of g^δ. Then it contains a subsequence converging weakly to ψ-minimizing solution. Further, if the ψ-minimizing solution $u^\dagger$ is unique, then the whole sequence converges weakly, and if the functional ψ satisfies the H-property, then the convergence is strong.

Proof. By the minimizing property of $u_n \equiv u_{\alpha(\delta_n)}^{\delta_n}$ and the identity $K(u^\dagger) = g^\dagger$, we have

$$\begin{aligned}\|K(u_n) - g^{\delta_n}\|^p + \alpha(\delta_n)\psi(u_n) &\le \|K(u^\dagger) - g^{\delta_n}\|^p + \alpha(\delta_n)\psi(u_n)\\ &\le \delta_n^p + \alpha(\delta_n)\psi(u^\dagger).\end{aligned}$$

Now the choice of the sequence $\{\alpha(\delta_n)\}$ yields that the sequences $\{\|K(u_n)-g^{\delta_n}\|^p\}$ and $\{\psi(u_n)\}$ are uniformly bounded. Further, it follows from the inequality $(a+b)^p \leq 2^{p-1}(a^p+b^p)$ for $a,b \geq 0$ and $p \geq 1$ that

$$\|K(u_n)-g^\dagger\|^p \leq 2^{p-1}(\|K(u_n)-g^{\delta_n}\|^p+\delta_n^p) < c.$$

Therefore, we have

$$\|K(u_n)-g^\dagger\|^p+\psi(u_n) \leq c.$$

By the coercivity of the functional J_α from Assumption 4.1(i), the sequence $\{u_n\}$ is uniformly bounded, and contains a subsequence, also denoted by $\{u_n\}$, and some $u^* \in \mathcal{C}$, such that $u_n \to u^*$ weakly. By the weak sequential closedness of K, cf. Assumption 4.1(ii), there holds $K(u_n) \to K(u^*)$ weakly, which together with the norm convergence $g^{\delta_n} \to g^\dagger$ implies $K(u_n)-g^\dagger \to K(u^*)-g^\dagger$. Now the weak lower semicontinuity of norms and the choice of $\alpha(\delta_n)$ yields

$$\begin{aligned} 0 \leq \|K(u^*)-g^\dagger\|^p &\leq \liminf_{n\to\infty} \|K(u_n)-g_n\|^p \\ &\leq \liminf_{n\to\infty} \delta_n^p+\alpha(\delta_n)\psi(u^\dagger) = 0, \end{aligned}$$

i.e., $K(u^*) = g^\dagger$. Furthermore, by the weak lower semicontinuity of the functional ψ, cf. Assumption 4.1(iii), we deduce

$$\psi(u^*) \leq \liminf_{n\to\infty} \psi(u_n) \leq \liminf_{n\to\infty}(\tfrac{\delta_n^p}{\alpha(\delta_n)}+\psi(u^\dagger)) = \psi(u^\dagger).$$

Hence u^* is a ψ-minimizing solution. The rest of the proof follows as before, and hence it is omitted. □

So far we have discussed basic properties of the Tikhonov model, i.e., the existence, stability and consistency of a Tikhonov minimizer u_α^δ. Naturally there is also the important question of uniqueness of the minimizer u_α^δ. In general, this is not guaranteed for nonlinear inverse problems. However, in the Hilbert space setting, we shall mention one such result in Section 4.2.

4.2 Classical convergence rate analysis

In this part, we discuss convergence rate analysis in a Hilbert space setting, i.e., both X and Y are Hilbert spaces, and the Tikhonov approach amounts to the following constrained optimization problem:

$$\min_{u\in C}\left\{J_\alpha(u) \equiv \|K(u)-g^\delta\|^2+\alpha\|u\|^2\right\}. \tag{4.7}$$

Like before, the minimizer of (4.7) is denoted by u_α^δ, and respectively, the minimizer for the exact data $g^\dagger$ by u_α. By convergence rate analysis, we aim at estimating the distance between the approximation u_α^δ and the ground truth $u^\dagger$ in terms of the noise level δ, and at developing rules for determining the scalar parameter $\alpha > 0$ automatically so as to obtain robust yet accurate approximations u_α^δ to the exact solution $u^\dagger$. In practice, one might have some preliminary knowledge about $u^\dagger$, e.g., in the neighborhood of u^*, then the functional J_α can be modified to $\|K(u)-g^\delta\|^2+\alpha\|u-u^*\|^2$ in order to accommodate such a priori information. The analysis below adapts straightforwardly to the case.

4.2.1 *A priori parameter choice*

In this part, we shall first recall the classical convergence theory due to Engl, Kunisch and Neubauer [89], and discuss several a posteriori choice rules and their convergence rates.

In order to derive convergence rates results, we require assumptions on the smoothness of the operator K, a restriction on the nonlinearity of K in a neighborhood of $u^\dagger$ and a source condition, which relates the penalty term to the operator K at $u^\dagger$.

Assumption 4.2. The operator K satisfies:

(i) K has a continuous Fréchet derivative;
(ii) there exists an $L > 0$ such that $\|K'(u^\dagger) - K'(u)\|_{L(X,Y)} \leq L\|u^\dagger - u\|$ for all $x \in \mathcal{C} \cap B_\rho(u^\dagger)$, with $\rho > 2\|u^\dagger\|$; and
(iii) there exists a $w \in Y$ such that $u^\dagger = K'(u^\dagger)^* w$ and $L\|w\| < 1$.

Now we can present the classical result in the pioneering work [89].

Theorem 4.6. *Let $K : X \to Y$ be continuous and weakly sequentially closed and Assumption 4.2 hold with $\rho = 2\|u^\dagger\| + \delta/\sqrt{\alpha}$. Then the minimizer u_α^δ to J_α satisfies*

$$\|u_\alpha^\delta - u^\dagger\| \leq \frac{\delta + \alpha\|w\|}{\sqrt{\alpha(1 - L\|w\|)}} \quad \textit{and} \quad \|K(u_\alpha^\delta) - g^\dagger\| \leq \delta + 2\alpha\|w\|.$$

Proof. By the minimizing property of u_α^δ to $J_\alpha(u)$, we obtain

$$\|K(u_\alpha^\delta) - g^\delta\|^2 + \alpha\|u_\alpha^\delta\|^2 \leq \delta^2 + \alpha\|u^\dagger\|^2,$$

which implies the boundedness of $\{J_\alpha(u_\alpha^\delta)\}$ and due to the choice $\alpha \sim \delta$ also the uniform boundedness of $\{u_\alpha^\delta\}$. A straightforward calculation shows

$$\|K(u_\alpha^\delta) - g^\delta\|^2 + \alpha\|u_\alpha^\delta - u^\dagger\|^2 \leq \delta^2 - 2\alpha\langle u_\alpha^\delta - u^\dagger, u^\dagger\rangle. \tag{4.8}$$

By the source condition from Assumption (4.2)(iii), we deduce

$$\|K(u_\alpha^\delta) - g^\delta\|^2 + \alpha\|u_\alpha^\delta - u^\dagger\|^2 \le \delta^2 - 2\alpha\langle K'(u^\dagger)(u_\alpha^\delta - u^\dagger), w\rangle. \tag{4.9}$$

To apply Assumption (4.2)(ii) to bound the inner product, we appeal again to inequality (4.8). Upon neglecting the term $\|K(u_\alpha^\delta) - g^\delta\|^2$ on the left hand side and applying the Cauchy-Schwarz inequality, we arrive at

$$\alpha\|u_\alpha^\delta - u^\dagger\|^2 \le \delta^2 + 2\alpha\|u^\dagger\|\|u^\dagger - u_\alpha^\delta\|,$$

and consequently,

$$(\|u_\alpha^\delta - u^\dagger\| - \|u^\dagger\|)^2 \le \|u^\dagger\|^2 + \frac{\delta^2}{\alpha}.$$

Hence, $u_\alpha^\delta \in B_\rho(u^\dagger)$ with $\rho = 2\|u^\dagger\| + \delta/\sqrt{\alpha}$. Now Assumptions 4.2(i) and (ii), i.e., Lipschitz continuity of the derivative and the convexity of the admissible set $\mathcal{C}$, imply

$$\begin{aligned} &\|K(u_\alpha^\delta) - K(u^\dagger) - K'(u^\dagger)(u_\alpha^\delta - u^\dagger)\| \\ =&\|\int_0^1 \left(K'(u^\dagger + t(u_\alpha^\delta - u^\dagger)) - K'(u^\dagger)\right)(u_\alpha^\delta - u^\dagger)dt\| \\ \le& \int_0^1 Lt\|u_\alpha^\delta - u^\dagger\|^2 dt = \frac{L}{2}\|u_\alpha^\delta - u^\dagger\|^2. \end{aligned} \tag{4.10}$$

By adding and subtracting $K(u_\alpha^\delta) - K(u^\dagger)$ in (4.9), and appealing to the Cauchy-Schwarz inequality, we obtain

$$\begin{aligned} \langle -K'(u^\dagger)(u^\dagger - u_\alpha^\delta), w\rangle \le& \tfrac{L}{2}\|w\|\|u_\alpha^\delta - u^\dagger\|^2 \\ &+ \|w\|\|K(u_\alpha^\delta) - g^\delta\| + \|w\|\delta. \end{aligned} \tag{4.11}$$

Inserting (4.11) in (4.9) and quadratic completion yields

$$\begin{aligned} (\|K(u_\alpha^\delta) - g^\delta\| - \alpha\|w\|)^2 + \alpha(1 - L\|w\|)\|u_\alpha^\delta - u^\dagger\|^2 \\ \le \delta^2 + 2\alpha\|w\|\delta + \alpha^2\|w\|^2 = (\delta + \alpha\|w\|)^2. \end{aligned}$$

The desired assertion follows from this and the condition $1 - L\|w\| > 0$ in Assumption 4.2(iii). □

Remark 4.3. The proof of Theorem 4.6 indicates the following estimates for exact data $g^\dagger$:

$$\|u_\alpha - u^\dagger\| \le \frac{\|w\|}{\sqrt{1 - L\|w\|}}\sqrt{\alpha} \quad \text{and} \quad \|K(u_\alpha) - g^\dagger\| \le 2\alpha\|w\|.$$

Further, with the choice $\alpha \sim \delta$, we have

$$\|u_\alpha^\delta - u^\dagger\| = \mathcal{O}(\sqrt{\delta}) \quad \text{and} \quad \|K(u_\alpha^\delta) - g^\dagger\| = \mathcal{O}(\delta).$$

Remark 4.4. The restriction on the nonlinearity of K, i.e., Assumption 4.2(ii) of Theorem 4.6, can be reformulated in several different ways. For example, the condition

$$\|K'(u)(u-\tilde{u})\| \leq C\|K(u)-K(\tilde{u})\|$$

has been studied. It leads to a more direct estimate for the critical inner product in (4.11):

$$\begin{aligned}|\langle -K'(u^\dagger)(u^\dagger - u_\alpha^\delta), w\rangle| &\leq \|K'(u^\dagger)(u^\dagger - u_\alpha^\delta)\|\|w\| \\ &\leq C\|K(u^\dagger) - K(u_\alpha^\delta)\|\|w\| \\ &\leq C(\|w\|\delta + \|w\|\|g^\delta - K(u_\alpha^\delta)\|)\end{aligned}$$

and the proof proceeds identically thereafter. We would like to note that the crucial inner product (4.11) generally is of indefinite sign and thus does not necessarily cause any trouble. For an analysis along this line of thought, we refer to Section 4.3.

Example 4.2. This example continues Example 4.1 on the differentiability of the operator K for the Robin inverse problem, and Lipschitz continuity of the derivative operator. In particular, we show that K is twice differentiable on $\mathcal{C}$: The mapping $u \mapsto y(u)$ is twice Fréchet differentiable from $L^2(\Gamma_i)$ to $H^1(\Omega)$, on $\mathcal{C}$ and for every $u \in \mathcal{C}$ and all directions $h_1, h_2 \in L^2(\Gamma_i)$, the derivatives are given by

(i) $y'(u)h_1 \in H^1(\Omega)$ is the solution z of

$$\int_\Omega \nabla z \cdot \nabla v dx + \int_{\Gamma_i} uzvds = -\int_{\Gamma_i} h_1 y(u) v ds$$

for all $v \in H^1(\Omega)$, and the following estimate holds

$$\|y'(u)h_1\|_{H^1(\Omega)} \leq C\|h_1\|_{L^2(\Gamma_i)}.$$

(ii) $y''(u)(h_1,h_2) \in H^1(\Omega)$ is the solution z of

$$\int_\Omega \nabla z \cdot \nabla v dx + \int_{\Gamma_i} uzvds = -\int_{\Gamma_i} h_1 y'(u) h_2 v ds - \int_{\Gamma_i} h_2 y'(u) h_1 v ds$$

for all $v \in H^1(\Omega)$, and the following estimate holds

$$\|y''(u)(h_1,h_2)\|_{H^1(\Omega)} \leq C\|h_1\|_{L^2(\Gamma_i)}\|h_2\|_{L^2(\Gamma_i)}.$$

In particular, (ii) implies that the derivative map $K'(u)$ is Lipschitz continuous. Here the characterization of the derivatives in (i) and (ii) follows from direct calculation. It remains to show the boundedness and continuity. By setting $v = y'(u)h_1$ in the weak formulation, Hölder's inequality, the trace theorem and the a priori estimate (4.4), we have

$$\begin{aligned}\|y'(u)h_1\|_{H^1(\Omega)}^2 &\leq C\|y'(u)h_1\|_{L^4(\Gamma_i)}\|h_1\|_{L^2(\Gamma_i)}\|y(u)\|_{L^4(\Gamma_i)} \\ &\leq C\|y'(u)h_1\|_{H^1(\Omega)}\|h_1\|_{L^2(\Gamma_i)}\|y(u)\|_{H^1(\Omega)} \\ &\leq C\|y'(u)h_1\|_{H^1(\Omega)}\|h_1\|_{L^2(\Gamma_i)},\end{aligned}$$

from which the first estimate follows. Analogously we deduce that

$$\|y(u+h_1) - y(u)\|_{H^1(\Omega)} \leq C\|h_1\|_{L^2(\Gamma_i)}.$$

Next let $w = y(u+h_1) - y(u) - y'(u)h_1$, which satisfies

$$\int_\Omega \nabla w \cdot \nabla v dx + \int_{\Gamma_i} uwvds = -\int_{\Gamma_i} h_1(y(u+h_1) - y(u))vds$$

for all $v \in H^1(\Omega)$. By repeating the proof of the preceding estimate, we deduce that

$$\|w\|_{H^1(\Omega)} \leq C\|h_1\|_{L^2(\Gamma_i)}\|y(u+h_1) - y(u)\|_{H^1(\Omega)},$$

from which it follows directly that $y'(u)h_1$ defined above is indeed the Fréchet derivative of $y(u)$ at u. By arguing similarly and using the first assertion, the second assertion follows.

Example 4.3. We illustrate the range inclusion condition $u^\dagger = K'(u^\dagger)^*w$ for Example 4.1. From Example 4.2, the derivative $K'(u^\dagger) : L^2(\Gamma_i) \to L^2(\Gamma_c)$ is given by $K'(u^\dagger)h = \gamma_{\Gamma_c}z$, where $z \in H^1(\Omega)$ solves

$$\int_\Omega \nabla z \cdot \nabla v dx + \int_{\Gamma_i} u^\dagger zvds = -\int_{\Gamma_i} hy(u^\dagger)vds.$$

The adjoint operator $K'(u^\dagger)^* : L^2(\Gamma_c) \to L^2(\Gamma_i)$ is given by

$$K'(u^\dagger)^*w = -\gamma_{\Gamma_i}(y(u^\dagger)\tilde{z}(u^\dagger)),$$

where $\tilde{z}(u^\dagger) \in H^1(\Omega)$ solves

$$\int_\Omega \nabla \tilde{z} \cdot \nabla v dx + \int_{\Gamma_i} u^\dagger \tilde{z}vds = \int_{\Gamma_c} wvds.$$

The adjoint representation $K'(u^\dagger)^*w$ follows from setting $v = \tilde{z}$ and $v = z$ in the weak formulations for z and $\tilde{z}$:

$$\begin{aligned}\langle K'(u^\dagger)h, w\rangle_{L^2(\Gamma_c)} &= \langle -hy(u^\dagger), \tilde{z}\rangle_{L^2(\Gamma_i)} \\ &= \langle h, -y(u^\dagger)\tilde{z}\rangle_{L^2(\Gamma_i)} \equiv \langle h, K'(u^\dagger)^*w\rangle.\end{aligned}$$

Consequently, the source condition reads: $u^\dagger = -\gamma_{\Gamma_i}(y(u^\dagger)\tilde{z}(u^\dagger))$. Since both $y(u^\dagger)$ and $\tilde{z}(u^\dagger)$ belongs to the space $H^1(\Omega)$, it follows from trace theorem that $-\gamma_{\Gamma_i}(y(u^\dagger)\tilde{z}(u^\dagger)) \in L^p(\Gamma_i)$, for any $p < \infty$. This is already much stronger than the condition $u^\dagger \in L^2(\Gamma_i)$ as was required by Tikhonov regularization. Alternatively, we can write

$$\tilde{z}(u^\dagger) = \frac{-u^\dagger}{y(u^\dagger)} \quad \text{on} \quad \Gamma_i.$$

Hence, for the existence of an element $w \in L^2(\Gamma_c)$, by the trace theorem, the right hand side $\frac{-u^\dagger}{\gamma_{\Gamma_i} y(u^\dagger)}$ must at least belong to the space $H^{\frac{1}{2}}(\Gamma_i)$, which imposes a certain smoothness condition on $u^\dagger$.

4.2.2 *A posteriori parameter choice*

Theorem 4.6 provides error estimates under a priori parameter choice. It is not necessarily convenient for practical purposes, since a reasonable value would necessarily require a knowledge of the source representer w, which is generally inaccessible. Therefore, there has been of immense interest in developing a posteriori rules with provable (optimal) convergence rates. One classical rule is the discrepancy principle, where the regularization parameter α is determined from the following nonlinear equation:

$$\|K(u_\alpha^\delta) - g^\delta\| = c_m \delta, \tag{4.12}$$

for some $c_m \geq 1$. The existence of a solution to the nonlinear equation (4.12) is highly nontrivial to verify, and we defer the issue to Section 4.2.3. We also refer to Section 3.4.1 for its efficient implementation.

Theorem 4.7. *Let Assumption 4.2 hold, and α be determined by the discrepancy principle* (4.12). *Then there holds*

$$\|u_\alpha^\delta - u^\dagger\| \leq \sqrt{\frac{2(1 + c_m)\|w\|}{1 - L\|w\|}}\sqrt{\delta}.$$

Proof. It follows from the minimizing property of u_α^δ and (4.12) that $\|u_\alpha^\delta\|^2 \leq \|u^\dagger\|^2$. Now the source condition and nonlinearity condition from Assumption 4.2 yields

$$\begin{aligned}
\|u_\alpha^\delta - u^\dagger\|^2 &\leq -2\langle u_\alpha^\delta - u^\dagger, u^\dagger\rangle \\
&\leq -2\langle K'(u^\dagger)(u_\alpha^\delta - u^\dagger), w\rangle \\
&= 2\langle K(u_\alpha^\delta) - K(u^\dagger) - K'(u^\dagger)(u_\alpha^\delta - u^\dagger), w\rangle \\
&\quad - 2\langle K(u_\alpha^\delta) - K(u^\dagger), w\rangle \\
&\leq L\|w\|\|u_\alpha^\delta - u^\dagger\|^2 + 2\|w\|\|K(u_\alpha^\delta) - K(u^\dagger)\|.
\end{aligned}$$

Meanwhile, the discrepancy principle (4.12) gives

$$\|K(u_\alpha^\delta) - K(u^\dagger)\| \leq \|K(u_\alpha^\delta) - g^\delta\| + \|g^\delta - K(u^\dagger)\| = (1 + c_m)\delta.$$

This concludes the proof of the theorem. □

One well-known drawback of the discrepancy principle is the suboptimality for higher-order source conditions, e.g., $u^\dagger = (K'(u^\dagger))^* K'(u^\dagger)w$. Hence it has been of interest to develop alternative a posteriori rules which achieve optimal convergence rates. We first give some basic estimates [284] on the approximation error $\|u_\alpha - u^\dagger\|$ and propagation error $\|u_\alpha^\delta - u_\alpha\|$, which represent the nonlinear analogues for Theorem 3.9 and are important for deriving convergence rates for a posteriori choice rules. In the proofs, one crucial ingredient is the optimality equation for u_α:

$$K'(u_\alpha)^*(K(u_\alpha) - g^\dagger) + \alpha u_\alpha = 0, \tag{4.13}$$

which is valid if u_α is an interior point of $\mathcal{C}$.

Theorem 4.8. *Let Assumption 4.2 hold with $\rho = 2\|x^\dagger\|$. If $3L\|w\| < 2$, then*

$$\|u_\alpha - u^\dagger\| \leq \frac{\sqrt{\alpha}\|w\|}{2\sqrt{1 - L\|w\|}} \quad \text{and} \quad \|K(u_\alpha) - g^\dagger\| \leq \alpha\|w\|.$$

Proof. Let $A_\alpha = K'(u_\alpha)$ and $A = K'(u^\dagger)$. Equation (4.13) implies

$$\begin{aligned}\alpha\|u_\alpha - u^\dagger\|^2 &= \alpha(u_\alpha - u^\dagger, -u^\dagger) + \alpha(u_\alpha - u^\dagger, u_\alpha)\\ &= -\alpha(u_\alpha - u^\dagger, u^\dagger) + (A_\alpha(u^\dagger - u_\alpha), K(u_\alpha) - g^\dagger).\end{aligned}$$

Upon adding $\|K(u_\alpha) - g^\dagger\|^2$ on both sides, and appealing to the source condition $u^\dagger = A^* w$ from Assumption 4.2 and estimate (4.10), we arrive at

$$\begin{aligned}&\|K(u_\alpha) - g^\dagger\|^2 + \alpha\|u_\alpha - u^\dagger\|^2\\ &\qquad= -\alpha(g^\dagger + A(u_\alpha - u^\dagger) - K(u_\alpha), w) - \alpha(K(u_\alpha) - g^\dagger, w)\\ &\qquad\quad+ (K(u_\alpha) + A_\alpha(u^\dagger - u_\alpha) - g^\dagger, K(u_\alpha) - g^\dagger)\\ &\qquad\leq \alpha\frac{L\|w\|}{2}\|u_\alpha - u^\dagger\|^2 + \alpha\|w\|\|K(u_\alpha) - g^\dagger\|\\ &\qquad\quad+ \frac{L}{2}\|u_\alpha - u^\dagger\|^2\|K(u_\alpha) - g^\dagger\|.\end{aligned}$$

It follows directly from the proof of Theorem 4.6, cf. Remark 4.3, that

$$\|K(u_\alpha) - g^\dagger\| \leq 2\alpha\|w\|,$$

and consequently

$$\begin{aligned}&\|K(u_\alpha)-g^\dagger\|^2+\alpha\|u_\alpha-u^\dagger\|^2\\&\leq \alpha\frac{3L\|w\|}{2}\|u_\alpha-u^\dagger\|^2+\alpha\|w\|\|K(u_\alpha)-g^\dagger\|.\end{aligned}$$

This together with the condition $3L\|w\|<2$ implies $\|K(u_\alpha)-g^\dagger\|^2\leq \alpha\|w\|\|K(u_\alpha)-g^\dagger\|$, i.e., the second estimate.

Now we use this estimate and Young's inequality to obtain

$$\begin{aligned}\|K(u_\alpha)-g^\dagger\|^2+&\alpha\|u_\alpha-u^\dagger\|^2\\&\leq \alpha L\|w\|\|u_\alpha-u^\dagger\|^2+\alpha\|w\|\|K(u_\alpha)-g^\dagger\|\\&\leq \alpha L\|w\|\|u_\alpha-u^\dagger\|^2+\frac{\alpha^2\|w\|^2}{4}+\|K(u_\alpha)-g^\dagger\|^2,\end{aligned}$$

which gives $(1-L\|w\|)\|u_\alpha-u^\dagger\|^2\leq \alpha\|w\|^2/4$, and hence also the first estimate. □

Now we estimate the propagation error $\|u_\alpha^\delta-u_\alpha\|$.

Theorem 4.9. *Let Assumption 4.2 hold with* $\rho=\delta/\sqrt{\alpha}+2\|x^\dagger\|$. *If* $3L\|w\|<2$, *then*

$$\|u_\alpha^\delta-u_\alpha\|\leq\frac{1}{\sqrt{1-L\|w\|}}\frac{\delta}{\sqrt{\alpha}}\quad\text{and}\quad\|K(u_\alpha^\delta)-g^\delta+g^\dagger-K(u_\alpha)\|\leq\delta.$$

Proof. By the minimizing property of u_α^δ, we obtain

$$\|K(u_\alpha^\delta)-g^\delta\|^2+\alpha\|u_\alpha^\delta\|^2\leq\|K(u_\alpha)-g^\delta\|^2+\alpha\|u_\alpha\|^2.$$

Adding both sides the expression

$$2(K(u_\alpha^\delta)-g^\delta,g^\dagger-K(u_\alpha))+\|K(u_\alpha)-g^\dagger\|^2+\alpha[-2(u_\alpha^\delta,u_\alpha)+\|u_\alpha\|^2],$$

completing the square, and equation (4.13) yield

$$\begin{aligned}&\|K(u_\alpha^\delta)-g^\delta+g^\dagger-K(u_\alpha)\|^2+\alpha\|u_\alpha^\delta-u_\alpha\|^2\\&\leq\|g^\dagger-g_\alpha^\delta\|^2+2(K(u_\alpha)-g^\dagger,K(u_\alpha)-K(u_\alpha^\delta))+2\alpha(u_\alpha,u_\alpha-u_\alpha^\delta)\\&=\|g^\dagger-g^\delta\|^2+2(K(u_\alpha)-g^\dagger,K(u_\alpha)+K'(u_\alpha)(u_\alpha^\delta-u_\alpha)-K(u_\alpha^\delta))\\&\leq\delta^2+L\|K(u_\alpha)-g^\dagger\|\|u_\alpha^\delta-u_\alpha\|^2.\end{aligned}$$

In view of Theorem 4.8, we deduce

$$\|K(u_\alpha^\delta)-g^\delta+g^\dagger-K(u_\alpha)\|^2+\alpha\|u_\alpha^\delta-u_\alpha\|^2\leq\delta^2+L\alpha\|w\|\|u_\alpha^\delta-u_\alpha\|^2.$$

Now the desired estimates follow directly from the condition $L\|w\|<1$. □

Now we give two a posteriori rules circumventing the early saturation issue of the discrepancy principle:

$$\Upsilon_1(\alpha) = \|(R_\alpha^\delta)^{1/2}[K(u_\alpha^\delta) - g^\delta]\| = c\delta, \tag{4.14}$$

$$\Upsilon_2(\alpha) = \frac{\|(R_\alpha^\delta)^{1/2}[K(u_\alpha^\delta) - g^\delta]\|^2}{\|R_\alpha^\delta[K(u_\alpha^\delta) - g^\delta]\|} = c\delta, \tag{4.15}$$

where $c > 1$ is a fixed constant. In the rules Υ_1 and Υ_2, the operator R_α^δ is given by $R_\alpha^\delta = \alpha(A_\alpha^\delta(A_\alpha^\delta)^* + \alpha I)^{-1}$, and $A_\alpha^\delta = K'(u_\alpha^\delta)$. The first rule Υ_1 was proposed by Scherzer et al [269], and it represents a nonlinear extension of the rule due to Gfrerer [104]. The rule Υ_2 is an extension of the monotone error rule [283] for linear inverse problems to the nonlinear case [284].

Both rules arc based on error estimation. We first briefly derive the rules. First, we note that the necessary optimality condition for u_α:

$$K'(u_\alpha)^*(K(u_\alpha) - g^\dagger) + \alpha u_\alpha = 0,$$

which upon formal differentiation with respect to α yields

$$K'(u_\alpha)^*K'(u_\alpha)\frac{du_\alpha}{d\alpha} + (K'(u_\alpha)^*)'(K(u_\alpha) - g^\dagger) + \alpha\frac{du_\alpha}{d\alpha} = -u_\alpha.$$

By ignoring the term involving $(K'(u_\alpha)^*)'$, we get

$$\frac{du_\alpha}{d\alpha} \approx -(\alpha I + K'(u_\alpha)^*K'(u_\alpha))^{-1}u_\alpha, \tag{4.16}$$

Similarly, the weak form with $u_\alpha - u^\dagger$ as a test function can be recast into

$$\begin{aligned}(K(u_\alpha) - g^\dagger, K'(u_\alpha)(u_\alpha - u^\dagger)) &\approx \alpha^2\left(\frac{du_\alpha}{d\alpha}, u_\alpha - u^\dagger\right)\\ &\quad + \alpha\left(K'(u_\alpha)\frac{du_\alpha}{d\alpha}, K'(u_\alpha)(u_\alpha - u^\dagger)\right).\end{aligned} \tag{4.17}$$

By the above splitting procedure, the total error can be bounded from above by $\frac{1}{2}\|u_\alpha - u^\dagger\|^2 + c\frac{\delta^2}{\alpha}$, cf. Theorem 4.9. Hence it is natural to choose α such that

$$\min \tfrac{1}{2}\|u_\alpha - u^\dagger\|^2 + c\frac{\delta^2}{\alpha}$$

where c is a fixed positive constant. The necessary optimality condition for this function is given by

$$c\delta^2 = \alpha^2\left(\frac{du_\alpha}{d\alpha}, u_\alpha - u^\dagger\right).$$

Inserting this into (4.17) yields

$$(K(u_\alpha)-g^\dagger, K'(u_\alpha)(u_\alpha-u^\dagger)) \approx c\delta^2 + \alpha\left(K'(u_\alpha)\frac{du_\alpha}{d\alpha}, K'(u_\alpha)(u_\alpha-u^\dagger)\right).$$

Next we approximate $K'(u_\alpha)(u_\alpha-u^\dagger)$ by $K(u_\alpha)-K(u^\dagger)$, and obtain from the approximation for $\frac{du_\alpha}{d\alpha}$ that

$$\|K(u_\alpha)-g^\dagger\|^2+\alpha(K(u_\alpha)-g^\dagger, K'(u_\alpha)(\alpha I+K'(u_\alpha)^*K'(u_\alpha))^{-1}u_\alpha)\approx c\delta^2,$$

which upon rearrangements yields

$$\alpha(K(u_\alpha)-g^\dagger,(\alpha I+K'(u_\alpha)K'(u_\alpha)^*)^{-1}(K(u_\alpha)-g^\dagger))\approx c\delta^2.$$

Finally, to arrive at a computable rule, we replace $g^\dagger$ by g^δ, and consequently u_α by u_α^δ, we arrive at the rule Υ_1:

$$\alpha(K(u_\alpha^\delta)-g^\delta,(\alpha I+K'(u_\alpha^\delta)K'(u_\alpha^\delta)^*)^{-1}(K(u_\alpha^\delta)-g^\delta))=c\delta^2.$$

Next we derive the monotone error rule. It looks for a computable regularization parameter α^* which is as small as possible and guarantees the property $\frac{d}{d\alpha}\|u_\alpha^\delta-u^\dagger\|^2>0$ for all $\alpha\in(\alpha^*,\infty)$. With

$$A_\alpha^\delta = K'(u_\alpha^\delta) \quad \text{and} \quad R_\alpha^\delta=\alpha(\alpha I+A_\alpha^\delta(A_\alpha^\delta)^*)^{-1},$$

we have

$$\begin{aligned}
&\tfrac{1}{2}\frac{d}{d\alpha}\|u_\alpha^\delta-u^\dagger\|^2\\
&=\left(u_\alpha^\delta-u^\dagger,\frac{du_\alpha^\delta}{d\alpha}\right)\\
&\approx(u_\alpha^\delta-u^\dagger,-(\alpha I+(A_\alpha^\delta)^*A_\alpha^\delta)^{-1}u_\alpha^\delta)\\
&=(u_\alpha^\delta-u^\dagger,\alpha^{-1}(\alpha I+(A_\alpha^\delta)^*A_\alpha^\delta)^{-1}(A_\alpha^\delta)^*(K(u_\alpha^\delta)-g^\delta))\\
&=\alpha^{-2}(A_\alpha^\delta(u_\alpha^\delta-u^\dagger),R_\alpha^\delta(K(u_\alpha^\delta)-g^\delta))\\
&\approx\alpha^{-2}(K(u_\alpha^\delta)-g^\delta+g^\delta-g^\dagger,R_\alpha^\delta(K(u_\alpha^\delta)-g^\delta))\\
&\geq\alpha^{-2}\|R_\alpha^\delta(K(u_\alpha^\delta)-g^\delta)\|\left(\frac{\|(R_\alpha^\delta)^{\frac{1}{2}}(K(u_\alpha^\delta)-g^\delta)\|^2}{\|R_\alpha^\delta(K(u_\alpha^\delta)-g^\delta)\|}-\delta\right).
\end{aligned}$$

In the derivations, we have employed (4.16) and the approximation $K'(u_\alpha^\delta)(u_\alpha^\delta-u^\dagger)\approx K(u_\alpha^\delta)-K(u^\dagger)$. This suggests the following monotone error rule for choosing a suitable α:

$$\frac{\|(R_\alpha^\delta)^{\frac{1}{2}}(K(u_\alpha^\delta)-g^\delta)\|^2}{\|R_\alpha^\delta(K(u_\alpha^\delta)-g^\delta)\|}=c\delta$$

for some $c\geq 1$, which (approximately) ensures the desired relation $\frac{d}{d\alpha}\|u_\alpha^\delta-u^\dagger\|^2\geq 0$. This gives the rule Υ_2.

The next result gives a lower bound for $\alpha(\delta)$ determined by rules Υ_1 and Υ_2.

Lemma 4.1. *Let α be chosen by rule Υ_1 or Υ_2 with $c > 1$, and Assumption 4.2 hold with $\rho = \delta/\sqrt{\alpha} + 2\|u^\dagger\|$. If $L\|w\| < 1$, then*

$$\alpha \geq \frac{c-1}{\|w\|}\delta.$$

Proof. For the rule Υ_1, we deduce from $\|R_\alpha^\delta\| \leq 1$ and Theorems 4.8 and 4.9 that

$$c\delta = \|(R_\alpha^\delta)^{1/2}[K(u_\alpha^\delta) - g^\delta]\| \leq \|K(u_\alpha^\delta) - g^\delta\| \leq \delta + \alpha\|w\|$$

which gives the desired assertion for rule Υ_1. For rule Υ_2, we use the estimate

$$c\delta = \frac{\|(R_\alpha^\delta)^{1/2}[K(u_\alpha^\delta) - g^\delta]\|^2}{\|R_\alpha^\delta[K(u_\alpha^\delta) - g^\delta]\|} \leq \|K(u_\alpha^\delta) - g^\delta\|,$$

and the rest follows as above. □

The next result shows the convergence rate [284] for the rules Υ_1 and Υ_2 under Assumption 4.2.

Theorem 4.10. *Let α be chosen by rule Υ_1 or Υ_2 with $c > 1$, and Assumption 4.2 hold with $\rho = \delta/\sqrt{\alpha} + 2\|u^\dagger\|$. Further, assume that $L\|w\|$ is sufficiently small such that $3L\|w\| < 2$ and*

$$\epsilon_1 := \sqrt{\frac{L\|w\|}{2}} + L\|w\| + \frac{L\|w\|}{8\sqrt{1 - L\|w\|}}\frac{c+1}{c-1} < 1$$

hold. Then

$$\|u_\alpha^\delta - u^\dagger\| \leq \frac{1}{1-\epsilon_1}\left\{(c+1)^{1/2} + \tfrac{1}{2}(c-1)^{-1/2}\right\}\|w\|^{1/2}\delta^{1/2}.$$

Proof. For convenience, we introduce the following notation

$$\begin{aligned} A_\alpha &= K'(u_\alpha), \quad A_\alpha^\delta = K'(u_\alpha^\delta), \\ R_\alpha^\delta &= \alpha(A_\alpha^\delta(A_\alpha^\delta)^* + \alpha I)^{-1}, \\ \widehat{R}_\alpha^\delta &= \alpha((A_\alpha^\delta)^* A_\alpha^\delta + \alpha I)^{-1}. \end{aligned}$$

First we show the result for the rule Υ_1. By the optimality condition for u_α^δ, we obtain

$$\begin{aligned} &((A_\alpha^\delta)^* A_\alpha^\delta + \alpha I)(u_\alpha^\delta - u^\dagger) \\ &= (A_\alpha^\delta)^* A_\alpha^\delta(u_\alpha^\delta - u^\dagger) + \alpha(u_\alpha^\delta - u^\dagger) \\ &= (A_\alpha^\delta)^* A_\alpha^\delta(u_\alpha^\delta - u^\dagger) + (A_\alpha^\delta)^*(g^\delta - K(u_\alpha^\delta)) - \alpha u^\dagger \\ &= -\alpha u^\dagger + (A_\alpha^\delta)^*(g^\delta + A_\alpha^\delta(u_\alpha^\delta - u^\dagger) - K(u_\alpha^\delta)). \end{aligned}$$

Multiplying both sides of the identity by $((A_\alpha^\delta)^* A_\alpha^\delta + \alpha I)^{-1}$ yields the following error representation

$$u_\alpha^\delta - u^\dagger = -\widehat{R}_\alpha^\delta u^\dagger + ((A_\alpha^\delta)^* A_\alpha^\delta + \alpha I)^{-1}(A_\alpha^\delta)^*(g^\delta + A_\alpha^\delta(u_\alpha^\delta - u^\dagger) - K(u_\alpha^\delta)).$$

Now by the inequality

$$\|((A_\alpha^\delta)^* A_\alpha^\delta + \alpha I)^{-1}(A_\alpha^\delta)^*\| \leq \frac{1}{2\sqrt{\alpha}},$$

(4.10), and the triangle inequality, we deduce

$$\begin{aligned}
\|u_\alpha^\delta - u^\dagger\| &\leq \|\widehat{R}_\alpha^\delta u^\dagger\| + \frac{1}{2\sqrt{\alpha}}\left(\delta + \frac{L}{2}\|u_\alpha^\delta - u^\dagger\|^2\right) \\
&\leq \|\widehat{R}_\alpha^\delta u^\dagger\| + \frac{1}{2\sqrt{\alpha}}\left(\delta + \frac{L}{2}\frac{\delta + \alpha\|w\|/2}{\sqrt{\alpha}\sqrt{1 - L\|w\|}}\|u_\alpha^\delta - u^\dagger\|\right) \\
&\leq \|\widehat{R}_\alpha^\delta u^\dagger\| + \frac{1}{2}\sqrt{\frac{\|w\|}{c-1}}\sqrt{\delta} + \frac{L\|w\|}{8\sqrt{1 - L\|w\|}}\frac{c+1}{c-1}\|u_\alpha^\delta - u^\dagger\|,
\end{aligned}$$

where the last line follows from the inequality $\delta/\alpha \leq \|w\|/(c-1)$ in Lemma 4.1. Now by the Lipschitz continuity of $K'(u)$ from Assumption 4.2, source condition and the estimate $\|\widehat{R}_\alpha^\delta\| \leq 1$, we obtain

$$\begin{aligned}
\|\widehat{R}_\alpha^\delta u^\dagger\|^2 &= (\widehat{R}_\alpha^\delta u^\dagger, \widehat{R}_\alpha^\delta (A_\alpha^\delta)^* w) + (\widehat{R}_\alpha^\delta u^\dagger, \widehat{R}_\alpha^\delta (A^* - (A_\alpha^\delta)^*) w) \\
&\leq (\widehat{R}_\alpha^\delta u^\dagger, \widehat{R}_\alpha^\delta (A_\alpha^\delta)^* w) + \|\widehat{R}_\alpha^\delta u^\dagger\| L\|w\|\|u_\alpha^\delta - u^\dagger\|,
\end{aligned}$$

which together with the implication $c^2 \leq a + bc \Rightarrow c \leq \sqrt{a} + b$ for $a, b, c \geq 0$ and the estimate $\|(\widehat{R}_\alpha^\delta)^{\frac{1}{2}}\| \leq 1$ gives

$$\begin{aligned}
\|\widehat{R}_\alpha^\delta u^\dagger\| &\leq \sqrt{(\widehat{R}_\alpha^\delta u^\dagger, \widehat{R}_\alpha^\delta (A_\alpha^\delta)^* w)} + L\|w\|\|u_\alpha^\delta - u^\dagger\| \\
&\leq \|(R_\alpha^\delta)^{3/2} A_\alpha^\delta u^\dagger\|^{1/2}\|w\|^{1/2} + L\|w\|\|u_\alpha^\delta - u^\dagger\|.
\end{aligned}$$

To estimate the term $\|(R_\alpha^\delta)^{3/2} A_\alpha^\delta u^\dagger\|^{1/2}$, we appeal again to the optimality condition for u_α^δ to obtain

$$A_\alpha^\delta (A_\alpha^\delta)^*(K(u_\alpha^\delta) - g^\delta) = -\alpha A_\alpha^\delta u_\alpha^\delta.$$

Adding the expression $\alpha(K(u_\alpha^\delta) - g^\delta)$ to both sides and then multiplying by $(A_\alpha^\delta (A_\alpha^\delta)^* + \alpha I)^{-1}$ yields

$$\begin{aligned}
K(u_\alpha^\delta) - g^\delta &= R_\alpha^\delta (K(u_\alpha^\delta) - A_\alpha^\delta u_\alpha^\delta - g^\delta) \\
&= R_\alpha^\delta (K(u_\alpha^\delta) + A_\alpha^\delta (u^\dagger - u_\alpha^\delta) - g^\delta) - R_\alpha^\delta A_\alpha^\delta u^\dagger.
\end{aligned}$$

Multiplying by $(R_\alpha^\delta)^{1/2}$ gives

$$-(R_\alpha^\delta)^{3/2} A_\alpha^\delta u^\dagger = (R_\alpha^\delta)^{1/2}(K(u_\alpha^\delta) - g^\delta) - (R_\alpha^\delta)^{3/2}(K(u_\alpha^\delta) + A_\alpha^\delta (u^\dagger - u_\alpha^\delta) - g^\delta).$$

Now by the definition of rule Υ_1, we obtain

$$\|(R_\alpha^\delta)^{3/2}A_\alpha^\delta u^\dagger\|^{1/2} \leq \sqrt{c\delta + \delta + \frac{L}{2}\|u_\alpha^\delta - u^\dagger\|^2}$$
$$\leq \sqrt{c+1}\sqrt{\delta} + \sqrt{\frac{L}{2}}\|u_\alpha^\delta - u^\dagger\|.$$

Now the assertion for Υ_1 follows from the preceding estimates.

To prove the assertion for Υ_2, we note that the definition of Υ_1 has only been exploited in Lemma 4.1 and the last step. However, due to Lemma 4.1 and the relation $\Upsilon_1(\alpha) \leq \Upsilon_2(\alpha)$, these estimates are also valid for Υ_2. This completes the proof of the theorem. □

4.2.3 *Structural properties*

Lastly, we mention some structural properties of the functional J_α. In general, the functional J_α is nonconvex, and have multiple local minima, which represents one of the major computational inconveniences of the approach for nonlinear inverse problems. The following result shows its local convexity [203]. We denote by $M = \sup_{u\in\mathcal{C}} \|K'(u)\|$.

Theorem 4.11. *Let Assumption 4.2 hold, and $\delta \leq \alpha/6L$, $12L\|w\| \leq 1$. Then the functional J_α is strongly convex on the ball $B_r(u_\alpha^\delta)$, with $r = \alpha/(12ML)$, with the constant of strong convexity $\theta = \alpha/4$. Further, it is convex on the ball $B_{\alpha/(3LM)}(u_\alpha^\delta)$.*

Proof. It is straightforward to show that

$$\langle J_\alpha'(u) - J_\alpha'(v), u - v\rangle = \alpha\|u - v\|^2 + \langle K'(u)^*K(u) - K'(u)^*K(v), u - v\rangle$$
$$+ \langle K'(u)^*(K(v) - g^\delta) - K'(v)(K(v) - g^\delta), u - v\rangle.$$

By the mean value theorem, we obtain

$$\langle K'(u)^*K(u) - K'(u)K(v), u - v\rangle$$
$$= \langle K'(u)^* \int_0^1 K'(v + t(u - v))(u - v)dt, u - v\rangle$$
$$= \langle K'(u)^* \int_0^1 [K'(v + t(u - v)) - K'(u)](u - v)dt, u - v\rangle$$
$$+ \|K'(u)(u - v)\|^2 \geq -\tfrac{1}{2}LM\|u - v\|^3 \quad \forall u, v \in \mathcal{C}.$$

Further, there holds

$$\langle K'(u)^*(K(v) - g^\delta) - K'(v)^*(K(v) - g^\delta), u - v\rangle$$
$$\leq \|K'(u)^* - K'(v)^*\|\|K(v) - g^\delta\|\|u - v\|.$$

We note that

$$\|K'(u)^* - K'(v)^*\| = \|K'(u) - K'(v)\| \leq L\|u - v\|$$

and by Theorem 4.6,

$$\begin{aligned}\|K(v) - g^\delta\| &\leq \|K(v) - K(u_\alpha^\delta)\| + \|K(u_\alpha^\delta) - g^\delta\| \\ &\leq \|\int_0^1 K'(u_\alpha^\delta + t(v - u_\alpha^\delta))(v - u_\alpha^\delta)dt\| + \delta + 2\alpha\|w\| \\ &\leq M\|u_\alpha^\delta - v\| + \delta + 2\alpha\|w\|.\end{aligned}$$

Consequently

$$\begin{aligned}|\langle K'(u)^*(K(v)-g^\delta) &- K'(v)^*(K(v) - g^\delta), u - v\rangle| \\ &\leq L\|u - v\|^2(M\|u_\alpha^\delta - v\| + \delta + 2\alpha\|w\|).\end{aligned}$$

Combing the preceding estimates yields

$$\begin{aligned}\langle J_\alpha'(u) - J_\alpha'(v), u - v\rangle \geq \|u - v\|^2 &\left(\alpha - \tfrac{1}{2}LM\|u - v\|\right. \\ &\left.-LM\|u_\alpha^\delta - v\| - L(\delta + 2\alpha\|w\|)\right) \quad \forall u, v \in \mathcal{C}.\end{aligned}$$

Now for all $u, v \in B_r(u_\alpha^\delta)$, there holds $\|u - v\| < 2r$ and $\|u_\alpha^\delta - v\| < r$, and hence

$$\begin{aligned}\langle J_\alpha'(u) - J_\alpha'(v), u - v\rangle &\geq \|u - v\|^2(\alpha - 2LMr - L(\delta + 2\alpha\|w\|)) \\ &\geq \tfrac{\alpha}{2}\|u - v\|^2 \quad \forall u, v \in B_r(u_\alpha^\delta).\end{aligned}$$

Hence J_α is strongly convex on the ball $B_r(u_\alpha^\delta)$ with a constant of strong convexity $\alpha/4$. □

Corollary 4.1. *The global minimizers to J_α are isolated.*

Theorem 4.12. *Let Assumption 4.2 hold, $\delta \leq \alpha/6L$, $12L\|w\| \leq 1$. The gradient J_α' has the Fejér property with parameter $\eta = \alpha/2$ on $B(u_\alpha^\delta)$, $r = \alpha/(3LM)$, i.e.,*

$$\langle J_\alpha'(u), u - u_\alpha^\delta\rangle \geq \eta\|u - u_\alpha^\delta\|^2 \quad \forall u \in B_r(u_\alpha^\delta).$$

Proof. We set $v = u_\alpha^\delta$ in the proof of Theorem 4.11. Accordingly we estimate

$$|\langle K'(u)^*(K(u_\alpha^\delta)-g^\delta)-K'(u_\alpha^\delta)^*(K(u_\alpha^\delta)-g^\delta), u-u_\alpha^\delta\rangle| \leq L\|u-u_\alpha^\delta\|^2(\delta+2\alpha\|w\|).$$

Now with the equality $J_\alpha'(u_\alpha^\delta) = 0$, we arrive at

$$\begin{aligned}\langle J_\alpha'(u), u - u_\alpha^\delta\rangle &\geq \|u - u_\alpha^\delta\|^2(\alpha - \tfrac{1}{2}LMr - L(\delta + 2\alpha\|w\|)) \\ &\geq \tfrac{\alpha}{2}\|u - u_\alpha^\delta\|^2 > 0 \quad \forall u \in B_r(u_\alpha^\delta).\end{aligned}$$

This completes the proof of the theorem. □

The uniqueness of the Tikhonov minimizer to J_α is important for defining a posteriori rules, e.g., discrepancy principle. Hence, we shall briefly discuss relevant sufficient conditions. It is known that the uniqueness is guaranteed for a class of weakly nonlinear operators [59]. Below we describe an approach developed in [176], which imposes the following assumption on the minimum-norm solution $u^\dagger$ and the operator K.

Assumption 4.3. The true solution $u^\dagger$ and the operator K satisfy

(i) There is a number $r > 4\|u^\dagger\|$ such that $B_r(u^\dagger) \subset \mathcal{C}$.
(ii) There is a constant K_2 such that for every $(u, v, z) \in B_r(u^\dagger) \times B_r(u^\dagger) \times X$ there is an element $k(u, v, z) \in X$ such that

$$\begin{aligned}(K'(u) - K'(v))z &= K'(v)k(u, v, z),\\ \|k(u, v, z)\| &\leq K_2\|z\|\|u - v\|.\end{aligned}$$

Theorem 4.13. *Let Assumption 4.3 hold, $c_1 > 1$ and $\alpha \geq (c_1\delta)^2$ be chosen such that $c_1 r > 2$, and $\tau_0 := \frac{K_2}{c_1} < 1$. If $\|u^\dagger\|$ is sufficiently small such that $K_2\|u^\dagger\| < 1 - \tau_0$, then for $\delta > 0$ sufficiently small, the functional J_α has a unique solution.*

Proof. Let $\tilde{u}_\alpha^\delta$ be another solution of J_α. It follows directly from the comparison $J_\alpha(\tilde{u}_\alpha^\delta) \leq J_\alpha(u^\dagger)$ that $\|\tilde{u}_\alpha^\delta\| \leq \frac{\delta}{\sqrt{\alpha}} + \|u^\dagger\|$. Hence for the choice $\alpha \geq (c_1\delta)^2$,

$$\|\tilde{u}_\alpha^\delta\| \leq \tfrac{1}{c_1} + \|u^\dagger\|, \tag{4.18}$$

since $c_1 r \geq 2$, we have $\frac{1}{c_1} + 2\|u^\dagger\| < r$, i.e., $\tilde{u}_\alpha^\delta$ is an interior point of $\mathcal{C}$. Hence the following first-order optimality condition holds:

$$K'(\tilde{u}_\alpha^\delta)^*(K(\tilde{u}_\alpha^\delta) - g^\delta) + \alpha\tilde{u}_\alpha^\delta = 0.$$

This together with the definition of u_α^δ yields

$$\begin{aligned}&\|K(u_\alpha^\delta) - K(\tilde{u}_\alpha^\delta)\|^2 + \alpha\|u_\alpha^\delta - \tilde{u}_\alpha^\delta\|^2\\ &= 2(K(u_\alpha^\delta) - K(\tilde{u}_\alpha^\delta), g^\delta - K(\tilde{u}_\alpha^\delta)) + 2\alpha(u_\alpha^\delta - \tilde{u}_\alpha^\delta, -\tilde{u}_\alpha^\delta)\\ &= 2(K(u_\alpha^\delta) - K(\tilde{u}_\alpha^\delta) - K'(\tilde{u}_\alpha^\delta)(u_\alpha^\delta - \tilde{u}_\alpha^\delta), g^\delta - K(\tilde{u}_\alpha^\delta)).\end{aligned}$$

Now by Assumption 4.3(ii), there holds

$$K(u_\alpha^\delta) - K(\tilde{u}_\alpha^\delta) - K'(\tilde{u}_\alpha^\delta)(u_\alpha^\delta - \tilde{u}_\alpha^\delta) = K'(\tilde{u}_\alpha^\delta)\int_0^1 \tilde{k}_t dt,$$

with $\tilde{k}_t = k(\tilde{u}_\alpha^\delta + t(u_\alpha^\delta - \tilde{u}_\alpha^\delta), \tilde{u}_\alpha^\delta, u_\alpha^\delta - \tilde{u}_\alpha^\delta)$ and $\| \int_0^1 \tilde{k}_t dt\| \leq \frac{K_2}{2}\|u_\alpha^\delta - \tilde{u}_\alpha^\delta\|^2$. Using the optimality condition again, we obtain

$$\begin{aligned}
&\|K(u_\alpha^\delta) - K(\tilde{u}_\alpha^\delta)\|^2 + \alpha\|u_\alpha^\delta - \tilde{u}_\alpha^\delta\|^2 \\
&\leq 2\Big(\int_0^1 \tilde{k}_t dt, K'(\tilde{u}_\alpha^\delta)^*(g^\delta - K(u_\alpha^\delta))\Big) \\
&= 2\alpha\Big(\int_0^1 \tilde{k}_t dt, \tilde{u}_\alpha^\delta\Big) \leq \alpha K_2 \|\tilde{u}_\alpha^\delta\|\|u_\alpha^\delta - \tilde{u}_\alpha^\delta\|^2.
\end{aligned}$$

Hence it suffices to show $1 - K_2\|\tilde{u}_\alpha^\delta\| > 0$, which however follows directly from the assumption and (4.18). □

Corollary 4.2. *Let the assumptions in Theorem 4.13 hold. Then the mapping $\alpha \mapsto u_\alpha^\delta$ is continuous, as is the mapping $\alpha \mapsto \|K(u_\alpha^\delta) - g^\delta\|$.*

Proof. By Theorem 4.13, the minimizer to J_α is unique. Let $\{\alpha_k\}$ be any sequence converging to α, and $\{u_k\}$ be the sequence of minimizers $\{J_{\alpha_k}\}$. Then by repeating the arguments in the proof of Theorem 4.2, the whole sequence $\{u_k\}$ converges weakly to u_α^δ, and further $\|u_k\| \to \|u_\alpha^\delta\|$. This completes the proof of the corollary. □

The last result gives an upper bound on the regularization parameter α selected by the discrepancy principle [300], generalizing Theorem 3.17.

Theorem 4.14. *Let the operator K be radially differentiable, i.e.,*

$$\lim_{t\to 0} \frac{K(u_\alpha^\delta + tu_\alpha^\delta) - K(u_\alpha^\delta)}{t} = \widetilde{K}u_\alpha^\delta$$

for some $\widetilde{K} \in L(X,Y)$. Then there holds

$$\alpha \leq \frac{\|\widetilde{K}u_\alpha^\delta\|}{\|u_\alpha^\delta\|^2}\|K(u_\alpha^\delta) - g^\delta\|.$$

Proof. By the minimizing property of u_α^δ, for sufficiently small t, $|t| \leq \gamma$, $\gamma > 0$, we have

$$J_\alpha(u_\alpha^\delta + tu_\alpha^\delta) - J_\alpha(u_\alpha^\delta) \geq 0.$$

Further,

$$\|(1+t)u_\alpha^\delta\|^2 - \|u_\alpha^\delta\|^2 = 2t\|u_\alpha^\delta\|^2 + t^2\|u_\alpha^\delta\|^2,$$

and

$$\|K(u_\alpha^\delta + tu_\alpha^\delta) - g^\delta\|^2 - \|K(u_\alpha^\delta) - g^\delta\|^2 = 2\|K(u_\alpha^\delta) - g^\delta\|\zeta + \zeta^2,$$

with

$$\zeta = \|K(u_\alpha^\delta + tu_\alpha^\delta) - g^\delta\| - \|K(u_\alpha^\delta) - g^\delta\|.$$

By the triangle inequality, $|\zeta| \leq \|K(u_\alpha^\delta + tu_\alpha^\delta) - K(u_\alpha^\delta)\|$ and thus $\lim_{t\to 0} \frac{|\zeta|}{|t|} \leq \|\widetilde{K}u_\alpha^\delta\|$. By choosing $t < 0$ sufficiently small, we get

$$2\|K(u_\alpha^\delta) - g^\delta\|\zeta + \zeta^2 \geq \alpha(-2t + t^2)\|u_\alpha^\delta\|^2 \geq 0,$$

and hence

$$\alpha \leq \frac{2\|K(u_\alpha^\delta) - g^\delta\||\zeta| + \zeta^2}{(-2t + t^2)\|u_\alpha^\delta\|^2}.$$

Now letting $t \to 0^-$, we arrive at the desired assertion. □

Remark 4.5. The radial differentiability is weaker than Gâteaux differentiability. In the latter case, the estimate can be deduced directly from the optimality condition.

4.3 A new convergence rate analysis

In this section, we develop a new approach to convergence rate analysis in the presence of convex constraints from the viewpoint of optimization theory, along the line of [58, 243]. We begin with a second-order necessary condition for the minimizer u_α. Then we propose using a second-order sufficient condition as a nonlinearity condition, show its connection with classical conditions, establish its role in deriving basic error estimates and convergence rate analysis. In Section 4.4, we shall illustrate the approach on a class of nonlinear parameter identifications.

4.3.1 *Necessary optimality condition*

Consider the following generic constrained Tikhonov regularization formulation

$$\min_{u \in \mathcal{C}} \phi(u, g^\dagger) + \alpha\psi(u),$$

where the fidelity $\phi(u, g)$ is differentiable in the first argument, the penalty $\psi(u)$ is convex and (weakly) lower semi-continuous, and the constraint set $\mathcal{C}$ is convex and closed.

Let u_α be a minimizer of the problem, i.e.,

$$\phi(u_\alpha, g^\dagger) + \alpha\psi(u_\alpha) \leq \phi(v, g^\dagger) + \alpha\psi(v) \quad \forall v \in \mathcal{C}.$$

Since ψ is convex, for $v = u_\alpha + t(u - u_\alpha) \in \mathcal{C}$ with $u \in \mathcal{C}, 0 < t \leq 1$

$$\frac{\phi(v, g^\dagger) - \phi(u_\alpha, g^\dagger)}{t} \geq -\alpha \frac{\psi(v) - \psi(u_\alpha)}{t} \geq -\alpha\,(\psi(u) - \psi(u_\alpha)).$$

By letting $t \to 0^+$, we obtain the necessary optimality condition

$$\langle \phi'(u_\alpha, g^\delta), u - u_\alpha\rangle + \alpha(\psi(u) - \psi(u_\alpha)) \geq 0 \quad \forall u \in \mathcal{C}.$$

Consequently, it follows from the convexity of ψ that if there exists an element $\xi_\alpha \in \partial\psi(u_\alpha)$ and let

$$\mu_\alpha = \frac{1}{\alpha}\left(\phi'(u_\alpha, g^\dagger) + \alpha\xi_\alpha\right),$$

then we have

$$\begin{cases} \phi'(u_\alpha, g^\dagger) + \alpha\xi_\alpha - \alpha\mu_\alpha = 0, \\ \langle \mu_\alpha, u - u_\alpha\rangle \geq 0 \ \ \forall u \in \mathcal{C}, \\ \xi_\alpha \in \partial\psi(u_\alpha). \end{cases} \tag{4.19}$$

Thus, $\mu_\alpha \in X^*$ serves as a Lagrange multiplier for the constraint $\mathcal{C}$, cf. Theorem 3.2 of [231]. If $\psi'(u_\alpha) \in X^*$ exists, then $\xi_\alpha = \psi'(u_\alpha)$ and thus $\mu_\alpha = \frac{1}{\alpha}\phi'(u_\alpha, g^\dagger) + \psi'(u_\alpha)$. In a more general constrained optimization, the existence of a Lagrange multiplier $\mu_\alpha \in X^*$ is guaranteed by the regular point condition [231, 163]. The inequality (4.19) is the first-order optimality condition. We refer to [231, 163] for a general theory of second-order conditions.

4.3.2 *Source and nonlinearity conditions*

Now we turn to problem (4.7). We will propose a new nonlinearity condition based on a second-order sufficient optimality condition. To this end, we first introduce the second-order error $E(u, \tilde{u})$ of the operator K defined by

$$E(u, \tilde{u}) = K(u) - K(\tilde{u}) - K'(\tilde{u})(u - \tilde{u}).$$

$E(u, \tilde{u})$ quantitatively measures the degree of nonlinearity, or pointwise linearization error, of the operator K, and will be used in deriving a nonlinearity condition. We also recall the first-order necessary optimality condition for u_α, cf., (4.19)

$$\begin{cases} K'(u_\alpha)^*(K(u_\alpha) - g^\dagger) + \alpha u_\alpha - \alpha\mu_\alpha = 0, \\ \langle \mu_\alpha, u - u_\alpha\rangle \geq 0 \ \ \forall u \in \mathcal{C}, \end{cases} \tag{4.20}$$

where μ_α is a Lagrange multiplier for the constraint $\mathcal{C}$. In view of the differentiability of the penalty, the Lagrange multiplier μ_α is explicitly given by $\mu_\alpha = u_\alpha + \frac{1}{\alpha}K'(u_\alpha)^*(K(u_\alpha) - g^\dagger)$.

Now we derive a second-order necessary optimality condition for problem (4.7).

Lemma 4.2. *The necessary optimality condition of a minimizer u_α to the functional J_α with the exact data $g^\dagger$ is given by: for any $u \in \mathcal{C}$*

$$\begin{aligned} \tfrac{1}{2}\|K(u_\alpha) - K(u)\|^2 + \tfrac{\alpha}{2}\|u_\alpha - u\|^2 + \alpha\langle \mu_\alpha, u - u_\alpha\rangle \\ + \langle K(u_\alpha) - g^\dagger, E(u, u_\alpha)\rangle \geq 0, \end{aligned} \tag{4.21}$$

where μ_α is a Lagrange multiplier associated with the constraint $\mathcal{C}$.

Proof. By the minimizing property of u_α, we have that for any $u \in \mathcal{C}$

$$\tfrac{1}{2}\|K(u_\alpha) - g^\dagger\|^2 + \tfrac{\alpha}{2}\|u_\alpha\|^2 \leq \tfrac{1}{2}\|K(u) - g^\delta\|^2 + \tfrac{\alpha}{2}\|u\|^2.$$

Straightforward computations show the following two elementary identities

$$\begin{aligned} \tfrac{1}{2}\|u_\alpha\|^2 - \tfrac{1}{2}\|u\|^2 &= -\tfrac{1}{2}\|u_\alpha - u\|^2 - \langle u_\alpha, u - u_\alpha\rangle, \\ \tfrac{1}{2}\|K(u_\alpha) - g^\dagger\|^2 - \tfrac{1}{2}\|K(u) - g^\dagger\|^2 &= -\tfrac{1}{2}\|K(u_\alpha) - K(u)\|^2 \\ &\quad - \langle K(u_\alpha) - g^\dagger, K(u) - K(u_\alpha)\rangle. \end{aligned}$$

Upon substituting these two identities, we arrive at

$$\begin{aligned} -\tfrac{1}{2}\|K(u_\alpha) - K(u)\|^2 - \tfrac{\alpha}{2}\|u_\alpha - u\|^2 - \alpha\langle u_\alpha, u - u_\alpha\rangle \\ - \langle K(u_\alpha) - g^\dagger, K(u) - K(u_\alpha)\rangle \leq 0. \end{aligned} \tag{4.22}$$

Now, the optimality condition for the minimizer u_α (cf. (4.20)) is given by

$$K'(u_\alpha)^*(K(u_\alpha) - g^\dagger) + \alpha u_\alpha - \alpha\mu_\alpha = 0,$$

where μ_α is a Lagrange multiplier for the constraint $\mathcal{C}$. Consequently,

$$\alpha\langle u_\alpha, u - u_\alpha\rangle + \langle K(u_\alpha) - g^\dagger, K'(u_\alpha)(u - u_\alpha)\rangle - \alpha\langle \mu_\alpha, u - u_\alpha\rangle = 0.$$

Together with this identity and the second-order error $E(u, u_\alpha)$, inequality (4.22) yields immediately the desired assertion. □

Remark 4.6. The case of a general convex ψ can be handled similarly using the generalized Bregman distance. In particular, repeating the proof in Lemma 4.2 gives the following necessary optimality condition

$$\tfrac{1}{2}\|K(u_\alpha) - K(u)\|^2 + \eta d_{\xi_\alpha, \mu_\alpha}(u, u_\alpha) + \langle K(u_\alpha) - g^\dagger, E(u, u_\alpha)\rangle \geq 0$$

where $\xi_\alpha \in \psi(u_\alpha)$ and μ_α is a Lagrangian multiplier. Then all the results in this section can be adapted to a general penalty ψ by replacing $\frac{1}{2}\|u - \tilde{u}\|^2$ with the Bregman distance $d_{\xi,\mu}(u, \tilde{u})$.

One salient feature of the optimality condition (4.21) is that it is true for any $u \in \mathcal{C}$ and thus it is a global one. Also, the term $\langle \mu_\alpha, u - u_\alpha \rangle$ is always nonnegative. The necessary condition (4.21) may be strengthened as follows: there exist some $c_s \in [0,1)$ and $\epsilon' > 0$ such that

$$\begin{aligned} &\tfrac{1}{2}\|K(u_\alpha) - K(u)\|^2 + \tfrac{\alpha}{2}\|u_\alpha - u\|^2 + \langle K(u_\alpha) - g^\dagger, E(u, u_\alpha)\rangle \\ &+ \alpha\langle \mu_\alpha, u - u_\alpha\rangle \geq \tfrac{c_s}{2}\|K(u_\alpha) - K(u)\|^2 + \tfrac{\epsilon'\alpha}{2}\|u - u_\alpha\|^2 \quad \forall u \in \mathcal{C}. \end{aligned} \tag{4.23}$$

That is, the left hand side of (4.23) is coercive in the sense that it is bounded from below by the positive term $\frac{c_s}{2}\|K(u_\alpha) - K(u)\|^2 + \frac{\epsilon'\alpha}{2}\|u - u_\alpha\|^2$. This condition is analogous to, but not identical with, the positive definiteness requirement on the Hessian in classical second-order conditions in optimization theory [231, 163]. Nonetheless, we shall call condition (4.21)/(4.23) a second-order necessary/sufficient optimality condition.

Note that there always holds $u_\alpha \to u^\dagger$ subsequentially as $\alpha \to 0^\dagger$, cf. Theorem 4.5. Now we further assume that $\frac{g^\dagger - K(u_\alpha)}{\alpha} \to w$ weakly and $\mu_\alpha \to \mu^\dagger$ weakly in suitable spaces as $\alpha \to 0^+$. Then by taking limit in equation (4.20) as $\alpha \to 0^+$, we arrive at a source condition: There exists a $w \in H$ and $\mu^\dagger \in X^*$ such that

$$\begin{cases} -K'(u^\dagger)^* w + u^\dagger - \mu^\dagger = 0, \\ \langle \mu^\dagger, u - u^\dagger \rangle \geq 0 \quad \forall u \in \mathcal{C}. \end{cases} \tag{4.24}$$

Now if the source representer w does satisfy $w = \lim_{\alpha\to 0} \frac{g^\dagger - K(u_\alpha)}{\alpha}$ (weakly), then asymptotically, we may replace $K(u_\alpha) - g^\dagger$ in (4.23) with $-\alpha w$, divide (4.23) by α and take $\alpha \to 0$ to obtain the following nonlinearity condition: There exists some $\epsilon > 0$ and $c_r \geq 0$ such that

$$\begin{aligned} &\tfrac{c_r}{2}\|K(u) - K(u^\dagger)\|^2 + \tfrac{1}{2}\|u - u^\dagger\|^2 - \langle w, E(u, u^\dagger)\rangle \\ &\qquad + \langle \mu^\dagger, u - u^\dagger \rangle \geq \tfrac{\epsilon}{2}\|u - u^\dagger\|^2 \quad \forall u \in \mathcal{C}, \end{aligned} \tag{4.25}$$

upon assuming the convergence of $\frac{1-c_s}{\alpha}$ to a finite constant c_r. Here the constant c_r may be made very large to accommodate the nonlinearity of the operator K. The only possibly indefinite term is $\langle w, E(u, u^\dagger)\rangle$. Hence, the analysis of $\langle w, E(u, u^\dagger)\rangle$ is key to demonstrating (4.25) for concrete problems.

Remark 4.7. The source condition (4.24) is equivalent to assuming the existence of a Lagrange multiplier w (for the equality constraint $K(u) = g^\dagger$) for the minimum-norm problem

$$\min \quad \|u\| \quad \text{subject to } K(u) = g^\dagger \text{ and } u \in \mathcal{C},$$

and, hence, the source condition (4.24) represents a necessary optimality condition for the minimum-norm solution $u^\dagger$.

Remark 4.8. On the nonlinearity condition (4.25), we have the following two remarks.

(1) In case of constrained Tikhonov regularization, we may have $w = 0$, which results in $\langle w, E(u, u^\dagger)\rangle = 0$ and thus the nonlinearity condition (4.25) automatically holds. For example, if $\mathcal{C} = \{u : u \geq c\}$, with c being a positive constant, and $u^\dagger = c$ is the exact solution (i.e., $g^\dagger = K(u^\dagger)$), then $w = 0$ and $\mu^\dagger = u^\dagger$ satisfy the source condition (4.24).

(2) One classical nonlinearity condition is from Assumption 4.2, which requires that $K'(u)$ is (locally) Lipschitz continuous with a Lipschitz constant L and further

$$L\|w\| < 1. \tag{4.26}$$

There are several other nonlinearity conditions. A very similar condition [133] is given by $\|E(u, \tilde{u})\| \leq \frac{L}{2}\|u - \tilde{u}\|^2$ with $L\|w\| < 1$, which clearly implies condition (4.25). Another popular condition reads

$$\|E(u, \tilde{u})\| \leq c_E\|K(u) - K(\tilde{u})\|\|u - \tilde{u}\|. \tag{4.27}$$

It has been used for analyzing iterative regularization methods. Clearly, it implies (4.25) for $c_r > (c_E\|w\|)^2$. We note that it also implies (4.23) after applying Young's inequality.

The following lemma shows that condition (4.25) is weaker than (4.26). Similarly one can show this for condition (4.27). Therefore, the proposed approach does cover the classical results.

Lemma 4.3. *Condition* (4.26) *implies condition* (4.25).

Proof. A direct estimate shows that under condition (4.26), we have

$$|\langle w, E(u, u^\dagger)\rangle| \leq \|w\|\|E(u, u^\dagger)\| \leq \|w\| \cdot \tfrac{L}{2}\|u - u^\dagger\|^2$$

by $\|E(u, u^\dagger)\| \leq \frac{L}{2}\|u - u^\dagger\|^2$ from the Lipschitz continuity of the operator $K'(u)$. Consequently,

$$\begin{aligned}
&\tfrac{1}{2}\|u - u^\dagger\|^2 - \langle w, E(u, u^\dagger)\rangle + \langle \mu^\dagger, u - u^\dagger\rangle \\
&\geq \tfrac{1}{2}\|u - u^\dagger\|^2 - \tfrac{1}{2}L\|w\|\|u - u^\dagger\|^2 + \langle \mu^\dagger, u - u^\dagger\rangle \\
&\geq \tfrac{1 - L\|w\|}{2}\|u - u^\dagger\|^2,
\end{aligned}$$

by noting the relation $\langle \mu^\dagger, u - u^\dagger\rangle \geq 0$ for any $u \in \mathcal{C}$. This shows that condition (4.25) holds with $\epsilon = 1 - L\|w\| > 0$ and $c_r = 0$. ☐

Remark 4.9. The condition $L\|w\| < 1$ is used for bounding the nonlinearity term $\langle w, E(u_\alpha, u^\dagger)\rangle$ from above. This is achieved by the Cauchy-Schwarz inequality, and thus the estimate might be too pessimistic since in general $\langle w, E(u, u^\dagger)\rangle$ can be either indefinite or negative. This might explain the effectiveness of Tikhonov regularization in practice even though assumption (4.26) on the solution $u^\dagger$ and the operator $K(u)$ may be not verified.

The drastic difference from the classical source condition is further illuminated by the following one-dimensional example: the smallness assumption ($L\|w\| < 1$) in the classical condition (4.26) is violated while (4.25) is always true.

Example 4.4. Let the nonlinear operator $K : \mathbb{R} \to \mathbb{R}$ be given by

$$K(u) = \epsilon u(1-u),$$

where $\epsilon > 0$. The solution depends sensitively on u if ϵ is very small, hence it mimics the ill-posed behavior of inverse problems. Let the exact data $g^\dagger$ (necessarily smaller than $\frac{\epsilon}{4}$) be given, then the minimum-norm solution $u^\dagger$ is given by

$$u^\dagger = \tfrac{1}{2}\left(1 - \sqrt{1 - 4\epsilon^{-1} g^\dagger}\right).$$

It is easy to verify that

$$K'(u^\dagger) = \epsilon(1 - 2u^\dagger),$$

and

$$\begin{aligned} E(u, u^\dagger) &= K(u) - K(u^\dagger) - K'(u^\dagger)(u - u^\dagger) \\ &= \epsilon u(1-u) - \epsilon u^\dagger(1-u^\dagger) - \epsilon(1-2u^\dagger)(u-u^\dagger) \\ &= -\epsilon(u - u^\dagger)^2. \end{aligned}$$

Now the source condition $K'(u^\dagger)^* w = u^\dagger$ implies that the source representer w is given by $w = \frac{u^\dagger}{\epsilon(1-2u^\dagger)}$. Therefore, the nonlinearity term $\langle w, E(u, u^\dagger)\rangle$ is given by

$$\langle w, E(u, u^\dagger)\rangle = \frac{-u^\dagger}{1 - 2u^\dagger}(u - u^\dagger)^2$$

which is smaller than zero for a fixed but sufficiently small $g^\dagger > 0$. Moreover, the prefactor $\left|\frac{u^\dagger}{1-2u^\dagger}\right|$ can be made arbitrarily large, thereby indicating that the smallness assumption ($L\|w\| < 1$) can never be satisfied then. Actually, the Lipschitz constant L of $K'(u)$ is $L = 2\epsilon$, and $L|w| = \frac{2u^\dagger}{|1-2u^\dagger|}$, which can be arbitrarily large (if $u^\dagger$ is sufficiently close to $\frac{1}{2}$), and thus (4.26) is violated. Hence (4.25) is indeed weaker.

This simple example also sheds light into the structure of the second-order sufficient condition (4.23). A direct calculation shows

$$\langle K(u_\alpha) - g^\dagger, E(u, u_\alpha)\rangle = \alpha \frac{u_\alpha}{1 - 2u_\alpha}(u - u_\alpha)^2.$$

Observe that the form of $\frac{1}{\alpha}\langle K(u_\alpha) - g^\dagger, E(u, u_\alpha)\rangle$ coincides with that of $-\langle w, E(u, u^\dagger)\rangle$. With this explicit representation at hand, the nonnegativity of the term $\langle K(u_\alpha) - g^\dagger, E(u, u_\alpha)\rangle$, and thus the second-order sufficient condition (4.23), can be numerically verified for a given $g^\dagger$ and every possible α since the Tikhonov minimizer u_α can be found by solving a cubic equation.

4.3.3 *Convergence rate analysis*

Now we illustrate conditions (4.24) and (4.25) for convergence rate analysis. First, we give a convergence rates result for a priori parameter choices.

Theorem 4.15. *Under conditions* (4.24) *and* (4.25), *we have the following estimates*

$$\begin{aligned}
\|u_\alpha^\delta - u^\dagger\| &\le \epsilon^{-\frac{1}{2}} \left(\sqrt{\tfrac{1+2c_r\alpha}{\alpha}}\delta + \sqrt{\tfrac{\alpha}{1-2c_r\alpha}}\|w\| + \sqrt{2\|w\|\delta} \right), \\
\|K(u_\alpha^\delta) - g^\delta\| &\le \tfrac{1}{\sqrt{1-2c_r\alpha}} \left(\sqrt{1+2c_r\alpha}\delta + \tfrac{2}{\sqrt{1-2c_r\alpha}}\alpha\|w\| + \sqrt{2\|w\|\alpha\delta} \right).
\end{aligned}$$

Proof. In view of the optimality of the minimizer u_α^δ and the source condition (4.24), we have

$$\begin{aligned}
&\tfrac{1}{2}\|K(u_\alpha^\delta) - g^\delta\|^2 + \tfrac{\alpha}{2}\|u_\alpha^\delta - u^\dagger\|^2 \\
&\le \tfrac{1}{2}\|K(u^\dagger) - g^\delta\|^2 - \alpha\langle u^\dagger, u_\alpha^\delta - u^\dagger\rangle \\
&= \tfrac{1}{2}\|K(u^\dagger) - g^\delta\|^2 - \alpha\langle w, K'(u^\dagger)(u_\alpha^\delta - u^\dagger)\rangle - \alpha\langle \mu^\dagger, u_\alpha^\delta - u^\dagger\rangle.
\end{aligned}$$

With the help of the second-order error $E(u, \tilde{u})$, we deduce

$$\begin{aligned}
\tfrac{1}{2}\|K(u_\alpha^\delta) - g^\delta\|^2 + \tfrac{\alpha}{2}\|u_\alpha^\delta - u^\dagger\|^2 - \alpha\langle w, E(u_\alpha^\delta, u^\dagger)\rangle + \alpha\langle \mu^\dagger, u_\alpha^\delta - u^\dagger\rangle& \\
\le \tfrac{1}{2}\|K(u^\dagger) - g^\delta\|^2 - \alpha\langle w, K(u_\alpha^\delta) - K(u^\dagger)\rangle.&
\end{aligned}$$

Now from the nonlinearity condition (4.25) and the Cauchy-Schwarz and Young's inequalities, we obtain

$$\begin{aligned}
&\frac{1-2c_r\alpha}{2}\|K(u_\alpha^\delta)-g^\delta\|^2+\frac{\epsilon\alpha}{2}\|u_\alpha^\delta-u^\dagger\|^2\\
\le\ &\frac{1+2c_r\alpha}{2}\|K(u^\dagger)-g^\delta\|^2-\alpha\langle w,K(u_\alpha^\delta)-K(u^\dagger)\rangle\\
\le\ &\frac{1+2c_r\alpha}{2}\delta^2+\alpha\|w\|\left(\|K(u_\alpha^\delta)-g^\delta\|+\|g^\dagger-g^\delta\|\right)\\
\le\ &\frac{1+2c_r\alpha}{2}\delta^2+\alpha\|w\|\|K(u_\alpha^\delta)-g^\delta\|+\alpha\delta\|w\|,
\end{aligned}$$

where we have made use of the inequality

$$\|K(u_\alpha^\delta)-K(u^\dagger)\|^2\le 2\|K(u_\alpha^\delta)-g^\delta\|^2+2\|K(u^\dagger)-g^\delta\|^2.$$

Using Young's inequality again and the implication $c^2\le a^2+b^2(a,b,c\ge 0)\Rightarrow c\le a+b$ gives the estimate for $\|u_\alpha^\delta-u^\dagger\|$. Meanwhile, by ignoring the term $\frac{\epsilon}{2}\|u_\alpha^\delta-u^\dagger\|^2$, we deduce from the implication $c^2\le a^2+bc(a,b,c\ge 0)\Rightarrow c\le\sqrt{a}+b$ the estimate on $\|K(u_\alpha^\delta)-g^\delta\|$. □

Hence the a priori choice $\alpha\sim\delta$ achieves a convergence rate $\mathcal{O}(\delta^{\frac{1}{2}})$ and $\mathcal{O}(\delta)$ for the error $\|u_\alpha^\delta-u^\dagger\|$ and the discrepancy $\|K(u_\alpha^\delta)-g^\dagger\|$, respectively, which coincide with that for the classical approach, cf. Theorem 4.6.

Next we illustrate the proposed approach on the discrepancy principle (4.12). The next result shows that the new approach recovers the canonical convergence rate $\mathcal{O}(\delta^{\frac{1}{2}})$, cf. Theorem 4.7.

Theorem 4.16. *Let conditions* (4.24) *and* (4.25) *be fulfilled, and* α *be determined by the discrepancy principle* (4.12). *Then the solution* u_α^δ *satisfies*

$$\|u_\alpha^\delta-u^\dagger\|\le\frac{1}{\sqrt{\epsilon}}\left(\sqrt{2(1+c)\|w\|}\delta^{\frac{1}{2}}+(1+c)\sqrt{c_r}\delta\right).$$

Proof. The minimizing property of u_α^δ and the defining relation (4.12) imply $\|u_\alpha^\delta\|^2\le\|u^\dagger\|^2$. Upon utilizing the source condition (4.24) and the second-order error $E(u_\alpha^\delta,u^\dagger)$, we deduce

$$\begin{aligned}
\tfrac{1}{2}\|u_\alpha^\delta-u^\dagger\|^2&\le-\langle u^\dagger,u_\alpha^\delta-u^\dagger\rangle\\
&=-\langle K'(u^\dagger)^*w+\mu^\dagger,u_\alpha^\delta-u^\dagger\rangle\\
&=-\langle w,K'(u^\dagger)(u_\alpha^\delta-u^\dagger)\rangle-\langle\mu^\dagger,u_\alpha^\delta-u^\dagger\rangle\\
&=-\langle w,K(u_\alpha^\delta)-K(u^\dagger)\rangle+\langle w,E(u_\alpha^\delta,u^\dagger)\rangle-\langle\mu^\dagger,u_\alpha^\delta-u^\dagger\rangle.
\end{aligned}$$

Now the nonlinearity condition (4.25) yields

$$\begin{aligned}
\tfrac{\epsilon}{2}\|u_\alpha^\delta-u^\dagger\|^2&\le\|w\|\|K(u_\alpha^\delta)-K(u^\dagger)\|+\tfrac{c_r}{2}\|K(u_\alpha^\delta)-K(u^\dagger)\|^2\\
&\le(c+1)\|w\|\delta+\tfrac{c_r}{2}(1+c)^2\delta^2,
\end{aligned}$$

where we have used the triangle inequality and (4.12) as follows

$$\|K(u_\alpha^\delta) - K(u^\dagger)\| \leq \|K(u_\alpha^\delta) - g^\delta\| + \|K(u^\dagger) - g^\delta\| \leq (1+c)\delta.$$

The desired estimate follows immediately from these inequalities. □

We derive two basic estimates under the source condition (4.24) and the nonlinearity condition (4.25): the approximation error $\|u_\alpha - u^\dagger\|$ due to the use of regularization and the propagation error $\|u_\alpha^\delta - u_\alpha\|$ due to the presence of data noise. These estimates are useful for a posteriori convergence rate analysis.

Lemma 4.4. *Assume that conditions* (4.24) *and* (4.25) *hold. Then the approximation error* $\|u_\alpha - u^\dagger\|$ *satisfies*

$$\|u_\alpha - u^\dagger\| \leq \epsilon^{-\frac{1}{2}}\|w\|\frac{\sqrt{\alpha}}{\sqrt{1-c_r\alpha}} \quad \text{and} \quad \|K(u_\alpha) - g^\dagger\| \leq \frac{2\alpha}{1-c_r\alpha}\|w\|.$$

Moreover, if there exists some $\epsilon' > 0$ *independent of* α *such that condition* (4.23) *holds for all* $u \in \mathcal{C}$, *then the propagation error* $\|u_\alpha^\delta - u_\alpha\|$ *satisfies*

$$\|u_\alpha^\delta - u_\alpha\| \leq \frac{1}{\sqrt{\epsilon' c_s}}\frac{\delta}{\sqrt{\alpha}} \quad \text{and} \quad \|K(u_\alpha^\delta) - K(u_\alpha)\| \leq \frac{2\delta}{c_s}.$$

Proof. The minimizing property of u_α and the relation $g^\dagger = K(u^\dagger)$ imply

$$\tfrac{1}{2}\|K(u_\alpha) - g^\dagger\|^2 + \tfrac{\alpha}{2}\|u_\alpha\|^2 \leq \tfrac{\alpha}{2}\|u^\dagger\|^2.$$

The source condition (4.24) and Cauchy-Schwarz inequality give

$$\begin{aligned}
&\tfrac{1}{2}\|K(u_\alpha) - g^\dagger\|^2 + \tfrac{\alpha}{2}\|u_\alpha - u^\dagger\|^2 \\
\leq& -\alpha\langle u^\dagger, u_\alpha - u^\dagger\rangle \\
=& -\alpha\langle w, K'(u^\dagger)(u_\alpha - u^\dagger)\rangle - \alpha\langle \mu^\dagger, u_\alpha - u^\dagger\rangle \\
=& -\alpha\langle w, K(u_\alpha) - g^\dagger\rangle + \alpha\langle w, E(u_\alpha, u^\dagger)\rangle - \alpha\langle \mu^\dagger, u_\alpha - u^\dagger\rangle \\
\leq& \,\alpha\|w\|\|K(u_\alpha) - g^\dagger\| + \alpha\langle w, E(u_\alpha, u^\dagger)\rangle - \alpha\langle \mu^\dagger, u_\alpha - u^\dagger\rangle.
\end{aligned}$$

By appealing to the nonlinearity condition (4.25), we arrive at

$$\frac{1-c_r\alpha}{2}\|K(u_\alpha) - g^\dagger\|^2 + \frac{\epsilon\alpha}{2}\|u_\alpha - u^\dagger\|^2 \leq \alpha\|w\|\|K(u_\alpha) - g^\dagger\|.$$

Consequently, by ignoring the term $\frac{\epsilon\alpha}{2}\|u_\alpha - u^\dagger\|^2$, we derive the estimate

$$\|K(u_\alpha) - g^\dagger\| \leq \frac{2\alpha}{1-c_r\alpha}\|w\|,$$

and meanwhile, by Young's inequality, we have

$$\begin{aligned}
&\frac{1-c_r\alpha}{2}\|K(u_\alpha) - g^\dagger\|^2 + \frac{\epsilon\alpha}{2}\|u_\alpha - u^\dagger\|^2 \\
\leq& \frac{1}{2(1-c_r\alpha)}\alpha^2\|w\|^2 + \frac{1-c_r\alpha}{2}\|K(u_\alpha) - g^\dagger\|^2.
\end{aligned}$$

This shows the first assertion.

Next we turn to the propagation error $\|u_\alpha^\delta - u_\alpha\|$. We use the optimality of the minimizer u_α^δ to get

$$\tfrac{1}{2}\|K(u_\alpha^\delta)-g^\delta\|^2+\tfrac{\alpha}{2}\|u_\alpha^\delta-u_\alpha\|^2 \leq \tfrac{1}{2}\|K(u_\alpha)-g^\delta\|^2-\alpha\langle u_\alpha, u_\alpha^\delta-u_\alpha\rangle. \quad (4.28)$$

Upon substituting the optimality condition of u_α (cf. (4.20)), i.e.,

$$\alpha u_\alpha = -K'(u_\alpha)^*(K(u_\alpha)-g^\dagger)+\alpha\mu_\alpha,$$

into (4.28), we arrive at

$$\begin{aligned}
&\tfrac{1}{2}\|K(u_\alpha^\delta)-g^\delta\|^2+\tfrac{\alpha}{2}\|u_\alpha^\delta-u_\alpha\|^2 \\
&\leq \tfrac{1}{2}\|K(u_\alpha)-g^\delta\|^2+\langle K(u_\alpha)-g^\dagger, K'(u_\alpha)(u_\alpha^\delta-u_\alpha)\rangle-\alpha\langle \mu_\alpha, u_\alpha^\delta-u_\alpha\rangle \\
&= \tfrac{1}{2}\|K(u_\alpha^\delta)-g^\delta\|^2+\tfrac{1}{2}\|K(u_\alpha^\delta)-K(u_\alpha)\|^2-\langle K(u_\alpha^\delta)-g^\delta, K(u_\alpha^\delta)-K(u_\alpha)\rangle \\
&\quad +\langle K(u_\alpha)-g^\dagger, K'(u_\alpha)(u_\alpha^\delta-u_\alpha)\rangle-\alpha\langle \mu_\alpha, u_\alpha^\delta-u_\alpha\rangle.
\end{aligned}$$

Now the second-order error $E(u_\alpha^\delta, u_\alpha)$ and the Cauchy-Schwarz inequality yield

$$\begin{aligned}
&\tfrac{1}{2}\|K(u_\alpha^\delta)-K(u_\alpha)\|^2+\tfrac{\alpha}{2}\|u_\alpha^\delta-u_\alpha\|^2+\alpha\langle \mu_\alpha, u_\alpha^\delta-u_\alpha\rangle \\
&\leq -\langle K(u_\alpha)-g^\delta, K(u_\alpha^\delta)-K(u_\alpha)\rangle+\langle K(u_\alpha)-g^\dagger, K'(u_\alpha)(u_\alpha^\delta-u_\alpha)\rangle \\
&= \langle g^\delta-g^\dagger, K(u_\alpha^\delta)-K(u_\alpha)\rangle-\langle K(u_\alpha)-g^\dagger, E(u_\alpha^\delta, u_\alpha)\rangle \\
&\leq \|g^\delta-g^\dagger\|\|K(u_\alpha^\delta)-K(u_\alpha)\|-\langle K(u_\alpha)-g^\dagger, E(u_\alpha^\delta, u_\alpha)\rangle.
\end{aligned}$$

Consequently, we have

$$\begin{aligned}
\tfrac{1}{2}\|K(u_\alpha^\delta)-K(u_\alpha)\|^2+\tfrac{\alpha}{2}\|u_\alpha^\delta-u_\alpha\|^2+\langle K(u_\alpha)-g^\dagger, E(u_\alpha^\delta, u_\alpha)\rangle& \\
+\alpha\langle \mu_\alpha, u_\alpha^\delta-u_\alpha\rangle \leq \delta\|K(u_\alpha^\delta)-K(u_\alpha)\|&.
\end{aligned}$$

Finally, the second-order sufficient optimality condition (4.23) implies

$$\|K(u_\alpha^\delta)-K(u_\alpha)\| \leq \tfrac{2\delta}{c_s} \quad \text{and} \quad \|u_\alpha^\delta-u_\alpha\| \leq \tfrac{1}{\sqrt{\epsilon' c_s}}\tfrac{\delta}{\sqrt{\alpha}}.$$

This completes the proof of the lemma. □

Now we can derive convergence rates for a posteriori choice rules, i.e. the balancing principle and Hanke-Raus rule. The balancing principle, introduced in [154] and discussed in Section 3.5, chooses an optimal regularization parameter α by minimizing

$$\alpha = \arg \min_{\alpha\in[0,\|K\|^2]} \left\{\Phi(\alpha) := \frac{F^{1+\gamma}(\alpha)}{\alpha}\right\}, \quad (4.29)$$

where $F(\alpha) = J_\alpha(u_\alpha^\delta)$ is the value function, and $\gamma > 0$ is a fixed constant. First, we give an a posteriori error estimate for the approximation u_α^δ.

Theorem 4.17. *Let the conditions in Lemma 4.4 hold, α be determined by rule (4.29), and $\delta_* = \|K(u_\alpha^\delta) - g^\delta\|$ be the realized residual. Then the following estimate holds*

$$\|u_\alpha^\delta - u^\dagger\| \leq c\left(\epsilon^{-\frac{1}{2}} + \epsilon'^{-\frac{1}{2}}\frac{F^{\frac{\gamma+1}{2}}(\delta)}{F^{\frac{\gamma+1}{2}}(\alpha)}\right)\max(\delta,\delta_*)^{\frac{1}{2}}.$$

Proof. By the triangle inequality, we have the error decomposition

$$\|u_\alpha^\delta - u^\dagger\| \leq \|u_\alpha^\delta - u_\alpha\| + \|u_\alpha - u^\dagger\|.$$

It suffices to bound the approximation error $\|u_\alpha - u^\dagger\|$ and the propagation error $\|u_\alpha^\delta - u_\alpha\|$. For the former, we have from the source condition (4.24) (cf. the proof of Lemma 4.4) that

$$\begin{aligned}\tfrac{1}{2}\|u_\alpha - u^\dagger\|^2 &\leq -\langle w, K'(u^\dagger)(u_\alpha - u^\dagger)\rangle - \langle \mu^\dagger, u_\alpha - u^\dagger\rangle \\ &= -\langle w, K(u_\alpha) - K(u^\dagger)\rangle + \langle w, E(u_\alpha, u^\dagger)\rangle - \langle \mu^\dagger, u_\alpha - u^\dagger\rangle.\end{aligned}$$

Consequently, we get

$$\tfrac{1}{2}\|u_\alpha - u^\dagger\|^2 - \langle w, E(u_\alpha, u^\dagger)\rangle + \langle \mu^\dagger, u_\alpha - u^\dagger\rangle \leq \|w\|\|K(u_\alpha) - g^\dagger\|.$$

However, by the triangle inequality and Lemma 4.4, the term $\|K(u_\alpha) - g^\dagger\|$ can be estimated by

$$\begin{aligned}\|K(u_\alpha) - g^\dagger\| &\leq \|K(u_\alpha) - K(u_\alpha^\delta)\| + \|K(u_\alpha^\delta) - g^\delta\| + \|g^\delta - g^\dagger\| \\ &\leq \frac{2\delta}{c_s} + \delta_* + \delta \leq \frac{2 + 2c_s}{c_s}\max(\delta, \delta_*).\end{aligned}$$

This together with the nonlinearity condition (4.25) yields

$$\tfrac{\epsilon}{2}\|u_\alpha - u^\dagger\|^2 \leq \|w\|\|K(u_\alpha) - g^\dagger\| + \tfrac{c_r}{2}\|K(u_\alpha) - g^\dagger\|^2,$$

i.e.,

$$\|u_\alpha - u^\dagger\| \leq \frac{1}{\sqrt{\epsilon}}\left(2\frac{\sqrt{1+c_s}}{\sqrt{c_s}}\sqrt{\|w\|}\max(\delta,\delta_*)^{\frac{1}{2}} + \frac{2+2c_s}{c_s}\sqrt{c_r}\max(\delta,\delta_*)\right).$$

Next we estimate the propagation error $\|u_\alpha^\delta - u_\alpha\|$. The minimizing property of the selected parameter α implies $\frac{1}{\alpha} \leq \frac{F^{\gamma+1}(\delta)}{F^{\gamma+1}(\alpha)}\frac{1}{\delta}$. By Lemma 4.4, we have

$$\begin{aligned}\|u_\alpha^\delta - u_\alpha\| &\leq \frac{1}{\sqrt{\epsilon'}}\frac{1}{\sqrt{c_s\alpha}}\delta \leq \frac{1}{\sqrt{c_s\epsilon'}}\frac{F^{\frac{\gamma+1}{2}}(\delta)}{F^{\frac{\gamma+1}{2}}(\alpha)}\delta^{\frac{1}{2}} \\ &\leq \frac{1}{\sqrt{c_s\epsilon'}}\frac{F^{\frac{\gamma+1}{2}}(\delta)}{F^{\frac{\gamma+1}{2}}(\alpha)}\max(\delta,\delta_*)^{\frac{1}{2}}.\end{aligned}$$

The desired estimate follows from this. □

We note that both δ and δ_* are naturally bounded, so the constant c in Theorem 4.17 can be made independent of $\max(\delta, \delta_*)$. The estimate provides an a posteriori check of the selected parameter. Nonetheless, its consistency remains unaddressed, even for linear inverse problems. We make a first attempt to this issue. First, we show a result on the realized residual δ_*.

Lemma 4.5. *Let the minimizer $\alpha \equiv \alpha(\delta)$ of rule (4.29) be realized in $(0, \|K\|^2)$. Then there holds*

$$\|K(u_\alpha^\delta) - g^\delta\| \to 0 \qquad \text{as } \delta \to 0.$$

Proof. By virtue of Theorem 3.23, the following balancing equation

$$\gamma\alpha\|u_\alpha^\delta\|^2 = \|K(u_\alpha^\delta) - g^\delta\|^2$$

holds at the local minimizer α. This together with the optimality of the selected parameter α that

$$\frac{[(1+\gamma^{-1})\|K(u_\alpha^\delta) - g^\delta\|^2]^{\gamma+1}}{\alpha} = \frac{F^{\gamma+1}(\alpha)}{\alpha} \leq \frac{F^{\gamma+1}(\tilde{\alpha})}{\tilde{\alpha}}, \tag{4.30}$$

for any $\tilde{\alpha} \in [0, \|K\|^2]$. However, with the choice $\tilde{\alpha} = \delta$ and by the optimality of the minimizer $u_{\tilde{\alpha}}^\delta$, we have

$$\begin{aligned} F(\tilde{\alpha}) &\equiv \tfrac{1}{2}\|K(u_{\tilde{\alpha}}^\delta) - g^\delta\|^2 + \tfrac{\tilde{\alpha}}{2}\|u_{\tilde{\alpha}}^\delta\|^2 \\ &\leq \tfrac{1}{2}\|K(u^\dagger) - g^\delta\|^2 + \tfrac{\tilde{\alpha}}{2}\|u^\dagger\|^2 \\ &\leq \tfrac{\delta^2}{2} + \delta\|u^\dagger\|^2 \sim \delta. \end{aligned}$$

Hence, by noting the condition $\gamma > 0$ and the a priori bound $\alpha \in [0, \|K\|^2]$, we deduce that the rightmost term in (4.30) tends to zero as $\delta \to 0$. This shows the desired assertion. □

We can now state a consistency result.

Theorem 4.18. *Let there exist some $M > 0$ such that $\|u^\dagger\| \leq M$, and the assumption in Lemma 4.5 be fulfilled under the constraint $\mathcal{C} = \{u \in X : \|u\| \leq M\}$. If the operator K is weakly closed and injective, then the sequence $\{u_{\alpha(\delta)}^\delta\}_\delta$ of solutions converges weakly to $u^\dagger$.*

Proof. Lemma 4.5 implies $\|K(u_\alpha^\delta) - g^\delta\| \to 0$ as $\delta \to 0$. The a priori bound $\|u_\alpha^\delta\| \leq M$ from the constraint $\mathcal{C}$ implies the existence of a subsequence of $\{u_\alpha^\delta\}$, also denoted by $\{u_\alpha^\delta\}$, and some u^* such that $u_\alpha^\delta \to u^*$ weakly. However, the sequential weak closedness of K and weak lower semicontinuity of norms yield $\|K(u^*) - g^\dagger\| = 0$. Hence, $K(u^*) = g^\dagger$, which together with the injectivity of the operator K implies $u^* = u^\dagger$. Since every subsequence has a subsequence converging weakly to $u^\dagger$, the whole sequence converges weakly to $u^\dagger$. □

Remark 4.10. Hence, the balancing principle is consistent provided that there exists a known upper bound on the solution $u^\dagger$, which is often available from physical considerations. This provides partial justification of its encouraging empirical results. In view of the uniform bound on the sequence $\{\alpha(\delta)\}$ in the defining relation (4.29), $\{\alpha(\delta)\}$ naturally contains a convergent subsequence. However, it remains unclear whether the (sub)sequence $\{\alpha(\delta)\}$ will also tend to zero as $\delta \to 0$.

The Hanke-Raus rule [125] is based on error estimation: the squared residual $\|K(u_\alpha^\delta) - g^\delta\|^2$ divided by the regularization parameter α behaves like an estimate for the total error (cf. Theorem 4.15). Hence, it chooses an optimal regularization parameter α by

$$\alpha^* = \arg \min_{\alpha \in [0, \|K\|^2]} \frac{\|K(u_\alpha^\delta) - g^\delta\|^2}{\alpha}. \tag{4.31}$$

We have the following a posteriori error estimate for the rule (4.31).

Theorem 4.19. *Let the conditions in Lemma 4.4 hold, α be determined by rule (4.31), and $\delta_* = \|K(u_\alpha^\delta) - g^\delta\| \neq 0$ be the realized residual. Then for any small noise level δ, there holds*

$$\|u_\alpha^\delta - u^\dagger\| \leq c\left(\epsilon^{-\frac{1}{2}}\|w\|^{\frac{1}{2}} + \epsilon'^{-\frac{1}{2}}\frac{\delta}{\delta_*}\right)\max(\delta, \delta^*)^{\frac{1}{2}}.$$

Proof. By the proof of Theorem 4.17, it suffices to estimate the error $\|u_\alpha^\delta - u_\alpha\|$. The definitions of α and δ_* indicate $\frac{\delta_*^2}{\alpha} \leq \frac{\|K(u_{\tilde\alpha}^\delta) - g^\delta\|^2}{\tilde\alpha}$ for any $\tilde\alpha \in [0, \|K\|^2]$. By taking $\tilde\alpha = \delta$ in the inequality and noting Lemma 4.4, we deduce

$$\begin{aligned}
\alpha^{-1} &\leq \delta_*^{-2}\delta^{-1}\|K(u_\delta^\delta) - g^\delta\|^2 \\
&\leq \delta_*^{-2}\delta^{-1}\left(\|K(u_\delta^\delta) - K(u_\delta)\| + \|K(u_\delta) - g^\dagger\| + \|g^\dagger - g^\delta\|\right)^2 \\
&\leq \delta_*^{-2}\delta^{-1}\left(\frac{2\delta}{c_s} + \frac{2\delta}{1 - 2c_r\delta}\|w\| + \delta\right)^2 \\
&= \left(\frac{2 + c_s}{c_s} + \frac{2}{1 - 2c_r\delta}\|w\|\right)^2 \delta\delta_*^{-2}.
\end{aligned}$$

Using again Lemma 4.4, we arrive at the following estimate

$$\begin{aligned}\|u_\alpha^\delta - u_\alpha\| &\leq \frac{1}{\sqrt{c_s\epsilon'}}\frac{1}{\sqrt{\alpha}}\delta \\ &\leq \frac{1}{\sqrt{c_s\epsilon'}}\frac{\delta}{\delta_*}\left(\frac{2+c_s}{c_s} + \frac{2}{1-2c_r\delta}\|w\|\right)\delta^{\frac{1}{2}} \\ &\leq \frac{1}{\sqrt{c_s\epsilon'}}\frac{\delta}{\delta_*}\left(\frac{2+c_s}{c_s} + \frac{2}{1-2c_r\delta}\|w\|\right)\max(\delta,\delta_*)^{\frac{1}{2}}.\end{aligned}$$

After setting $c = \max(2\frac{\sqrt{1+c_s}}{\sqrt{c_s}}\sqrt{\|w\|} + \frac{2+2c_s}{c_s}\sqrt{c_r}\max(\delta,\delta_*)^{\frac{1}{2}}, \frac{1}{\sqrt{c_s}}(\frac{2+c_s}{c_s} + \frac{2}{1-2c_r\delta}\|w\|))$, the desired assertion follows from the triangle inequality. □

4.4 A class of parameter identification problems

Now we revisit conditions (4.24) and (4.25) for a general class of nonlinear parameter identification problems, and illustrate their features by using the structure of the adjoint operator $K'(u^\dagger)^*$. Then we specialize to problems with bilinear structures, and show the unnecessity of the source representer w for (numerically) evaluating the nonlinearity term $\langle w, E(u,u^\dagger)\rangle$. Finally, we describe several concrete examples.

4.4.1 *A general class of nonlinear inverse problems*

Generically, parameter identification problems can be described by

$$\begin{cases} e(u,y) = 0, \\ K(u) = Cy(u), \end{cases}$$

where $e(u,y) : X \times Y \to Y^*$ denotes a (differential) operator which is differentiable with respect to both arguments u and y, and the derivative e_y is assumed to be invertible. The notation $y(u) \in Y$ refers to the unique solution to the operator equation $e(u,y) = 0$ for a given u, and the operator C is linear and bounded. Typically, the operator C represents an embedding or trace operator.

To make the source condition (4.24) more precise and tangible, we compute the derivative $K'(u)\delta u$ (with the help of the implicit function theorem) and the adjoint operator $K'(u)^*$. Observe that the derivative $y'(u)\delta u$ of the solution $y(u)$ with respect to u in the direction δu satisfies

$$e_u(u,y(u))\delta u + e_y(u,y(u))y'(u)\delta u = 0,$$

from which follows the derivative formula

$$y'(u)\delta u = -(e_y(u,y(u)))^{-1}e_u(u,y(u))\delta u.$$

Consequently, we arrive at the following explicit representation

$$K'(u)\delta u = -C(e_y(u,y(u)))^{-1}e_u(u,y(u))\delta u.$$

Obviously, the adjoint operator $K'(u)^*$ is given by

$$K'(u)^*w = -e_u(u,y(u))^*(e_y(u,y(u)))^{-*}C^*w.$$

With the expression for the adjoint operator $K'(u)^*$, the source condition (4.24), i.e., $K'(u^\dagger)^*w = u^\dagger - \mu^\dagger$, can be expressed explicitly as

$$-e_u(u^\dagger,y(u^\dagger))^*(e_y(u^\dagger,y(u^\dagger)))^{-*}C^*w = u^\dagger - \mu^\dagger.$$

This identity remains valid by setting $\rho = -(e_y(u^\dagger,y(u^\dagger)))^{-*}C^*w$. In other words, instead of the source condition (4.24), we require the existence of $\rho \in Y$ and $\mu^\dagger$ such that

$$e_u(u^\dagger,y(u^\dagger))^*\rho = u^\dagger - \mu^\dagger, \tag{4.32}$$

and $\langle \mu^\dagger, u - u^\dagger\rangle \geq 0$ for any $u \in \mathcal{C}$. This identity represents an alternative (new) source condition. A distinct feature of this approach is that potentially less regularity is imposed on ρ, instead of on w. This follows from the observation that the existence of $\rho \in Y$ does not necessarily guarantee the existence of an element $w \in H$ satisfying $\rho = -(e_y(u^\dagger,y(u^\dagger)))^{-*}C^*w$ due to possibly extra smoothing property of the operator $(e_y(u^\dagger,y(u^\dagger)))^{-*}$; see Example 4.5 below for an illustration. Conversely, the existence of w always implies the existence of ρ satisfying the new source condition (4.32). Therefore, it opens an avenue to relax the regularity requirement of the source representer. Such a source condition underlies the main idea of the interesting approach in [90] for a parabolic inverse problem.

Under the source condition (4.32), we have

$$\begin{aligned}
\langle w, E(u,u^\dagger)\rangle &= \langle w, Cy(u) - Cy(u^\dagger) - Cy'(u^\dagger)(u-u^\dagger)\rangle \\
&= \langle C^*w, y(u) - y(u^\dagger) - y'(u^\dagger)(u-u^\dagger)\rangle \\
&= -\langle \rho, e_y(u^\dagger,y(u^\dagger))(y(u) - y(u^\dagger) - y'(u^\dagger)(u-u^\dagger))\rangle.
\end{aligned}$$

Accordingly, the nonlinearity condition (4.25) can be expressed by

$$\begin{aligned}
&\tfrac{c_r}{2}\|K(u) - K(u^\dagger)\|^2 + \tfrac{1}{2}\|u-u^\dagger\|^2 + \langle \mu^\dagger, u-u^\dagger\rangle \\
&\quad + \langle \rho, e_y(u^\dagger,y(u^\dagger))(y(u) - y(u^\dagger) - y'(u^\dagger)(u-u^\dagger))\rangle \\
&\quad \geq \tfrac{\epsilon}{2}\|u-u^\dagger\|^2 \quad \forall u \in \mathcal{C}.
\end{aligned} \tag{4.33}$$

Therefore, the term $\langle \rho, e_y(u^\dagger,y(u^\dagger))(y(u)-y(u^\dagger)-y'(u^\dagger)(u-u^\dagger))\rangle$ will play an important role in studying the degree of nonlinearity of the operator K, and in analyzing related Tikhonov regularization methods. We shall illustrate its usage in Example 4.7. We would like to point out that the nonlinearity condition (4.33) can be regarded as the (weak) limit of the second-order sufficient condition (4.23), which in the current context reads

$$\begin{aligned}
&\tfrac{1}{2}\|K(u_\alpha)-K(u)\|^2+\tfrac{\alpha}{2}\|u_\alpha-u\|^2+\alpha\langle\mu_\alpha,u-u_\alpha\rangle\\
&\qquad+\langle e_y(u_\alpha,y(u_\alpha))^{-*}C^*(K(u_\alpha)-g^\dagger),\\
&\qquad e_y(u_\alpha,y(u_\alpha))(y(u)-y(u_\alpha)-y'(u_\alpha)(u-u_\alpha))\rangle\\
&\qquad\geq\tfrac{c_s}{2}\|K(u_\alpha)-K(u)\|^2+\tfrac{\epsilon'\alpha}{2}\|u-u_\alpha\|^2.
\end{aligned}$$

Conditions (4.32) and (4.33) can yield identical convergence rates for nonlinear Tikhonov models as (4.24) and (4.25), since the former is exactly the representations of the latter in the context of parameter identifications. The main changes to the proofs are the following two key identities

$$\begin{aligned}
\langle w,K'(u^\dagger)(u-u^\dagger)\rangle&=\langle\rho,e_u(u^\dagger,y(u^\dagger))(u-u^\dagger)\rangle(=\langle u^\dagger-\mu^\dagger,u-u^\dagger\rangle),\\
\langle w,K(u)-K(u^\dagger)\rangle&=-\langle\rho,e_y(u^\dagger,y(u^\dagger))(y(u)-y(u^\dagger))\rangle,
\end{aligned}$$

and the remaining steps proceed identically.

In the rest, we further specialize to the case where the operator equation $e(u,y)=0$ assumes the form

$$A(u)y-f=0.$$

A lot of parameter identification problems for linear partial differential equations (systems) can be cast into this abstract model, e.g., the second-order elliptic operator $A(u)y=-\nabla\cdot(a(x)\nabla y)+\mathbf{b}(x)\cdot\nabla y+c(x)y$ with the parameter u being one or some combinations of $a(x)$, $\mathbf{b}(x)$ and $c(x)$. Then upon denoting the derivative of $A(u)$ with respect to u by $A'(u)$, we have

$$e_u(u,y(u))\delta u=A'(u)\delta u y(u)$$

and

$$e_y(u,y(u))=A(u).$$

The derivative $A'(u)\delta u y(u)$ can be either local (separable) or nonlocal. For example, in the former category, $A(u)y=(-\Delta+u)y$ with $A'(u)\delta u y(u)=y(u)\delta u$. The case $A(u)y=-\nabla\cdot(u\nabla y)$ with $A'(u)\delta u y(u)=-\nabla\cdot(\delta u\nabla y(u))$ belongs to the latter category. The local case will be further discussed in Section 4.4.2. Consequently, the (new) source and nonlinearity conditions respectively simplify to

$$e_u(u^\dagger,y(u^\dagger))^*\rho=u^\dagger-\mu^\dagger$$

and

$$\begin{aligned}
&\tfrac{c_r}{2}\|K(u)-K(u^\dagger)\|^2+\langle\rho,A(u^\dagger)(y(u)-y(u^\dagger)-y'(u^\dagger)(u-u^\dagger))\rangle\\
&\qquad+\tfrac{1}{2}\|u-u^\dagger\|^2+\langle\mu^\dagger,u-u^\dagger\rangle\geq\tfrac{\epsilon}{2}\|u-u^\dagger\|^2\quad\forall u\in\mathcal{C}.
\end{aligned}$$

4.4.2 *Bilinear problems*

Here we elaborate the structure of $\langle w, E(u,u^\dagger)\rangle$ in (4.25). Interestingly, it admits a representation without resorting to the source representer w for bilinear problems. Specifically, the following class of inverse problems is considered. Let the operator $e(u,y)$ be (affine) bilinear with respect to the arguments u and y for fixed y and u, respectively, and for a given u, $e_u(u,y)$ is defined pointwise (local/separable).

We begin with the second-order error $E(u,u^\dagger)$ for bilinear problems. The bilinear structure of the operator $e(u,y)$ implies

$$\begin{aligned}0 &= e(u,y(u)) - e(u^\dagger, y(u^\dagger))\\ &= e_y(u^\dagger, y(u^\dagger))(y(u) - y(u^\dagger)) + e_u(u,y(u))(u-u^\dagger),\end{aligned}$$

i.e.,

$$y(u) - y(u^\dagger) = -(e_y(u^\dagger, y(u^\dagger)))^{-1} e_u(u,y(u))(u-u^\dagger).$$

Therefore, we deduce that

$$\begin{aligned}E(u,u^\dagger) &= K(u) - K(u^\dagger) - K'(u^\dagger)(u-u^\dagger)\\ &= Cy(u) - Cy(u^\dagger) + C(e_y(u^\dagger,y(u^\dagger)))^{-1}e_u(u^\dagger,y(u^\dagger))(u-u^\dagger),\\ &= -C(e_y(u^\dagger,y(u^\dagger)))^{-1}e_u(u,y(u^\dagger))(u-u^\dagger)\\ &\qquad + C(e_y(u^\dagger,y(u^\dagger)))^{-1}e_u(u^\dagger,y(u^\dagger))(u-u^\dagger)\\ &= -C(e_y(u^\dagger,y(u^\dagger)))^{-1}(e_u(u,y(u)) - e_u(u^\dagger,y(u^\dagger)))(u-u^\dagger).\end{aligned}$$

With the help of the preceding three relations, the source condition $K'(u^\dagger)^*w = u^\dagger - \mu^\dagger$ and locality (separability) of $e_u(u,y(u))$, we get

$$\begin{aligned}\langle w, E(u,u^\dagger)\rangle &= \langle w, -C(e_y(u^\dagger,y(u^\dagger)))^{-1}(e_u(u,y(u)) - e_u(u^\dagger,y(u^\dagger)))(u-u^\dagger)\rangle\\ &= \Big\langle -e_u(u^\dagger,y(u^\dagger))^*(e_y(u^\dagger,y(u^\dagger)))^{-*}C^*w,\\ &\qquad \frac{e_u(u,y(u)) - e_u(u^\dagger,y(u^\dagger))}{e_u(u^\dagger,y(u^\dagger))}(u-u^\dagger)\Big\rangle\\ &= \left\langle K'(u^\dagger)^*w, \frac{e_u(u,y(u)) - e_u(u^\dagger,y(u^\dagger))}{e_u(u^\dagger,y(u^\dagger))}(u-u^\dagger)\right\rangle\\ &= \left\langle u^\dagger - \mu^\dagger, \frac{e_u(u,y(u)) - e_u(u^\dagger,y(u^\dagger))}{e_u(u^\dagger,y(u^\dagger))}(u-u^\dagger)\right\rangle.\end{aligned}$$

Therefore, we have arrived at the following concise representation

$$\langle w, E(u,u^\dagger)\rangle = \left\langle u^\dagger - \mu^\dagger, \frac{e_u(u,y(u)) - e_u(u^\dagger,y(u^\dagger))}{e_u(u^\dagger,y(u^\dagger))}(u-u^\dagger)\right\rangle. \tag{4.34}$$

Remark 4.11. The derivations indicate that the source representer w actually is not needed for evaluating $\langle w, E(u,u^\dagger)\rangle$, which enables possible numerical verification of the nonlinearity condition (4.25). Note that even if we do know the exact solution $u^\dagger$, the representer w is still not accessible since the operator equation $K'(u^\dagger)^* w = u^\dagger$ is generally also ill-posed. Hence, the representation (4.34) is of much practical significance.

Remark 4.12. Another important consequence is that it may enable estimates of form (4.27), thereby validating condition (4.25). This can be achieved by applying Hölder-type inequality if the image of $K(u)$ and the coefficient u share the domain of definition, e.g., in recovering the potential/leading coefficient in an elliptic equation from distributed measurements in the domain; see Example 4.7 below for an illustration.

Finally, we point out that formally the representation (4.34) can be regarded as the limit of

$$\left\langle u_\alpha - \mu_\alpha, \frac{e_u(u, y(u)) - e_u(u_\alpha, y(u_\alpha))}{e_u(u_\alpha, y(u_\alpha))}(u - u_\alpha) \right\rangle$$

as α goes to zero, which might be computationally amenable, and hence enable possible numerical verification of (4.23).

4.4.3 *Three elliptic examples*

In this part, we illuminate the nonlinearity condition (4.25) and the structure of the term $\langle w, E(u, u^\dagger)\rangle$ with examples, and discuss the usage of the source and nonlinearity conditions (4.32) and (4.33).

First, we consider an elliptic parameter identification problem to show that the smallness assumption $L\|w\| < 1$ of the classical nonlinearity condition (4.26) is unnecessary by deriving an explicit representation of the nonlinearity term $\langle w, E(u, u^\dagger)\rangle$. The derivations also illustrate clearly structural properties developed in Section 4.4.2.

Example 4.5. We continue Example 4.1 with the structure of the term $\langle w, E(u, u^\dagger)\rangle$. We know from Example 4.2 that $K : L^2(\Gamma_i) \mapsto L^2(\Gamma_c)$ is Fréchet differentiable with a Lipschitz continuous derivative, and

$$\begin{aligned} K'(u^\dagger)\delta u &= \gamma_{\Gamma_c} \tilde{z}(u^\dagger), \\ E(u, u^\dagger) &= \gamma_{\Gamma_c} v(u, u^\dagger), \\ K'(u^\dagger)^* w &= -\gamma_{\Gamma_i}(y(u^\dagger) z(u^\dagger)), \end{aligned}$$

where the functions $\tilde{z}(u^\dagger)$, $v(u,u^\dagger)$ and $z(u^\dagger) \in H^1(\Omega)$ satisfy

$$\int_\Omega \nabla\tilde{z}\cdot\nabla\tilde{v}dx + \int_{\Gamma_i} u^\dagger \tilde{z}\tilde{v}ds = -\int_{\Gamma_i} \delta u y(u)\tilde{v}ds \quad \forall \tilde{v}\in H^1(\Omega),$$

$$\int_\Omega \nabla v\cdot\nabla\tilde{v}dx + \int_{\Gamma_i} u^\dagger v\tilde{v}ds = -\int_{\Gamma_i} (u-u^\dagger)(y(u)-y(u^\dagger))\tilde{v}ds \ \forall \tilde{v}\in H^1(\Omega),$$

$$\int_\Omega \nabla z\cdot\nabla\tilde{v}dx + \int_{\Gamma_i} u^\dagger z\tilde{v}ds = \int_{\Gamma_c} w\tilde{v}ds \quad \forall \tilde{v}\in H^1(\Omega).$$

Assume that the source condition (4.25) holds with the representer $w \in L^2(\Gamma_c)$. Then by setting $\tilde{v} = z(u^\dagger)$ and $\tilde{v} = v(u,u^\dagger)$ respectively in the their weak formulations, it follows that

$$\langle w, E(u,u^\dagger)\rangle_{L^2(\Gamma_c)} = \left\langle u^\dagger - \mu^\dagger, (u-u^\dagger)\frac{y(u)-y(u^\dagger)}{y(u^\dagger)}\right\rangle_{L^2(\Gamma_i)}.$$

Hence, the term $\langle w, E(u,u^\dagger)\rangle$ exhibits the desired structure, cf. (4.34). Next, by maximum principle, the solution $y(u)$ is positive for positive f. Moreover there holds

$$\int_\Omega |\nabla(y(u)-y(u^\dagger))|^2dx + \int_{\Gamma_i} u(y(u)-y(u^\dagger))^2ds$$
$$+ \int_{\Gamma_i} (u-u^\dagger)y(u^\dagger)(y(u)-y(u^\dagger))ds = 0.$$

Note also the monotonicity relation, i.e., if $u \geq u^\dagger$, then $y(u) \leq y(u^\dagger)$. It follows from the above two relations that if $u^\dagger - \mu^\dagger \geq 0$, then

$$-\langle w, E(u,u^\dagger)\rangle \geq 0.$$

This shows that the nonlinearity condition (4.25) holds without resorting to the smallness condition $L\|w\| < 1$ in the classical nonlinearity condition (4.26) under the designated circumstance.

Next we contrast the source condition (4.32) with the conventional one (4.24). The operator $e(u,y)$ is bilinear, and $e_u(u,y(u)) = \gamma_{\Gamma_i}y(u)$, $e_u(u,y(u))^*\rho = \gamma_{\Gamma_i}(\rho y(u))$. Hence the new source condition (4.32) requires the existence of some element $\rho \in H^1(\Omega)$ such that

$$\gamma_{\Gamma_i}(\rho y(u^\dagger)) = u^\dagger - \mu^\dagger.$$

This admits an easy interpretation: for $\gamma_{\Gamma_i}\rho$ to be fully determined, $\gamma_{\Gamma_i}y(u^\dagger)$ cannot vanish, which is exactly the identifiability condition. The representers ρ and w are related by (in weak form)

$$\int_\Omega \nabla\rho\cdot\nabla\tilde{v}dx + \int_{\Gamma_i} u^\dagger\rho\tilde{v}ds = -\int_{\Gamma_c} w\tilde{v}ds \quad \forall \tilde{v}\in H^1(\Omega).$$

This relation shows clearly the different regularity assumptions on ρ and w: the existence of $\rho \in H^1(\Omega)$ does not actually guarantee the existence of $w \in L^2(\Gamma_c)$. To ensure the existence of $w \in L^2(\Gamma_c)$, one necessarily needs higher regularity on ρ than $H^1(\Omega)$, presumably $\rho \in H^{\frac{3}{2}}(\Omega)$. Conversely, the existence of $w \in L^2(\Gamma_c)$ automatically ensures the existence of $\rho \in H^1(\Omega)$.

Next we give an example of inverse medium scattering to show the same structure of the nonlinearity term $\langle w, E(u, u^\dagger)\rangle$ but with less definitiveness.

Example 4.6. Here we consider the two-dimensional time-harmonic inverse scattering problem, cf. Section 2.2.3, of determining the index of refraction n^2 from near-field scattered field data u^s, given one incident field y^i. Let $y = y(x)$ denote the transverse mode wave and satisfy

$$\Delta y + n^2 k^2 y = 0.$$

Let the incident plane wave be $y^i = e^{kx\cdot d}$ with $d = (d_1, d_2) \in \mathbb{S}^1$ being the incident direction. Then for the complex coefficient $u = (n^2 - 1)k^2$ with its support within $\Omega \subset \mathbb{R}^2$, the total field $y = y^{tot}$ satisfies

$$y = y^i + \int_\Omega G(x,z)u(z)y(z)dz,$$

where $G(x, z)$ is the free space fundamental solution, i.e., $G(x, z) = \frac{i}{4}H_0^1(k|x - z|)$, the Hankel function of the first kind and zeroth order. The inverse problem is to determine the refraction coefficient u from the scattered field

$$y^s(x) = \int_\Omega G(x,z)u(z)y(z)dz$$

measured on a near-field boundary Γ. Consequently, we have $K(u) = \gamma_\Gamma y^s(x) \in L^2(\Gamma)$. The Tikhonov approach for recovering u takes the form

$$\min_{u\in\mathcal{C}} \int_\Gamma |K(u) - g^\delta|^2 ds + \alpha \int_\Omega |u|^2 dx.$$

Here g^δ denotes the measured scattered field, and the constraint set $\mathcal{C}$ is taken to be $\mathcal{C} = \{u \in L^\infty(\Omega) : \Re(u) \geq 0, \mathrm{supp}(u) \subset\subset \Omega\}$. It can be shown that the forward operator $K : L^2(\Omega) \mapsto L^2(\Gamma)$ is Fréchet differentiable, and the derivative is Lipschitz continuous on $\mathcal{C}$. Now let $G^\dagger(x, z)$ be the fundamental solution to the elliptic operator $\Delta + k^2 + u^\dagger$. Then we can deduce

$$K'(u^\dagger)\delta u = -\int_\Omega G^\dagger(x,z)\delta u y(u^\dagger)dz \quad x \in \Gamma,$$

$$E(u, u^\dagger) = -\int_\Omega G^\dagger(x,z)(u - u^\dagger)(y(u) - y(u^\dagger))dz \quad x \in \Gamma,$$

$$K'(u^\dagger)^* w = -\overline{y(u^\dagger)} \int_\Gamma \overline{G^\dagger(x,z)} w(x) dx,$$

where the overbar refers to complex conjugate. Then, by the source condition $K'(u^\dagger)^*w = u^\dagger - \mu^\dagger$, we get

$$\begin{aligned}
&\langle w, E(u,u^\dagger)\rangle_{L^2(\Gamma)} \\
&= -\int_\Gamma w(x)\overline{\int_\Omega G^\dagger(x,z)(u(z)-u^\dagger(z))(y(u)(z)-y(u^\dagger)(z))dzdx} \\
&= -\int_\Omega \overline{y(u^\dagger)}\int_\Gamma \overline{G^\dagger(x,z)}w(x)dx(\overline{u(z)}-\overline{u^\dagger(z)})\frac{\overline{y(u)(z)}-\overline{y(u^\dagger)(z)}}{\overline{y(u^\dagger)(z)}}dz \\
&= \left\langle u^\dagger-\mu^\dagger, (u-u^\dagger)\frac{y(u)-y(u^\dagger)}{y(u^\dagger)}\right\rangle_{L^2(\Omega)}.
\end{aligned}$$

Note that the structure of $\langle w, E(u,u^\dagger)\rangle_{L^2(\Gamma)}$ coincides with that in Example 4.5, which further corroborates the theory for bilinear problems in Section 4.4.2. However, an analogous argument for definitive sign is missing since the maximum principle does not hold for the Helmholtz equation. Nonetheless, one might still expect some norm estimate of the form (4.27), which remains open. In particular, then a small $u^\dagger - \mu^\dagger$ would imply the nonlinearity condition (4.25).

The last example shows the use of the source condition (4.32) and nonlinearity condition (4.33).

Example 4.7. Let $\Omega \subset \mathbb{R}^2$ be an open bounded domain with a smooth boundary Γ. We consider the following elliptic equation

$$\begin{cases} -\nabla\cdot(u\nabla y) = f & \text{in } \Omega, \\ y = 0 & \text{on } \Gamma. \end{cases}$$

Let $y(u) \in H_0^1(\Omega)$ be the solution. We measure y (denoted by $g^\delta \in H_0^1(\Omega)$) in the domain Ω with $\|\nabla(g^\delta - y(u^\dagger))\|_{L^2(\Omega)} \le \delta$, i.e., $K(u) = y(u)$, and are interested in recovering the conductivity $u \in \mathcal{C} = \{u \in H^1(\Omega) : c_0 \le u \le c_1\}$ for some finite $c_0, c_1 > 0$ by means of Tikhonov regularization

$$\min_{u\in\mathcal{C}} \int_\Omega |\nabla(K(u)-g^\delta)|^2dx + \alpha\int_\Omega |u|^2 + |\nabla u|^2dx.$$

It follows from Meyers' theorem [234] that the operator $K : H^1(\Omega) \mapsto H_0^1(\Omega)$ is Fréchet differentiable and the derivative is Lipschitz continuous. The operator $A(u)$ is given by $-\nabla\cdot(u\nabla\cdot)$. It is easy to see that

$$\begin{aligned}
A'(u)\delta u y(u) &= -\nabla\cdot(\delta u\nabla y(u)), \\
e_u(u,y(u))^*\rho &= \nabla y(u)\cdot\nabla\rho.
\end{aligned}$$

Consequently, the source condition (4.32) reads: there exists some $\rho \in H_0^1(\Omega)$ such that

$$\nabla y(u^\dagger) \cdot \nabla \rho = (I - \Delta)u^\dagger - \mu^\dagger,$$

which amounts to the solvability condition $\nabla y(u^\dagger) \neq 0$ (cf., e.g., [261, 161]). The nonlinearity condition (4.33) is given by

$$\begin{aligned} &\tfrac{c_r}{2}\|\nabla(y(u) - y(u^\dagger))\|^2_{L^2(\Omega)} - \langle u^\dagger \nabla \rho, \nabla E(u, u^\dagger)\rangle \\ &\qquad + \tfrac{1}{2}\|u - u^\dagger\|^2_{H^1(\Omega)} + \langle \mu^\dagger, u - u^\dagger\rangle \geq \tfrac{\epsilon}{2}\|u - u^\dagger\|^2_{H^1(\Omega)} \quad \forall u \in \mathcal{C}, \end{aligned}$$

where $E(u, u^\dagger) = K(u) - K(u^\dagger) - K'(u^\dagger)(u - u^\dagger)$ is the second-order error. By setting $\tilde{v} = \rho$ in the weak formulation of $E(u, u^\dagger)$, i.e.,

$$\int u^\dagger \nabla E(u, u^\dagger) \cdot \nabla \tilde{v} dx = -\int_\Omega (u - u^\dagger)\nabla(y(u) - y(u^\dagger)) \cdot \nabla \tilde{v} dx \quad \forall \tilde{v} \in H_0^1(\Omega),$$

and applying the generalized Hölder's inequality and Sobolev embedding theorem, we get

$$\begin{aligned} |\langle u^\dagger \nabla \rho, \nabla E(u, u^\dagger)\rangle| &\leq \|\nabla(y(u) - y(u^\dagger))\|_{L^2(\Omega)}\|u - u^\dagger\|_{L^q(\Omega)}\|\nabla \rho\|_{L^p(\Omega)} \\ &\leq C\|\nabla \rho\|_{L^p(\Omega)}\|\nabla(y(u) - (y^\dagger))\|_{L^2(\Omega)}\|u - u^\dagger\|_{H^1(\Omega)}, \end{aligned}$$

where the exponents $p, q > 2$ satisfy $\frac{1}{p} + \frac{1}{q} = \frac{1}{2}$ (the exponent p can be any number greater than 2). Therefore, we have established condition (4.27) for the inverse conductivity problem, and the nonlinearity condition (4.33) holds provided that the source representer $\rho \in W_0^{1,p}(\Omega)$ for some $p > 2$. We note that the smallness of the representer ρ is not required for the nonlinearity condition (4.33) for this example. The convergence theory in Section 4.3 implies a convergence rate $\|u_\alpha^\delta - u^\dagger\|_{H^1(\Omega)} \leq C\sqrt{\delta}$ for the Tikhonov model with the a priori choice $\alpha \sim \delta$.

The classical source condition (4.24) reads: there exists some $w \in H_0^1(\Omega)$ such that

$$K'(u^\dagger)^* w = (I - \Delta)u^\dagger - \mu^\dagger,$$

or equivalently in the weak formulation

$$\langle \nabla K'(u^\dagger)h, \nabla w\rangle = \langle u^\dagger, h\rangle_{H^1(\Omega)} - \langle \mu^\dagger, h\rangle \quad \forall h \in H^1(\Omega).$$

This source condition is difficult to interpret due to the lack of an explicit characterization of the range of the adjoint operator $K'(u^\dagger)^*$. Also the weak formulation of $K'(u^\dagger)h \in H_0^1(\Omega)$, i.e.,

$$\langle u^\dagger \nabla K'(u^\dagger)h, \nabla v\rangle = \langle h \nabla y(u^\dagger), \nabla v\rangle \quad \forall v \in H_0^1(\Omega),$$

does not directly help due to subtle differences in the relevant bilinear forms. Nonetheless, the representers ρ and w are closely related by

$$\rho = (A(u^\dagger))^{-1}(-\Delta)w.$$

This relation indicates that the operator $(A(u^\dagger))^{-1}(-\Delta)$ renormalizes the standard inner product $\langle\nabla\cdot,\nabla\cdot\rangle$ on $H_0^1(\Omega)$ to a problem-adapted weighted inner product $\langle u^\dagger\nabla\cdot,\nabla\cdot\rangle$, and thus facilitates the interpretation of the resulting source condition. This shows clearly the advantage of the source condition (4.32).

Remark 4.13. With stronger regularity on the representer w, e.g., $w \in W_0^{1,\infty}(\Omega) \cap W^{2,p}(\Omega)$ $(p > 2)$, one might also establish the nonlinearity condition for $L^2(\Omega)$ data, i.e., $g^\delta \in L^2(\Omega)$.

4.5 Convergence rate analysis in Banach spaces

Now we turn to convergence rate analysis for general convex variational penalty, i.e.,

$$J_\alpha(u) = \tfrac{1}{2}\|K(u) - g^\delta\|^2 + \alpha\psi(u).$$

Like in the linear case, the main tool for the analysis is Bregman distance. In order to obtain convergence rates, one assumes source condition on the true solution $u^\dagger$ and nonlinearity condition on the operator K. In this part, we discuss several extensions of the classical approach in Section 4.2, including generalizing the source condition and the nonlinearity condition, especially variational inequalities.

4.5.1 *Extensions of the classical approach*

We observe that the starting point of the basic convergence rate analysis is a comparison of the functional values $J_\alpha(u_\alpha^\delta)$ with $J_\alpha(u^\dagger)$, which yields after reordering

$$\tfrac{1}{2}\|K(u_\alpha^\delta) - g^\delta\|^2 + \alpha\left(\psi(u_\alpha^\delta) - \psi(u^\dagger)\right) \le \tfrac{1}{2}\|K(u^\dagger) - g^\delta\|^2.$$

Analogous to quadratic completion, we add $-\alpha\langle\xi^\dagger, u_\alpha^\delta - u^\dagger\rangle$ to both sides, and obtain a Bregman distance $d_{\xi^\dagger}(u_\alpha^\delta, u^\dagger)$ with $\xi^\dagger \in \partial\psi(u^\dagger)$:

$$\tfrac{1}{2}\|K(u_\alpha^\delta) - g^\delta\|^2 + \alpha d_{\xi^\dagger}(u_\alpha^\delta, u^\dagger) \le \tfrac{1}{2}\|K(u^\dagger) - g^\delta\|^2 - \alpha\langle\xi^\dagger, u_\alpha^\delta - u^\dagger\rangle. \quad (4.35)$$

The constraint can be incorporated by using instead the generalized Bregman distance $d_{\xi^\dagger,\mu^\dagger}(u_\alpha^\delta, u^\dagger)$. The final steps of the proof in Theorem 4.6 need two essential ingredients:

(a) A source condition on $u^\dagger$, which allows us to estimate $-\langle \xi^\dagger, u_\alpha^\delta - u^\dagger\rangle$ and to transfer these quantities in X and X' to quantities in Y;
(b) An assumption on the operator K, typically a restriction on the nonlinearity of K, which allows to estimate this quantity in Y.

There are several viable approaches to these conditions. We begin with the most natural extension [260].

Theorem 4.20. *Let Assumption 4.1 hold, and further there hold:*

(i) *K is Gâteaux differentiable;*
(ii) *there exists a $w \in Y$ and a Lagrange multiplier $\mu^\dagger \in X^*$, i.e.,*
$$\langle \mu^\dagger, u - u^\dagger\rangle \geq 0 \quad \forall u \in \mathcal{C},$$
such that $K'(u^\dagger)^* w + \mu^\dagger = \xi^\dagger \in \partial\psi(u^\dagger)$.
(iii) *there exists an $L > 0$ with $L\|w\| < 1$, such that for all $u \in \mathcal{C}$*
$$\|E(u, u^\dagger)\| \leq L d_{\xi^\dagger,\mu^\dagger}(u, u^\dagger).$$

Then the minimizer u_α^δ to the functional J_α satisfies
$$d_{\xi^\dagger,\mu^\dagger}(u_\alpha^\delta, u^\dagger) \leq \frac{1}{2(1 - L\|w\|)}\left(\frac{\delta}{\sqrt{\alpha}} + \sqrt{\alpha}\|w\|\right)^2,$$
$$\|K(u_\alpha^\delta) - g^\delta\| \leq \delta + 2\alpha\|w\|.$$
In particular, with the choice $\alpha \sim \delta$, there hold
$$d_{\xi^\dagger,\mu^\dagger}(u_\alpha^\delta, u^\dagger) = \mathcal{O}(\delta) \quad \text{and} \quad \|K(u_\alpha^\delta) - g^\dagger\| = \mathcal{O}(\delta).$$

Proof. It follows from (4.35) and the source condition (ii) that
$$\begin{aligned}\tfrac{1}{2}\|K(u_\alpha^\delta) - g^\delta\|^2 &+ \alpha d_{\xi^\dagger,\mu^\dagger}(u_\alpha^\delta, u^\dagger)\\ &\leq \tfrac{1}{2}\|K(u^\dagger) - g^\delta\|^2 - \alpha\langle w, K'(u^\dagger)(u_\alpha^\delta - u^\dagger)\rangle.\end{aligned}$$
We rewrite the term $-\langle w, K'(u^\dagger)(u_\alpha^\delta - u^\dagger)\rangle$ into
$$-\langle w, K'(u^\dagger)(u_\alpha^\delta - u^\dagger)\rangle = -\langle w, K(u_\alpha^\delta) - K(u^\dagger)\rangle + \langle w, E(u_\alpha^\delta, u^\dagger)\rangle.$$
Applying the Cauchy-Schwarz inequality and (iii) gives
$$\begin{aligned}&\tfrac{1}{2}\|K(u_\alpha^\delta) - g^\delta\|^2 + \alpha d_{\xi^\dagger,\mu^\dagger}(u_\alpha^\delta, u^\dagger)\\ &\leq \tfrac{1}{2}\|K(u^\dagger) - g^\delta\|^2 - \alpha\langle w, K(u_\alpha^\delta) - K(u^\dagger)\rangle + \alpha\langle w, E(u_\alpha^\delta, u^\dagger)\rangle\\ &\leq \tfrac{1}{2}\|K(u^\dagger) - g^\delta\|^2 - \alpha\langle w, K(u_\alpha^\delta) - K(u^\dagger)\rangle + L\|w\| d_{\xi^\dagger,\mu^\dagger}(u_\alpha^\delta, u^\dagger).\end{aligned}$$
By completing squares, we deduce
$$\begin{aligned}\tfrac{1}{2}\|K(u_\alpha^\delta) - g^\delta - \alpha w\|^2 &+ \alpha(1 - L\|w\|) d_{\xi^\dagger,\mu^\dagger}(u_\alpha^\delta, u^\dagger)\\ &\leq \tfrac{1}{2}\|K(u^\dagger) - g^\delta - \alpha w\|^2,\end{aligned}$$
from which the desired assertions follow directly. □

Alternatively, one can control the term $-\langle E(u,u^\dagger),w\rangle$ directly [50].

Theorem 4.21. *Let Assumption 4.1 hold, and further there hold:*

(i) *K is Gâteaux differentiable;*
(ii) *there exists a $w \in Y$ and a Lagrange multiplier $\mu^\dagger \in X^*$, i.e.,*

$$\langle \mu^\dagger, u-u^\dagger\rangle \geq 0 \quad \forall u \in \mathcal{C},$$

such that $K'(u^\dagger)^* w + \mu^\dagger = \xi^\dagger \in \partial\psi(u^\dagger)$.
(iii) *there exists a $\gamma > 0$, such that for all $u \in \mathcal{C}$*

$$-\langle E(u,u^\dagger),w\rangle \leq \gamma \|K(u)-K(u^\dagger)\|\|w\|.$$

Then the minimizer u_α^δ to the functional J_α satisfies

$$\begin{aligned} d_{\xi^\dagger,\mu^\dagger}(u_\alpha^\delta,u^\dagger) &\leq \tfrac{1}{2\alpha}\left(\delta + (\gamma+1)\|w\|\alpha\right)^2, \\ \|K(u_\alpha^\delta)-g^\delta\| &\leq \delta + 2(\gamma+1)\alpha\|w\|. \end{aligned}$$

Proof. We proceed as in the proof of Theorem 4.20, and obtain

$$\begin{aligned} \tfrac{1}{2}\|K(u_\alpha^\delta)-g^\delta\|^2 + \alpha d_{\xi^\dagger,\mu^\dagger}(u_\alpha^\delta,u^\dagger) \leq{}& \tfrac{1}{2}\|K(u^\dagger)-g^\delta\|^2 \\ & - \alpha\langle w, K(u_\alpha^\delta)-K(u^\dagger)\rangle + \alpha\langle w, E(u_\alpha^\delta,u^\dagger)\rangle. \end{aligned}$$

Now condition (iii) yields

$$\tfrac{1}{2}\|K(u_\alpha^\delta)-g^\delta\|^2 + \alpha d_{\xi^\dagger,\mu^\dagger}(u_\alpha^\delta,u^\dagger) \leq \tfrac{1}{2}\delta^2 + (\gamma+1)\alpha\|w\|\|K(u_\alpha^\delta)-K(u^\dagger)\|.$$

From this and the implication $a^2 \leq b^2 + ac \Rightarrow a \leq b + c$ for $a,b,c \geq 0$ the desired assertions follow directly. □

4.5.2 *Variational inequalities*

Now we relax the assumptions on the source condition to variational inequalities, which was first introduced in [141], and further developed in [117, 35, 94, 143, 142]. Variational inequalities are tailored such that the estimates exploiting the source condition and the restriction on the nonlinearity of K are circumvented in one shot. Below we first give the main result from [141], and then discuss the case of a more general approach using the index function.

Theorem 4.22. *Let Assumption 4.1 hold, and further there exist $\beta_1, \beta_2 > 0$, $\beta_1 < 1$ and a $\xi^\dagger \in \partial\psi(u^\dagger)$ such that for all $u \in \mathcal{C}$*

$$-\langle \xi^\dagger, u-u^\dagger\rangle \leq \beta_1 d_{\xi^\dagger}(u,u^\dagger) + \beta_2\|K(u)-K(u^\dagger)\|.$$

Then for any minimizer u_α^δ to J_α, there hold

$$
\begin{aligned}
d_\xi(u_\alpha^\delta, u^\dagger) &\leq \frac{1}{2\alpha(1-\beta_1)}(\delta^2 + \alpha^2\beta_2^2 + 2\alpha\beta_2\delta), \\
\|K(u_\alpha^\delta) - g^\delta\| &\leq \delta + \sqrt{2\alpha\beta_2\delta} + 2\beta_2\alpha.
\end{aligned}
$$

In particular, the choice $\alpha \sim \delta$ yields

$$d_\xi(u_\alpha^\delta, u^\dagger) = \mathcal{O}(\delta) \quad \text{and} \quad \|K(u_\alpha^\delta) - g^\dagger\| = \mathcal{O}(\delta).$$

Proof. By (4.35) and the variational inequality, we deduce

$$
\begin{aligned}
&\tfrac{1}{2}\|K(u_\alpha^\delta) - g^\delta\|^2 + \alpha d_\xi(u_\alpha^\delta, u^\dagger) \\
&\quad \leq \tfrac{1}{2}\delta^2 + \alpha(\beta_1 d_\xi(u_\alpha^\delta, u^\dagger) + \beta_2\|K(u_\alpha^\delta) - K(u^\dagger)\|) \\
&\quad \leq \tfrac{1}{2}\delta^2 + \alpha\beta_1 d_\xi(u_\alpha^\delta, u^\dagger) + \alpha\beta_2\|K(u_\alpha^\delta) - g^\delta\| + \alpha\beta_2\delta.
\end{aligned}
$$

The result follows directly from an application of Young's inequality. □

Remark 4.14. The variational inequality in Theorem 4.22 holds under the conditions in Theorem 4.20 (in the absence of constraints):

$$
\begin{aligned}
|\langle \xi, u - u^\dagger\rangle| &= |\langle w, K'(u^\dagger)(u - u^\dagger)\rangle| \\
&\leq \|w\|\|K(u) - K(u^\dagger)\| + \|w\|\|K(u) - K(u^\dagger) - K'(u^\dagger)(u - u^\dagger)\| \\
&\leq \|w\|\|K(u) - K(u^\dagger)\| + L\|w\| d_\xi(u, u^\dagger).
\end{aligned}
$$

Hence one may take $\beta_2 = \|w\|$ and $\beta_1 = L\|w\| < 1$ so as to arrive at the desired variational inequality. Clearly, it remains valid if the equality $\xi = K'(u^\dagger)w$ is replaced with the inequality

$$|\langle \xi, u - u^\dagger\rangle| \leq |\langle w, K'(u^\dagger)(u - u^\dagger)\rangle| \quad \forall u \in \mathcal{C}.$$

Analogously, the condition in Theorem 4.21 can also be absorbed into a variational inequality with $\beta_1 = 0$ and $\beta_2 = (1+\gamma)\|w\|$:

$$
\begin{aligned}
\langle \xi, u - u^\dagger\rangle &\leq \langle K'(u^\dagger)^* w, u - u^\dagger\rangle \\
&= \langle w, K'(u^\dagger)(u - u^\dagger)\rangle \\
&= \langle w, K(u^\dagger) + K'(u^\dagger)(u - u^\dagger) - K(u)\rangle + \langle w, K(u) - K(u^\dagger)\rangle \\
&\leq (1+\gamma)\|w\|\|K(u) - K(u^\dagger)\|.
\end{aligned}
$$

The source condition of Theorem 4.22 can be further weakened. To motivate the source condition of [117] we start again with the basic comparison $J_\alpha(u_\alpha^\delta) \leq J_\alpha(u^\dagger)$, which upon rearranging yields

$$\tfrac{1}{2}\|K(u_\alpha^\delta) - g^\delta\|^2 + \alpha\left(\psi(u_\alpha^\delta) - \psi(u^\dagger)\right) \leq \tfrac{1}{2}\delta^2.$$

The source condition of Theorem 4.22 was obtained by adding the crucial term $-\alpha\langle\xi, u_\alpha^\delta - u^\dagger\rangle$ on both sides, which did yield the Bregman distance on the left hand side. Hence it requires a source condition that gives a bound of this term. The source condition there was given in terms of a Bregman distance. Grasmair et al [117] introduced a new source condition, which states an estimate directly for $\psi(u_\alpha^\delta) - \psi(u^\dagger)$.

Theorem 4.23. *Let Assumption 4.1 hold, and further there exist $\beta_1, \beta_2 > 0$ and $r > 0$ such that for all $u \in C$*

$$\psi(u) - \psi(u^\dagger) \geq \beta_1 \|u - u^\dagger\|^r - \beta_2 \|K(u) - K(u^\dagger)\|.$$

Then for any minimizer u_α^δ to J_α, there holds

$$\|u_\alpha^\delta - u^\dagger\|^r \leq \frac{\delta^2 + \alpha\beta_2\delta + (\alpha\beta_2)^2/2}{\alpha\beta_1},$$

$$\|K(u_\alpha^\delta) - g^\delta\|^2 \leq 2\delta^2 + 2\alpha\beta_2\delta + (\alpha\beta_2)^2.$$

Proof. The minimizing property of u_α^δ yields

$$\tfrac{1}{2}\|K(u_\alpha^\delta) - g^\delta\|^2 + \alpha\left(\psi(u_\alpha^\delta) - \psi(u^\dagger)\right) \leq \tfrac{1}{2}\delta^2,$$

which together with the variational inequality gives

$$\tfrac{1}{2}\|K(u_\alpha^\delta) - g^\delta\|^2 + \alpha(\beta_1\|u - u^\dagger\|^r - \beta_2\|K(u_\alpha^\delta) - K(u^\dagger)\|) \leq \tfrac{1}{2}\delta^2.$$

Upon rearranging the terms and using the Young's inequality, we get

$$\begin{aligned}\tfrac{1}{2}\|K(u_\alpha^\delta) - g^\delta\|^2 &+ \alpha\beta_1\|u - u^\dagger\|^r \\ &\leq \tfrac{1}{2}\delta^2 + \alpha\beta_2\|K(u_\alpha^\delta) - K(u^\dagger)\| \\ &\leq \tfrac{1}{2}\delta^2 + \alpha\beta_2\delta + \alpha\beta_2\|K(u_\alpha^\delta) - g^\delta\|.\end{aligned}$$

The desired assertion follows from this directly. □

Remark 4.15. Theorem 4.23 states a convergence rate $\mathcal{O}(\delta^{1/r})$ if the rather abstract source condition is fulfilled. In particular, it would yield a surprising convergence rate $\|u_\alpha^\delta - u^\dagger\| = \mathcal{O}(\delta)$ if one can establish the source condition for the case $r = 1$. Then one can achieve a reconstruction error on the same level as the data error δ, and thus the reconstruction problem is no longer ill-posed!

We illustrate the abstract source condition in Theorem 4.23 with ℓ^1 regularization for sparse signals, taken from [117], inspired by earlier works [218]. Sparsity constraints are very common in signal and image processing, and recently also in parameter identifications [172].

Example 4.8. Let X be a Hilbert space, and $\{\varphi_i\} \subset X$ be a given frame. We consider the ℓ^1-penalty, i.e., $\psi(u) = \sum_i |\langle\varphi_i, u\rangle|$. Then under the following assumption, the variational inequality in Theorem 4.23 holds.

(i) The operator equation $K(u) = g^\dagger$ has a ψ-minimizing solution $u^\dagger$ that is sparse with respect to $\{\varphi_i\}$.

(ii) K is Gâteaux differentiable at $u^\dagger$, and for every finite set $\mathbb{J} \subset \mathbb{N}$, the restriction of its derivative $K'(u^\dagger)$ to $\{\varphi_j : j \in \mathbb{J}\}$ is injective.

(iii) There exists $\gamma_1, \gamma_2 > 0$ such that for all $u \in \mathcal{C}$:
$$\psi(u) - \psi(u^\dagger) \geq \gamma_1 \|K(u) - K(u^\dagger) - K'(u^\dagger)(u - u^\dagger)\| - \gamma_2 \|K(u) - K(u^\dagger)\|.$$

(iv) There exists $\xi \in \partial\psi(u^\dagger)$ and $\gamma_3 > 0$ such that for all $u \in \mathcal{C}$:
$$\psi(u) - \psi(u^\dagger) \geq -\gamma_3 \langle \xi, u - u^\dagger \rangle - \gamma_2 \|K(u) - K(u^\dagger)\|.$$

To show the claim, we define $\mathbb{J} := \{j \in \mathbb{N} : |\langle \varphi_j, \xi \rangle| \geq 1/2\}$ and $W = \mathrm{span}\{\varphi_j : j \in \mathbb{J}\}$. Since $\xi \in X$, $\mathbb{J}$ is finite. Thus there exists $C > 0$ such that $C\|K'(u^\dagger)w\| \geq \|w\|$ for all $w \in W$. Next we denote $\pi_W, \pi_W^\perp : X \to X$ the projections
$$\pi_W u = \sum_{j \in \mathbb{J}} \langle \varphi_j, u \rangle \quad \text{and} \quad \pi_W^\perp u = \sum_{j \notin \mathbb{J}} \langle \varphi_j, u \rangle.$$
By assumption $\langle \varphi_j, u^\dagger \rangle = 0$, if $j \notin \mathbb{J}$, i.e., $u^\dagger = \pi_W u^\dagger$, and $\pi_W^\perp u^\dagger = 0$. Hence,
$$\begin{aligned}
\|u - u^\dagger\| &\leq \|\pi_W(u - u^\dagger)\| + \|\pi_W^\perp u\| \\
&\leq C\|K'(u^\dagger)(\pi_W(u - u^\dagger)\| + \|\pi_W^\perp u\| \\
&\leq C\|K'(u^\dagger)(u - u^\dagger)\| + (1 + C\|K'(u^\dagger)\|)\|\pi_W^\perp u\|.
\end{aligned}$$
Next we denote by $m = \max\{|\langle \varphi_j, \xi \rangle| : \xi \notin \mathbb{J}\}$, which, in view of the fact $\{\langle \xi, \varphi_j \rangle\} \in \ell^2$, is well defined. Using the inequality $0 \leq m < 1$ and $\langle \varphi_j, \xi \rangle \leq m$, the assumption $\xi \in \partial\psi(u^\dagger)$ and (iv), we can estimate with $c = \frac{1}{1-m}$
$$\begin{aligned}
\|\pi_W^\perp u\| &= \left(\sum_{j \notin \mathbb{J}} |\langle \varphi_j, u \rangle|^2 \right)^{1/2} \leq \sum_{j \notin \mathbb{J}} |\langle \varphi_j, u \rangle| \\
&\leq c \sum_{j \in \mathbb{J}} (1 - m) |\langle \varphi_j, u \rangle| \\
&\leq c \sum_{j \in \mathbb{J}} (|\langle \varphi_j, u \rangle| - \langle \varphi_j, \xi \rangle \langle \varphi_j, u \rangle) \\
&\leq c \sum_{j \in \mathbb{N}} (|\langle \varphi_j, u \rangle| - \langle \varphi_j, u^\dagger \rangle - \langle \varphi_j, \xi \rangle \langle \varphi_j, u - u^\dagger \rangle) \\
&= c(\psi(u) - \psi(u^\dagger) - \langle \xi, u - u^\dagger \rangle) \\
&\leq c((1 + \gamma_3^{-1})(\psi(u) - \psi(u^\dagger)) + \gamma_2/\gamma_3 \|K(u) - K(u^\dagger)\|).
\end{aligned}$$

Here the third to the last lines follow from the definition of a subgradient and the fact that $\langle \varphi_j, u^\dagger \rangle = 0$ for $j \notin \mathbb{J}$. For the term $\|K'(u^\dagger)(u - u^\dagger)\|$ we estimate from (iii) by

$$\begin{aligned}\|K'(u^\dagger)(u - u^\dagger)\| &\leq \|K(u) - K(u^\dagger) - K'(u^\dagger)(u - u^\dagger)\| + \|K(u) - K(u^\dagger)\| \\ &\leq \gamma_1^{-1}(\psi(u) - \psi(u^\dagger)) + (1 + \gamma_2/\gamma_1)\|K(u) - K(u^\dagger)\|.\end{aligned}$$

Now the desired claim follows by collecting terms.

Remark 4.16. The local injectivity assumption in Example 4.8 is crucial to the estimate. Usually it is difficult to verify, and holds only under restrictive assumptions on the unknown. The constant in the estimate is usually large for ill-posed problems. We refer interested readers to [171] for an application to electrical impedance tomography.

Last, we describe an approach of generalized variational inequalities using an index function [143, 95, 142].

Definition 4.2. A function $\varphi : (0, \infty) \to (0, \infty)$ is called an index function if it is continuous, strictly increasing, and satisfies the limit condition $\lim_{t \to 0^+} \varphi(t) = 0$.

Assumption 4.4. There exists a constant $0 < \beta \leq 1$ and a concave index function φ such that for some $\xi^\dagger \in \partial\psi(u^\dagger)$

$$\beta d_{\xi^\dagger}(u, u^\dagger) \leq \psi(u) - \psi(u^\dagger) + \varphi(\|K(u) - K(u^\dagger)\|). \tag{4.36}$$

The variational inequality directly gives several basic estimates.

Lemma 4.6. *Let Assumption 4.4 hold. Then for $\alpha > 0$, there hold*

$$\begin{aligned}\psi(u^\dagger) - \psi(u_\alpha^\delta) &\leq \varphi(\|K(u_\alpha^\delta) - K(u^\dagger)\|), \\ \psi(u_\alpha^\delta) - \psi(u^\dagger) &\leq \tfrac{\delta^2}{2\alpha}.\end{aligned}$$

Proof. The first assertion is an immediate consequence of (4.36), $\beta > 0$ and $d_\xi(u, u^\dagger) \geq 0$. By the minimizing property of u_α^δ, we derive

$$\begin{aligned}\tfrac{1}{2}\|K(u_\alpha^\delta) - g^\delta\|^2 + \alpha\psi(u_\alpha^\delta) &\leq \tfrac{1}{2}\|K(u^\dagger) - g^\delta\|^2 + \alpha\psi(u^\dagger) \\ &\leq \tfrac{1}{2}\delta^2 + \alpha\psi(u^\dagger),\end{aligned}$$

from which the second assertion follows. □

Lemma 4.7. *Under Assumption 4.4, for $\alpha > 0$, there holds*

$$\|K(u_\alpha^\delta) - K(u^\dagger)\|^2 \leq 4\delta^2 + 4\alpha\varphi(\|K(u_\alpha^\delta) - K(u^\dagger)\|).$$

Proof. It follows from (4.36) and the minimizing property of u_α^δ that

$$\begin{aligned}0 &\leq \psi(u_\alpha^\delta) - \psi(u^\dagger) + \varphi(\|K(u_\alpha^\delta) - K(u^\dagger)\|)\\ &\leq \alpha^{-1}\left(\tfrac{1}{2}\delta^2 - \tfrac{1}{2}\|K(u_\alpha^\delta) - g^\delta\|^2\right) + \varphi(\|K(u_\alpha^\delta) - K(u^\dagger)\|).\end{aligned}$$

Meanwhile, by Young's inequality,

$$\tfrac{1}{2}\|K(u_\alpha^\delta) - g^\delta\|^2 \geq \tfrac{1}{4}\|K(u_\alpha^\delta) - K(u^\dagger)\|^2 - \tfrac{1}{2}\delta^2.$$

Inserting it into the first inequality, we deduce

$$0 \leq \delta^2 - \tfrac{1}{4}\|K(u_\alpha^\delta) - K(u^\dagger)\|^2 + \alpha\varphi(\|K(u_\alpha^\delta) - K(u^\dagger)\|),$$

which completes the proof. □

There are several possible ways to use Lemma 4.7. First, given a specific value of the regularization parameter $\alpha > 0$, one can bound the fidelity from above. Second, the value of α can be bounded from below if the fidelity is larger than δ. To this end, we introduce the function

$$\tilde{\varphi} = \frac{t^2}{\varphi(t)} \quad t > 0,$$

where φ is an arbitrary concave index function. By the concavity of φ, $\tilde{\varphi}(t)$ is an index function.

Corollary 4.3. *Let α_* be determined by $\alpha_* = \tilde{\varphi}(\delta)$. Then for $\alpha \leq \alpha^*$, there holds*

$$\|K(u_\alpha^\delta) - K(u^\dagger)\| \leq 8\delta.$$

Proof. We treat the two cases $\|K(u_\alpha^\delta) - K(u^\dagger)\| > \delta$ and $\|K(u_\alpha^\delta) - K(u^\dagger)\| \leq \delta$ separately. If $\|K(u_\alpha^\delta) - K(u^\dagger)\| > \delta$, then by Lemma 4.7 and the value for α_*, we obtain

$$\begin{aligned}\tfrac{1}{2}\|K(u_\alpha^\delta) - K(u^\dagger)\|^2 &\leq 2\delta^2 + 2\alpha\varphi(\|K(u_\alpha^\delta) - K(u^\dagger)\|)\\ &\leq 2\delta^2 + 2\alpha_*\varphi(\|K(u_\alpha^\delta) - K(u^\dagger)\|).\end{aligned}$$

Now the choice of the value α_* gives

$$\alpha_*\varphi(\|K(u_\alpha^\delta) - K(u^\dagger)\|) = \delta^2\frac{\varphi(\|K(u_\alpha^\delta) - K(u^\dagger)\|)}{\varphi(\delta)}.$$

Combing this with the preceding inequality, together with the condition $\|K(u_\alpha^\delta) - K(u^\dagger)\| > \delta$ and the monotonicity of the index function φ, yields

$$\begin{aligned}\|K(u_\alpha^\delta) - K(u^\dagger)\|^2 &\leq 4\delta^2\left(1 + \frac{\varphi(\|K(u_\alpha^\delta) - K(u^\dagger)\|)}{\varphi(\delta)}\right)\\ &\leq 8\delta^2\frac{\varphi(\|K(u_\alpha^\delta) - K(u^\dagger)\|)}{\varphi(\delta)}.\end{aligned}$$

Now the concavity of the index function φ yields

$$\varphi(\|K(u_\alpha^\delta) - K(u^\dagger)\|) \leq \delta^{-1}\|K(u_\alpha^\delta) - K(u^\dagger)\|\varphi(\delta).$$

The desired assertion now follows immediately. Clearly, the bound is also valid for $\|K(u_\alpha^\delta) - K(u^\dagger)\| \leq \delta$. This completes the proof of the lemma. □

Corollary 4.4. *Let $\tau > 1$. Suppose that the parameter $\alpha > 0$ is chosen such that $\|K(u_\alpha^\delta) - g^\delta\| > \tau\delta$. Then there holds*

$$\alpha \geq \frac{1}{4}\frac{\tau^2 - 1}{\tau^2 + 1}\tilde{\varphi}((\tau - 1)\delta).$$

Proof. Using the minimizing property u_α^δ and the first assertion in Lemma 4.6, we have

$$\begin{aligned}
\tau^2\delta^2 &\leq \|K(u_\alpha^\delta) - g^\delta\|^2 \\
&\leq \delta^2 + 2\alpha(\psi(u^\dagger) - \psi(u_\alpha^\delta)) \\
&\leq \delta^2 + 2\alpha\varphi(\|K(u_\alpha^\delta) - K(u^\dagger)\|).
\end{aligned}$$

Therefore,

$$\delta^2 \leq \frac{2}{\tau^2 - 1}\alpha\varphi(\|K(u_\alpha^\delta) - K(u^\dagger)\|).$$

Upon plugging it into Lemma 4.7, we obtain (with $t_\alpha = \|K(u_\alpha^\delta) - K(u^\dagger)\|$) that

$$\begin{aligned}
t_\alpha^2 &\leq 4\delta^2 + 4\alpha\varphi(t_\alpha) \leq \frac{4}{\tau^2 - 1}\alpha\varphi(t_\alpha) + 4\alpha\varphi(t_\alpha) \\
&= 4\frac{\tau^2 + 1}{\tau^2 - 1}\alpha\varphi(t_\alpha).
\end{aligned}$$

Since $\tau\delta \leq t_\alpha + \delta$, we arrive, using the function $\tilde{\varphi}$, at

$$\tilde{\varphi}((\tau - 1)\delta) \leq \tilde{\varphi}(t_\alpha) \leq 4\frac{\tau^2 + 1}{\tau^2 - 1}\alpha.$$

□

Now we state a convergence rate result under Assumption 4.4.

Theorem 4.24. *Let Assumption 4.4 hold, and $\alpha_* = \alpha_*(\delta) = \tilde{\varphi}(\delta)$. Then for sufficiently small δ, there hold*

$$\begin{aligned}
d_{\xi^\dagger}(u_{\alpha_*}^\delta, u^\dagger) &= \mathcal{O}(\varphi(\delta)), \\
\|K(u_{\alpha_*}^\delta) - K(u^\dagger)\| &= \mathcal{O}(\delta), \\
|\psi(u_{\alpha_*}^\delta) - \psi(u^\dagger)| &= \mathcal{O}(\varphi(\delta)).
\end{aligned}$$

Proof. Corollary 4.3 yields a bound on the discrepancy, i.e.,

$$\|K(u_{\alpha_*}^\delta) - K(u^\dagger)\| \leq 8\delta.$$

In view of the first assertion of Lemma 4.6 and the concavity of the index function φ, this also bounds $\psi(u^\dagger) - \psi(u_{\alpha_*}^\delta)$ by

$$\begin{aligned}\psi(u^\dagger) - \psi(u_{\alpha_*}^\delta) &\leq \varphi(\|K(u_{\alpha_*}^\delta) - K(u^\dagger)\|)\\ &\leq \varphi(8\delta) \leq 8\varphi(\delta).\end{aligned}$$

Further, by the second assertion of Lemma 4.6 and the choice of α_*, we have

$$\psi(u_{\alpha_*}^\delta) - \psi(u^\dagger) \leq \tfrac{\delta^2}{2\alpha_*} \leq \tfrac{1}{2}\varphi(\delta).$$

These two estimates together give the desired bound on $|\psi(u_{\alpha_*}^\delta) - \psi(u^\dagger)|$. It follows from (4.36) that

$$d_{\xi^\dagger}(u_{\alpha_*}^\delta, u^\dagger) \leq \tfrac{1}{\beta}\left(\psi(u_{\alpha_*}^\delta) - \psi(u^\dagger) + \varphi(\|K(u_{\alpha_*}^\delta) - K(u^\dagger)\|)\right) \leq \tfrac{17}{2\beta}\varphi(\delta).$$

This completes the proof of the theorem. □

We end this part with an a posteriori parameter choice, the discrepancy principle. We restrict the selection of the regularization parameter to a discrete exponential grid. More precisely, we fix $0 < q < 1$, choose a largest α_0, and consider the set

$$\Delta_q = \{\alpha_j : \alpha_j = q^j\alpha_0,\ j = 1, 2, \ldots\}.$$

Theorem 4.25. *Let Assumption* (4.4) *hold. Let $\tau > 1$ be given, $\alpha_* \in \Delta_q$, $\alpha_* < \alpha_1$ be chosen by the discrepancy principle, as the largest parameter from Δ_q for which*

$$\|K(u_\alpha^\delta) - g^\delta\| \leq \tau\delta.$$

Then there hold

$$\begin{aligned} d_{\xi^\dagger}(u_{\alpha_*}^\delta, u^\dagger) &= \mathcal{O}(\varphi(\delta)),\\ \|K(u_{\alpha_*}^\delta) - K(u^\dagger)\| &= \mathcal{O}(\delta),\\ |\psi(u_{\alpha_*}^\delta) - \psi(u^\dagger)| &\leq \mathcal{O}(\varphi(\delta)).\end{aligned}$$

Proof. We first bound

$$\|K(u_{\alpha_*}^\delta) - K(u^\dagger)\| \leq \|K(u_{\alpha_*}^\delta) - g^\delta\| + \|g^\delta - K(u^\dagger)\| \leq (\tau + 1)\delta.$$

Under Assumption 4.4, this also gives

$$\psi(u^\dagger) - \psi(u_{\alpha_*}^\delta) \leq \varphi((\tau + 1)\delta) \leq (\tau + 1)\varphi(\delta),$$

cf. Lemma 4.6. Next we note that the parameter α_*/q fulfills the assumption of Corollary 4.4, and hence

$$\frac{\alpha_*}{q} \geq \frac{1}{4}\frac{\tau^2-1}{\tau^2+1}\tilde{\varphi}((\tau-1)\delta).$$

This together with the second assertion of Lemma 4.6 yields

$$\begin{aligned}\psi(u_{\alpha_*}^\delta) - \psi(u^\dagger) &\leq \frac{2\frac{\tau^2+1}{\tau^2-1}\delta^2}{q\tilde{\varphi}((\tau-1)\delta)} \\ &= \frac{1}{2q}\frac{4}{(\tau-1)^2}\frac{\tau^2+1}{\tau^2-1}\varphi((\tau-1)\delta) \\ &\leq \frac{2}{q(\tau-1)^2}\frac{\tau^2+1}{\tau^2-1}\varphi((\tau+1)\delta) \\ &\leq \frac{2}{q}\frac{\tau^2+1}{(\tau-1)^3}\varphi(\delta).\end{aligned}$$

The convergence rate in Bregman distance follows directly from these two estimates and (4.36). □

4.6 Conditional stability

In this section, we briefly discuss an alternative approach for obtaining error estimates, i.e., conditional stability due to Cheng and Yamamoto [65]. It connects two distinct research topics in inverse problems, regularization theory and conditional stability estimates. We note that many conditional stability estimates exist for inverse problems for partial differential equations. One general setting is as follows: Let K be a densely defined and injective operator from a Banach space X to a Banach space Y. Let $Z \subset X$ be another Banach space and let the embedding $Z \hookrightarrow X$ be continuous. We set $\mathcal{U}_M = \{u \in Z : \|u\|_Z \leq M\}$ for $M > 0$ and choose $Q \subset Z$ arbitrary.

Let the function $\omega = \omega(\epsilon)$ ($\epsilon \geq 0$) be nonnegative and monotonically increasing and satisfy

$$\lim_{\epsilon \downarrow 0} \omega(\epsilon) = 0 \quad \text{and} \quad \omega(k\epsilon) \leq B(k)\omega(\epsilon) \quad \text{for } k \in (0,\infty)$$

where $B(k)$ is a positive function defined in $(0,\infty)$.

Definition 4.3. The operator equation $K(u) = g$ is said to satisfy a conditional stability, if for a given $M > 0$, there exists a constant $c = c(M) > 0$ such that

$$\|u_1 - u_2\|_X \leq c(M)\omega(\|K(u_1) - K(u_2)\|_Y)$$

for all $u_1, u_2 \in \mathcal{U}_M \cap Q$. The function ω is called the modulus of the conditional stability.

Now we consider the Tikhonov functional

$$J_\alpha(u) = \|Ku - g^\delta\|_Y^2 + \alpha\|u\|_Z^2.$$

One basic result is a convergence rate for an a priori choice strategy.

Theorem 4.26. *Let the operator equation $K(u) = g$ satisfy the conditional stability and $K(u^\dagger) = g^\dagger$, $u^\dagger \in Q \cap Z$, and $\|g^\delta - g^\dagger\|_Y \leq \delta$. Let $u_\alpha^\delta \in Q \cap Z$ be an approximate minimizer in the sense*

$$J_\alpha(u_\alpha^\delta) \leq \inf_{u \in Q \cap Z} J_\alpha(u) + c_0\delta^2$$

where $c_0 > 0$ is a constant. Then with the choice rule $\alpha \sim \delta^2$, there holds

$$\|u_\alpha^\delta - u^\dagger\|_X \leq c\omega(\delta).$$

Proof. Let $M = \|u^\dagger\|_Z$. Since $J_\alpha(u_\alpha^\delta) \leq J_\alpha(u) + c_0\delta^2$ for any $u \in Q \cap Z$, we have

$$\begin{aligned} \|K(u_\alpha^\delta) - g^\delta\|_Y^2 + \alpha\|u_\alpha^\delta\|_Z^2 &\leq \|K(u^\dagger) - g^\delta\|_Y^2 + \alpha\|u^\dagger\|_Z^2 + c_0\delta^2 \\ &= \|g^\dagger - g^\delta\|_Y^2 + \alpha\|x^\dagger\|_Z^2 + c_0\delta^2 \\ &\leq (1 + c_0)\delta^2 + \alpha M^2. \end{aligned}$$

This together with the choice $c_1\delta^2 \leq \alpha \leq c_2\delta^2$ gives

$$\|K(u_\alpha^\delta) - g^\delta\|_Y \leq (1 + c_0 + c_2M^2)^{\frac{1}{2}}\delta$$

and

$$\|u_\alpha^\delta\|_Z \leq ((1 + c_0)c_1^{-1} + M^2)^{\frac{1}{2}} \equiv M_1.$$

Consequently,

$$\|K(u_\alpha^\delta) - K(u^\dagger)\|_Y \leq \|K(u_\alpha^\delta) - g^\delta\|_Y + \|g^\delta - g^\dagger\|_Y \leq (c + 1)\delta,$$

with $c = (1 + c_0 + c_2M^2)^{\frac{1}{2}}$. Finally, the conditional stability yields

$$\|u_\alpha^\delta - u^\dagger\|_X \leq c(M_1)\omega((c + 1)\delta) \leq c(M_1)B(c + 1)\omega(\delta),$$

which completes the proof of the theorem. □

Remark 4.17.

(i) In Theorem 4.26, the existence of a minimizer to the functional J_α is not required. However, this does not resolve the problem of finding an approximate minimizer. Further, in the Tikhonov functional J_α, the fidelity and penalty are both restricted to powered norms. It is unclear how to incorporate general convex variational penalties into the framework.

(ii) In Theorem 4.26, the choice $\alpha = \mathcal{O}(\delta^2)$ is different from other choice rules based on canonical source conditions. The goal here is to control the size of the discrepancy, while bounding uniformly the norm of the approximate minimizer.

Apart from the a priori choice $\alpha = \mathcal{O}(\delta^2)$, the classical discrepancy principle can reproduce an almost identical convergence rate.

Theorem 4.27. *Let the operator equation $K(u) = g$ satisfy the conditional stability and $K(u^\dagger) = g^\dagger$, $u^\dagger \in Q \cap Z$, and $\|g^\delta - g^\dagger\|_Y \leq \delta$. Let $u_\alpha^\delta \in Q \cap Z$ be an approximate minimizer in the sense*

$$J_\alpha(u_\alpha^\delta) \leq \inf_{u \in Q \cap Z} J_\alpha(u) + c_0\delta^2.$$

Let $\alpha(\delta)$ be determined by the discrepancy principle

$$c_1\delta \leq \|K(u_\alpha^\delta) - g^\delta\|_Y \leq c_2\delta,$$

where $c_2 > c_1 \geq \sqrt{1 + c_0}$. Then there holds the following estimate

$$\|u_\alpha^\delta - u^\dagger\|_X \leq c\omega(\delta).$$

Proof. Let $M = \|u^\dagger\|_Z$. Then by the minimizing property of u_α^δ we have

$$\begin{aligned}\|K(u_\alpha^\delta) - g^\delta\|_Y^2 + \alpha\|u_\alpha^\delta\|_Z^2 &\leq \|Ku^\dagger - g^\delta\|_Y^2 + \alpha\|u^\dagger\|^2 + c_0\delta^2 \\ &= (1 + c_0)\delta^2 + \alpha\|u^\dagger\|_Z^2.\end{aligned}$$

This together with the definition of the discrepancy principle yields

$$\|u_\alpha^\delta\|_Z^2 \leq \|u^\dagger\|_Z^2.$$

However,

$$\begin{aligned}\|K(u_\alpha^\delta) - K(u^\dagger)\|_Y &\leq \|K(u_\alpha^\delta) - g^\delta\|_Y + \|g^\delta - K(u^\dagger)\|_Y \\ &\leq (c_2 + 1)\delta.\end{aligned}$$

Then the conditional stability estimate yields

$$\begin{aligned}\|u_\alpha^\delta - u^\dagger\|_X &\leq C(M)\omega(\|K(u_\alpha^\delta) - K(u^\dagger)\|_Y) \\ &\leq C(M)\omega((c_2 + 1)\delta) \leq C(M)B(c_2 + 1)\omega(\delta).\end{aligned}$$

This completes the proof of the theorem. □

Remark 4.18. One distinct feature of the approach via conditional stability is that no source/ nonlinearity conditions are required for the convergence rate, by using conditional stability estimates instead. However, conditional stability estimates on computationally amenable Sobolev/Banach spaces can be very challenging to derive. This is practically important since otherwise faithfully minimizing the Tikhonov functional (respecting the smoothness assumption) might be very difficult, if the involved norms $\|\cdot\|_Y$ and $\|\cdot\|_Z$ are exotic.

Typically for PDE inverse problems, conditional stability is established via Carleman estimates. We refer interested readers to [48, 202, 310] for the applications of Carleman estimate in inverse problems. We conclude the section with one classical linear inverse problem, the backward heat problem, taken from [216].

Example 4.9. In this example, we consider the heat equation

$$\begin{aligned} y_t &= A(x)y \quad \text{in } \Omega \times (0,T), \\ y &= 0 \quad \text{on } \partial\Omega \times (0,T), \end{aligned} \tag{4.37}$$

where Ω is a bounded domain in $\mathbb{R}^d (d = 2, 3)$ with a smooth boundary $\partial\Omega$, and $A(x)$ is a second-order self-adjoint elliptic operator

$$A(x)y = \nabla \cdot (a(x)\nabla y) - c(x)y,$$

where the nonnegative function $c(x) \in L^\infty(\Omega)$ and $a(x) \in C^1(\overline{\Omega})$ with $a \geq \alpha_0$ for some $\alpha_0 > 0$. Let ω be an arbitrary subdomain of Ω, $\tau > 0$ be a fixed constant. The inverse problem is to reconstruct the initial temperature $y(\cdot, 0)$ from the measurement data of y in $\omega \times (\tau, T)$. To show a conditional stability estimate, we first introduce an admissible set $\mathcal{U}_M$ by: for any fixed $\epsilon \in (0,1)$ and $M > 0$

$$\mathcal{U}_M = \{a \in H^{2\epsilon}(\Omega) : \|a\|_{H^{2\epsilon}(\Omega)} \leq M\}.$$

Then for any $y(\cdot, 0) \in \mathcal{A}$, there exists a constant $\kappa = \kappa(M, \epsilon)$ such that

$$\|y(\cdot, 0)\|_{L^2(\Omega)} \leq C(M, \epsilon)(-\log \|y\|_{L^2(\omega\times(\tau,T))})^{-\kappa}. \tag{4.38}$$

In particular, with the help of this estimate (4.38), one can show that in Tikhonov regularization, if one chooses the setup as

$$\begin{aligned} X &= L^2(\Omega), \quad Y = L^2(\omega \times (\tau, T)), \\ Q &= Z = H^{2\epsilon}(\Omega), \quad \omega(\eta) = (-\log \eta)^{-\kappa}, \end{aligned}$$

then the (approximate) Tikhonov minimizer u_α^δ, with either the a priori parameter choice $\alpha \sim \delta^2$ or the discrepancy principle, satisfies a convergence rate $\mathcal{O}(\omega(\delta))$.

We shall assume that $\|y\|_{L^2(\omega\times(\tau,T))}$ is sufficiently small, and show the conditional stability estimate (4.38) in three steps.

step (i): application of a Carleman estimate. First we show that for any fixed $\theta \in (\tau, T)$, there exists constants $C = C(M) > 0$ and $\kappa_1 \in (0,1)$ such that

$$\|y(\cdot, \theta)\|_{L^2(\Omega)} \leq C\|y\|^{\kappa_1}_{L^2(\omega\times(\tau,T))}. \tag{4.39}$$

To show this, we first introduce two weight functions

$$\varphi(x,t) = \frac{e^{\lambda d(x)}}{(t-\tau)(T-t)},$$
$$\alpha(x,t) = \frac{e^{\lambda d(x)} - e^{2\lambda \|d\|_{C(\overline{\Omega})}}}{(t-\tau)(T-t)},$$

where $d(x)$ is a suitably chosen positive function in Ω [150, 310]. Then by the Carleman estimates (c.f Theorem 2.1 of [150] or [310]) for the solution $y(x,t)$ to (4.37), there exists some constant s_0 such that for all $s \geq s_0$ and $(x,t) \in \Omega \times (\tau, T)$ there holds

$$\int_\tau^T \int_\Omega \left(\frac{1}{s\varphi}|\nabla y|^2 + s\varphi y^2\right) e^{2s\alpha} dxdt \leq C \int_\tau^T \int_\omega s\varphi y^2 e^{2s\alpha} dxdt. \quad (4.40)$$

Further, for any two arbitrarily fixed θ_1, θ_2 with $\tau < \theta_1 < \theta_2 < T$, clearly we have

$$\varphi e^{2s\alpha} \leq \tilde{C} \quad \text{for } (x,t) \in \omega \times (\tau, T),$$
$$\varphi e^{2s\alpha} \geq \tilde{c}, \quad \varphi^{-1} e^{2s\alpha} \geq \tilde{c} \quad \text{for } (x,t) \in \Omega \times (\theta_1, \theta_2),$$

where the positive constants $\tilde{C}$ and $\tilde{c}$ depend only on τ, T, θ_1 and θ_2. Then it follows from (4.40) that

$$\|y\|_{L^2(\theta_1,\theta_2;H^1(\Omega))} \leq C_M \|y\|_{L^2(\omega\times(\tau,T))}. \quad (4.41)$$

Meanwhile, multiplying both sides of (4.37) by any $v \in H_0^1(\Omega)$ yields

$$\int_\Omega y_t v dx = -\int_\Omega a\nabla y \cdot \nabla v - \int_\Omega cyvdx,$$

which together with (4.41) gives

$$\begin{aligned} \|y_t\|_{L^2(\theta_1,\theta_2;H^{-1}(\Omega))} &\leq C\|y\|_{L^2(\theta_1,\theta_2;H^1(\Omega))} \\ &\leq C_M \|y\|_{L^2(\omega\times(\tau,T))}. \end{aligned}$$

Therefore, these two estimates and Sobolev embedding give

$$\|y\|_{C([\theta_1,\theta_2];H^{-1}(\Omega))} \leq C_M \|y\|_{L^2(\omega\times(\tau,T))}. \quad (4.42)$$

Furthermore, by the semigroup theory [249], we have $y(t) = e^{tA}y(\cdot,0)$, where the elliptic operator A is defined on the domain $\mathcal{D}(A) = H_0^1(\Omega) \cap H^2(\Omega)$. Then for any $\gamma > 0$, there exists a constant $C_\gamma > 0$ such that

$$\begin{aligned} \|(-A)^\gamma e^{tA}\| &\leq C_\gamma t^{-\gamma}, \\ \|a\|_{H^{2\gamma}(\Omega)} &\leq C_\gamma \|(-A)^\gamma a\|_{L^2(\Omega)}. \end{aligned} \quad (4.43)$$

This implies

$$\|y\|_{C([\theta_1,\theta_2];H^{2\gamma}(\Omega))} \leq C_\gamma \|(-A)^\gamma e^{tA} y(\cdot,0)\|_{C([\theta_1,\theta_2];L^2(\Omega))} \leq \frac{C_\gamma}{\theta_1^\gamma}\|y(\cdot,0)\|_{L^2(\Omega)} \leq \frac{C_\gamma M}{\theta_1^\gamma}.$$

From this and (4.42), by the interpolation theory, we deduce

$$\|y\|_{C([\theta_1,\theta_2];L^2(\Omega))} \leq \|y\|^{\frac{2\gamma}{2\gamma+1}}_{C([\theta_1,\theta_2];H^{-1}(\Omega))}\|y\|^{\frac{1}{2\gamma+1}}_{C([\theta_1,\theta_2];H^{2\gamma}(\Omega))} \leq C_M^{\frac{2\gamma}{2\gamma+1}}\left(\frac{C_\gamma M}{\theta_1^\gamma}\right)^{\frac{1}{2\gamma+1}}\|y\|^{\frac{2\gamma}{2\gamma+1}}_{L^2(\omega\times(\tau,T))},$$

which completes the proof of (4.39).
step (ii) logarithmic convexity inequality. Next we show an important logarithmic convexity inequality

$$\|y(\cdot,t)\|_{L^2(\Omega)} \leq \|y(\cdot,0)\|^{1-\frac{t}{\theta}}_{L^2(\Omega)}\|y(\cdot,\theta)\|^{\frac{t}{\theta}}_{L^2(\Omega)} \quad \forall 0\leq t\leq\theta. \tag{4.44}$$

To show this, we consider the function $V(t) = \|y(\cdot,t)\|^2_{L^2(\Omega)}$. Using (4.37) and integration by parts we deduce

$$V'(t) = 2\int_\Omega y(x,t)y_t(x,t)dx = 2\int_\Omega y(x,t)(Ay)(x,t)dx = -2\int_\Omega (a|\nabla y|^2 + cy^2)dx.$$

Further differentiation and integration by parts yields

$$V''(t) = -4\int_\Omega a\nabla y\cdot\nabla y_t + cyy_t dx = 4\int_\Omega y_t Ay dx = 4\int_\Omega y_t^2 dx.$$

Using these formulae for $V'(t)$ and $V''(t)$ and the Cauchy-Schwarz inequality, we have

$$V'(t)^2 - V''(t)V(t) = (2\int_\Omega yy_t dx)^2 - 4\int_\Omega y_t^2 dx\int_\Omega y^2 dx \leq 0,$$

which yields

$$(\log V(t))'' = \frac{V''(t)V(t) - V'(t)^2}{V(t)^2} \geq 0.$$

Therefore, we know that $\log V(t)$ is convex, i.e., $\log V(t) \leq (1-\frac{t}{\theta})\log V(0) + \frac{t}{\theta}\log V(\theta)$, from which the inequality (4.44) follows immediately. Upon squaring both sides of (4.44) and then integrating over $t\in(0,\theta)$, we obtain

$$\int_0^\theta \|y(\cdot,t)\|^2_{L^2(\Omega)}dt \leq C_M\int_0^\theta \|y(\cdot,\theta)\|^{\frac{2t}{\theta}}_{L^2(\Omega)}dt.$$

It follows from this and (4.39) that

$$\begin{aligned}\|y\|_{L^2(0,\theta;L^2(\Omega))} &\leq C_M(-\log \|y(\cdot,\theta)\|_{L^2(\Omega)})^{-\frac{1}{2}} \\ &\leq C_M(-\log \|y\|_{L^2(\omega\times(\tau,T))})^{-\frac{1}{2}}.\end{aligned} \tag{4.45}$$

step (iii) stability estimate. By the semigroup representation of the solution $y(\cdot,t)$, we have

$$y_t(\cdot,t) = Ae^{tA}y(\cdot,0) = -(-A)^{1-\epsilon}e^{tA}(-A)^{\epsilon}y(\cdot,0).$$

Now by appealing to (4.43), we obtain

$$\|y_t(\cdot,t)\|_{L^2(\Omega)} \leq Ct^{\epsilon-1}\|(-A)^{\epsilon}y(\cdot,0)\|_{L^2(\Omega)}.$$

Now for any $1 < p < 1/(1-\epsilon)$, upon noting $y(\cdot,0) \in \mathcal{A}$, we deduce

$$\begin{aligned}\int_0^\theta \|y_t(\cdot,t)\|^p_{L^2(\Omega)}dt &\leq C\int_0^\theta t^{p(\epsilon-1)}dt\|(-A)^{\epsilon}y(\cdot,0)\|^p_{L^2(\Omega)} \\ &\leq C\|y(\cdot,0)\|^p_{H^{2\epsilon}(\Omega)} \leq C(M),\end{aligned}$$

i.e.,

$$\|y\|_{W^{1,p}(0,\theta;L^2(\Omega))} \leq C(M). \tag{4.46}$$

We choose p such that $1 < p \leq 2$. Now by (4.45) and the inequality $\|\eta\|_{L^p(0,T)} \leq C\|\eta\|_{L^2(0,T)}$ for $p \leq 2$, we have

$$\|y\|_{L^p(0,\theta;L^2(\Omega))} \leq C(M)(-\log \|y\|_{L^2(\omega\times(\tau,T))})^{-\frac{1}{2}}. \tag{4.47}$$

Using (4.46), (4.47) and Sobolev interpolation, we derive that for $0 < s < 1$

$$\|y\|_{W^{1-s,p}(0,\theta;L^2(\Omega))} \leq C(M)(-\log \|y\|_{L^2(\omega\times(\tau,T))})^{-\frac{s}{2}}.$$

Now we choose $s \in (0,1)$ such that $(1-s)p > 1$ and thus the space $W^{1-s,p}(0,\theta;L^2(\Omega))$ continuously embeds into $C([0,\theta];L^2(\Omega))$. Consequently,

$$\begin{aligned}\|y\|_{C([0,\theta];L^2(\Omega))} &\leq C\|y\|_{W^{1-s,p}(0,\theta;L^2(\Omega))} \\ &\leq C(M)(-\log \|y\|_{L^2(\omega\times(\tau,T))})^{-\frac{s}{2}}.\end{aligned}$$

This shows the claimed conditional stability estimate (4.38).

Bibliographical notes

Our discussions in Section 4.2 on the convergence rate analysis in Hilbert space setting have exclusively focused on the classical source condition, i.e. $u^\dagger = K'(u^\dagger)^* w$. This is largely motivated by its easier interpretation as a necessary optimality condition, especially in the context of parameter identification for differential equations. However, there are more general source conditions. One is the power-type source condition

$$u^\dagger = (K'(u^\dagger)^* K'(u^\dagger))^\nu w,$$

where the parameter $0 < \nu \leq 1$ controls the smoothness of the solution $u^\dagger$. A more general case reads

$$u^\dagger = \varphi(K'(u^\dagger)^* K(u^\dagger))w,$$

where φ is a continuous nondecreasing index function defined on some interval $[0, \sigma]$ containing the spectrum of $K'(u^\dagger)^* K'(u^\dagger)$, and $\varphi(0) = 0$. The convergence rate analysis can be extended to these cases, provided certain further conditions on the operator K are met with. We shall not dwell into these interesting theoretical issues here, but refer interested readers to [284, 222]. The existence of a source representer w for parameter identification problems is nontrivial to verify. Nonetheless, there are a few studies, where sufficient conditions were provided. We refer to [90] for a parabolic inverse problem, and [128] for an elliptic inverse problem.

The variational inequality approach is fairly versatile for representing structural properties of nonlinear inverse problems. However, the application to concrete inverse problems remains fairly scarce. It is known that it is more general than the classical source condition; see [49] for the case of sparsity constraints. The approach of conditional stability was further studied in [184]; see also [79] for an application to the Landweber method.

Chapter 5

Nonsmooth Optimization

In this chapter we discuss optimization theory for a general class of variational problems arising in Tikhonov formulations of inverse problems. Throughout, we let X and Y be Banach spaces, and H be a Hilbert space lattice, the functional $F : X \to \mathbb{R}^+$ and the map $E : X \to Y$ be continuously differentiable. Further, we let $\mathcal{C} \subset X$ be a closed convex subset of X, and $\Lambda \in \mathcal{L}(X, H)$ be a bounded linear operator from X to H. Throughout, we denote by $\|\cdot\|$ the norm, and $|\cdot|$ the absolute value or the Euclidean norm on $\mathbb{R}^{\mathrm{d}}$, $\mathrm{d} = 2, 3$.

We shall consider minimization problems of the form

$$\min_{x \in \mathcal{C}} \; J(x) = F(x) + \alpha \psi(\Lambda x) \tag{5.1}$$

subject to the equality constraint

$$E(x) = 0.$$

Problem (5.1) encompasses a wide variety of optimization problems including variational inequalities of the first and second kind [112, 111]. Here ψ is a nonsmooth functional on H. One example is the L^1-type regularizations for inverse medium problems, which are often used to obtain geometrically sharp and enhanced minimizers. Then the functional $\psi(z)$ is given by

$$\psi(z) = \int_\Omega |z| \, d\omega,$$

and the linear operator Λ is the natural injection for the sparsity imaging method, $\Lambda = \nabla$ for the total variation regularization, and $\Lambda = \Delta$ for the biharmonic nonlinear filter. Alternatively, one may consider the following nonconvex functional

$$\psi(z) = \int_\Omega |z|^p \, d\omega, \quad 0 \le p < 1.$$

Then the minimizers feature sparsity structure, e.g., point-like distribution and clustered medium. Especially, $L^0(\Omega)$ with $p = 0$ represents the measure of support set $\{z \neq 0\}$ of the function z. As was mentioned in Chapter 3, it is very essential to select proper regularization functionals to obtain an enhanced yet robust reconstruction that captures distinct features and properties of the medium. In particular, one may use multiple regularization functionals $\{\psi_k\}$ and $\psi = \sum_k \alpha_k \psi_k$ to capture different features and properties of the sought-for medium. We note that the constraint $x \in \mathcal{C}$ can also be formulated by the nonsmooth functional ψ by

$$\psi(x) = I_{\mathcal{C}} \equiv \begin{cases} 0, & x \in \mathcal{C}, \\ \infty, & x \notin \mathcal{C}. \end{cases}$$

In a general class of inverse problems (control problems), the functional F denotes the fidelity or performance index, ψ represents the regularization, and $\alpha > 0$ is the regularization parameter. In many applications we have the natural decomposition of coordinate $x = (y, u) \in X_1 \times X_2$ and $E(x) = E(y, u) = 0$ represents the equality constraint for the state $y \in X_1$ and the control/design/medium variable $u \in \mathcal{C} \subset X_2$. So we have a constrained minimization problem of the form

$$\begin{aligned} &F(y) + \alpha\psi(u), \\ &\text{subject to } E(y, u) = 0. \end{aligned} \tag{5.2}$$

Throughout, we always assume that $E(y, u) = 0$ has a (locally) unique solution $y = y(u) \in X_1$ for any given $u \in \mathcal{C} \subset X_2$. Then Problem (5.2) is equivalent to Problem (5.1) with $F(u) = F(y(u), u)$. In general, the equality constraint $E(y, u) = 0$ is governed by partial differential equations.

Now we give two examples illustrating the abstract setup, one elliptic problem and one parabolic problem.

Example 5.1. This example is concerned with inverse medium problems for elliptic equations. For the inverse medium problem, we have

$$F(u) = \|Cy(u) - z\|_Y^2 \quad \text{and} \quad \psi = \psi(u),$$

where $C \in \mathcal{L}(X, Y)$ is an observation operator, typically a trace operator or restriction to a subdomain, and $z \in Y$ is the observational data. For example, for the inverse medium scattering problem, the equality constraint $E(y, u) = 0$ represents the Helmholtz equation

$$E(y, u) = \Delta y + n^2 k^2 y = 0,$$

where k is the wave number and $u = n^2$ (with $1 \leq u \leq u_{\max}$ almost everywhere) represents the refractive index of the medium. In the inverse potential problem, $E(y,u) = 0$ represents the Schrodinger equation

$$E(y,u) = -\Delta y + Vy = 0,$$

where $u = V$ is the spatially dependent potential function. For the inverse conductivity problem, e.g., electrical impedance tomography, it represents the conductivity equation

$$E(y,u) = -\nabla \cdot (\sigma \nabla y) = 0,$$

together with suitable boundary conditions, where $u = \sigma$ is the electrical conductivity of the medium. Due to physical constraints, we usually assume the medium function u belongs to the bilateral constraint set $\mathcal{C} = \{0 < u_{\min} \leq u \leq u_{\max}\}$.

For the control problem, the equality constraint is often given by

$$E_0 y + f(y,u) = 0, \tag{5.3}$$

where the operator $E_0 \in \mathcal{L}(X_0, X_0^*)$, with X_0 being a closed subspace of $L^2(\Omega)$, is boundedly invertible. We refer the existence, uniqueness, and regularity of weak solutions y to (5.3) for a given $u \in \mathcal{C}$ to classical PDE theory, cf. e.g., [294, 106].

Example 5.2. This example is concerned with optimal control of the heat equation. Let $Q = (0,T) \times \Omega$, with $\Omega \subset \mathbb{R}^{\mathrm{d}}$ being an open bounded domain. The functional F and equality constraint E are respectively given by

$$F(y,u) = \int_0^T (\ell(y(t)) + h(u(t)))\, dt + G(y(T)),$$

and

$$E_0 y = \frac{\partial}{\partial t} y - \operatorname{div}(\sigma \nabla y + b\, y).$$

Here the space X_0 is given by $X_0 = \{y \in L^2(0,T; H^1(\Omega)),\, y_t \in L^2(0,T; (H^1(\Omega))^*)\}$, and the functionals ℓ, h and G in the performance index F are nonnegative and Lipschitz continuous. Specifically, for the boundary control problem

$$\nu \cdot (\sigma \nabla y + by) = c(y)u \quad \text{on } \partial\Omega,$$

where ν is the unit outward normal to the boundary $\partial\Omega$, we have for a.e. $t \in (0,T)$ and all $\phi \in H^1(\Omega)$ such that

$$\begin{aligned}&\langle E_0 y + f(y,u), \phi \rangle_{(H^1(\Omega))^* \times H^1(\Omega)} \\ &= \langle \frac{\partial y}{\partial t}, \phi \rangle + \int_\Omega (\sigma \nabla y + by, \nabla \phi) d\omega + \int_{\partial\Omega} c(y) u \phi ds.\end{aligned}$$

In the formulation, σ, b and c are known medium parameters, and $u \in L^2(0,T; L^2(\partial\Omega))$ is the control function.

5.1 Existence and necessary optimality condition

Now we discuss the existence of minimizers to problem (5.1) and derive the necessary optimal condition.

5.1.1 *Existence of minimizers*

First we discuss the existence of minimizers to problem (5.1). We make the following assumption.

Assumption 5.1. Problem (5.1) satisfies the following assumptions:

(i) It is feasible, i.e., there exists an element $x_0 \in \mathcal{C}$ such that $E(x_0) = 0$.

(ii) Either $\mathcal{C}$ is bounded in X or J is coercive, i.e.,
$$J(x) \to \infty \text{ as } |x| \to \infty, \quad x \in \mathcal{C} \text{ satisfying } E(x) = 0.$$

(iii) The operator E is weakly (weakly star) continuous, i.e., for $x_n \to \bar{x}$ weakly/weakly $*$ in X, there holds
$$E(x_n) \to E(\bar{x}) \quad \text{weakly in } Y.$$

(iv) $J(x)$ is weakly (weakly $*$) lower sequentially semi-continuous, i.e. for any sequence $x_n \to \bar{x}$ weakly/weakly $*$ in X, there holds $J(\bar{x}) \leq \liminf_{n\to\infty} J(x_n)$.

Remark 5.1. If the functionals F and ψ are convex, then J is weakly (weakly star) lower sequentially semi-continuous [16].

Now we can state an existence result.

Theorem 5.1. *Let Assumption 5.1 hold. Then there exists a minimizer of problem* (5.1).

Proof. By Assumption 5.1(i), there exists a minimizing sequence $\{x_n\} \subset \mathcal{C}$ satisfying $E(x_n) = 0$ such that
$$\lim_{n\to\infty} J(x_n) = \inf_{x\in\mathcal{C},\ E(x)=0} J(x),$$
and further it is uniformly bounded by Assumption 5.1(ii). Hence, there exists a subsequence, also denoted by $\{x_n\}$, that converges weakly/weakly star to some $\bar{x} \in \mathcal{C}$ (by the convexity and closedness of $\mathcal{C}$). By Assumption 5.1(iii), $E(\bar{x}) = 0$. Now by the weak lower semicontinuity of J from Assumption 5.1(iv), there holds
$$J(\bar{x}) \leq \liminf_{n\to\infty} J(x_n) = \inf_{x\in\mathcal{C},\ E(x)=0} J(x),$$
i.e., $\bar{x}$ is a minimizer. □

5.1.2 *Necessary optimality*

Now we derive the necessary optimality condition for problem (5.1). First we consider the case in the absence of the equality constraint $E(x) = 0$. We shall assume that the functional F is C^1 continuous and ψ is convex, i.e., for all $x_1, x_2 \in X$ and $t \in [0,1]$:

$$\psi((1-t)x_1 + x_2) \leq (1-t)\psi(x_1) + t\psi(x_2).$$

Then we have a variational inequality as the necessary optimality.

Theorem 5.2. *Let $x^* \in \mathcal{C}$ be a minimizer to problem (5.1), F be C^1 continuous and ψ be convex. Then there holds for all $x \in \mathcal{C}$*

$$\langle F'(x^*), x - x^* \rangle + \alpha(\psi(\Lambda x) - \psi(\Lambda x^*)) \geq 0.$$

Proof. Let $x_t = x^* + t\,(x - x^*)$ for all $x \in \mathcal{C}$ and $0 \leq t \leq 1$. By the convexity of the set $\mathcal{C}$, $x_t \in \mathcal{C}$. The optimality of $x^* \in \mathcal{C}$ yields

$$F(x_t) - F(x^*) + \alpha(\psi(\Lambda x_t) - \psi(x^*)) \geq 0.$$

By the convexity of ψ, we have

$$\psi(\Lambda x_t) - \psi(\Lambda x^*) \leq t\,(\psi(\Lambda x) - \psi(\Lambda x^*)).$$

Meanwhile, by the differentiability of F, there holds

$$\frac{F(x_t) - F(x^*)}{t} \to F'(x^*)(x - x^*) \quad \text{as } t \to 0^+.$$

Now the desired variational inequality follows by letting $t \to 0^+$. □

We illustrate Theorem 5.2 with $L^1(\Omega)$ optimization.

Example 5.3. Let U be a closed convex subset in $\mathbb{R}$. We consider the following minimization problem on $X = L^2(\Omega)$ and $\mathcal{C} = \{u \in U \text{ a.e. in } \Omega\}$:

$$\min_{u \in \mathcal{C}} \tfrac{1}{2}\|Ku - b\|_Y^2 + \alpha \int_\Omega |u(\omega)|\, d\omega,$$

where the operator $K \in \mathcal{L}(X, Y)$. Then by Theorem 5.2, the optimality condition is given by:

$$(Ku^* - b, K(u - u^*))_Y + \alpha \int_\Omega (|u| - |u^*|)\, d\omega \geq 0 \quad \forall u \in \mathcal{C}. \tag{5.4}$$

Next we derive a pointwise representation of the optimality condition. To this end, we let $p = -K^*(Ku^* - b) \in X$ and

$$u = \begin{cases} v, & |\omega - \bar{\omega}| \leq \delta, \\ u^*, & \text{otherwise,} \end{cases}$$

where the scalar $v \in U$, $\delta > 0$ is a small number, and the point $\bar{\omega} \in \Omega$ is fixed. It follows from (5.4) that

$$\frac{1}{\delta} \int_{|\omega - \bar{\omega}| \leq \delta} (-p(v - u^*(\omega)) + \alpha (|v| - |u^*(\omega)|) \, d\omega \geq 0.$$

Now suppose $u^* \in L^1(\Omega)$. Then by letting $\delta \to 0^+$ we obtain from Lebesgue's theorem that

$$-p(v - u^*(\bar{\omega})) + \alpha (|v| - |u^*(\bar{\omega})|) \geq 0$$

holds almost everywhere $\bar{\omega} \in \Omega$ (Lebesgue points of $|u^*|$ and for all $v \in U$). That is, $u^*(\bar{\omega}) \in U$ minimizes

$$-pu + \alpha |u| \quad \text{over } u \in U. \tag{5.5}$$

First we consider the case $U = \mathbb{R}$. Then (5.5) implies that if $|p| < \alpha$, then $u^* = 0$ and otherwise $p = \alpha \frac{u^*}{|u^*|} = \alpha \operatorname{sign}(u^*)$. Thus, we obtain the following pointwise necessary optimality

$$(K^*(Ku^* - b))(\omega) + \alpha \, \partial |u^*|(\omega) = 0, \quad \text{a.e. } \omega \text{ in } \Omega,$$

where the subdifferential $\partial |u|$ of $|u|$ is defined by

$$\partial |u|(\omega) = \begin{cases} 1, & u(\omega) > 0, \\ [-1, 1], & u(\omega) = 0, \\ -1, & u(\omega) < 0. \end{cases}$$

Next, we consider the case $U = [-1, 1]$. Then (5.5) has the optimality condition $u^* \in \Psi(p)$ defined by

$$\Psi(p) = \begin{cases} 0, & |p| < \alpha, \\ [0, 1], & p = \alpha, \\ 1, & p > \alpha, \\ [-1, 0], & p = -\alpha, \\ -1, & p < -\alpha. \end{cases} \tag{5.6}$$

In general, the optimality condition (5.5) is a monotone graph $u^* \in \Psi(p)$.

Next we consider the constrained case, i.e., $x = (y, u)$, $E(y, u) = 0$ and $F(x) = F(y)$, where $y \in X_1$ denotes the state and $u \in \mathcal{C} \subset X_2$ represents the medium parameters/control/design variables, i.e.,

$$\min_{u \in \mathcal{C}} \; F(y) + \alpha \psi(\Lambda u) \text{ subject to } E(y, u) = 0. \tag{5.7}$$

We shall assume that at $x^* = (y^*, u^*) \in X_1 \times X_2$, the map

$$E_y(y^*, u^*) : X_1 \to Y = X_1^*$$

is bijective. Then by the implicit function theorem, there exists a unique solution graph $y = y(u)$ in a neighborhood of (y^*, u^*) for $E(y, u) = 0$. Consequently, one can rewrite problem (5.7) as

$$\min_{u \in \mathcal{C}} F(y(u)) + \alpha\psi(u).$$

Next we introduce the associated Lagrange functional for $\lambda \in X_1$

$$L(y, u, \lambda) = F(y) + \alpha\psi(\Lambda u) + \langle \lambda, E(y, u) \rangle.$$

Now we can state the fundamental Lagrange calculus.

Lemma 5.1. *Let $E_y(y, u)$ be boundedly invertible, and let $\widehat{F}(u) = F(y(u))$. Then for all $d \in X_2$, there holds*

$$\widehat{F}'(u)(d) = \langle \lambda, E_u(y, u)(d) \rangle,$$

where the adjoint $\lambda \in X_1$ satisfies

$$L_y(y, u, \lambda) = (E_y(y, u))^*\lambda + F'(y) = 0.$$

Proof. It follows from the implicit function theorem that

$$(\widehat{F}'(u), d) = (F'(y), \dot{y}),$$

where $\dot{y}$ satisfies $E_y(y, u)\dot{y} + E_u(y, u)d = 0$. Thus the claim follows from

$$(F'(y), \dot{y}) = -((E_y(y, u))^*\lambda, \dot{y}) = -\langle \lambda, E_y(y, u)\dot{y} \rangle = \langle \lambda, E_u(y, u)d \rangle. \qquad \square$$

Thus, we obtain the following necessary optimality condition.

Corollary 5.1. *Let $u^* \in \mathcal{C}$ be a minimizer of (5.7), then there holds for all $u \in \mathcal{C}$*

$$\langle E_u(y^*, u^*)^*\lambda, u - u^* \rangle + \alpha(\psi(\Lambda u) - \psi(\Lambda u^*)) \geq 0.$$

Now we illustrate the necessary optimality condition with two examples.

Example 5.4. This example illustrates the necessary optimality condition with one inverse medium problem, i.e., electrical impedance tomography. Let $\Omega \subset \mathbb{R}^d$, $d = 2, 3$, be an open bounded domain. We consider the following variational formulation:

$$\min \quad \frac{1}{2} \sum_{k=1}^{N} \int_{\partial\Omega} |f_k - u_k|^2 ds + \alpha\psi(\Lambda\sigma)$$

subject to

$$\begin{aligned} \nabla \cdot (\sigma \nabla u_k) &= 0, \quad \text{in } \Omega, \\ \sigma \tfrac{\partial}{\partial \nu} u_k &= g_k, \quad \text{on } \partial\Omega, \end{aligned}$$

and over the admissible set $\mathcal{C} = \{0 < \underline{\sigma} \le \sigma \le \bar{\sigma} < \infty, \ a.e.\}$. Here f_k is the (noisy) voltage measurement corresponding to the kth applied current g_k, $1 \le k \le N$. The necessary optimality can be written as

$$\begin{aligned}
&\sum_{k=1}^{N} \nabla u_k \cdot \nabla p_k + \alpha \Lambda^* \mu + \lambda = 0, \qquad \mu \in \partial\psi(\Lambda\sigma),\\
&\lambda = \max(0, \lambda + c\,(\sigma - \bar{\sigma})) + \min(0, \lambda + c\,(\sigma - \underline{\sigma})),\\
&\nabla \cdot (\sigma \nabla p_k) = 0 \text{ in } \Omega,\\
&\sigma \frac{\partial}{\partial \nu} p_k = f_k - u_k \text{ at } \partial\Omega,
\end{aligned}$$

where the constant $c > 0$ is arbitrary and the complementarily condition on (σ, λ) is for the bilateral constraint $\sigma \in \mathcal{C}$.

Example 5.5. This example is concerned with an inverse problem in heat conduction of determining the heat transfer coefficient/radiation coefficient γ in the heat equation for $u = u(t,x)$, $(t,x) \in (0,T) \times \Omega$:

$$\begin{cases}
\dfrac{\partial}{\partial t} u - \Delta u = 0, & \text{in } \Omega,\\[2mm]
\qquad \dfrac{\partial u}{\partial \nu} = \gamma u, & \text{on } \Gamma_1,\\[2mm]
\qquad \dfrac{\partial u}{\partial \nu} = 0, & \text{on } \Gamma_2,\\[2mm]
\qquad u(0) = u_0, & \text{in } \Omega,
\end{cases}$$

where $\Omega \subset \mathbb{R}^{\mathrm{d}}, \mathrm{d} = 1, 2, 3$, is an open bounded domain, Γ_1 and Γ_2 are two disjoint parts of the boundary $\partial\Omega$, and u_0 is the initial data. We formulate the following optimization problem

$$\min \ \tfrac{1}{2} \int_0^T \int_{\Gamma_2} |f - u|^2 ds dt + \alpha \psi(\Lambda\gamma)$$

subject to the above equation and the constraint $\gamma \in \mathcal{C} = \{\gamma \ge 0 \text{ a.e. in } \Gamma_1\}$, where f is a (noisy) measurement of the temperature $u(t,x)$ on the lateral boundary $(0,T) \times \Gamma_2$. The necessary optimality condition can be written as

$$\begin{cases}
up + \alpha \Lambda^* \mu + \lambda = 0,\\
\mu \in \partial\psi(\Lambda\gamma),\\
\lambda = \max(0, \lambda - c\,\gamma),
\end{cases}$$

where the adjoint variable p satisfies

$$\begin{cases} -\dfrac{\partial}{\partial t}p + \Delta p = 0, \quad \text{in } \Omega, \\ \dfrac{\partial p}{\partial \nu} = \gamma p, \quad \text{on } \Gamma_1, \\ \dfrac{\partial p}{\partial \nu} = u - f, \quad \text{on } \Gamma_2, \\ p(T) = 0, \quad \text{in } \Omega. \end{cases}$$

In general, we have the following Lagrange multiplier theorem [163].

Theorem 5.3. *Let $x^* \in \mathcal{C}$ be a minimizer of problem (5.1). Assume that the regular point condition holds at x^*, i.e., $0 \in int\{E'(x^*)(x-x^*) : x \in \mathcal{C}\}$ and that ψ is C^1 continuous. Then there exists a Lagrange multiplier $\lambda \in X$ such that for all $x \in \mathcal{C}$:*

$$\langle (E'(x^*))^*\lambda + F'(x^*) + \alpha\Lambda^*\psi'(\Lambda x^*), x - x^*\rangle \geq 0, \quad E(x^*) = 0.$$

Meanwhile, if ψ is convex and the regular point condition holds, then there exists a Lagrange multiplier $\lambda \in X$ such that for all $x \in \mathcal{C}$:

$$\begin{aligned} &\langle (E'(x^*))^*\lambda + F'(x^*) + \alpha\Lambda^*\mu, x - x^*\rangle \geq 0, \\ &\mu \in \partial\psi(\Lambda x^*), \quad E(x^*) = 0. \end{aligned}$$

5.2 Nonsmooth optimization algorithms

In this section, we describe several algorithms for nonsmooth optimization, including augmented Lagrangian method, exact penalty method, Gauss-Newton method and semismooth Newton method. The implementation of these algorithms will be discussed.

5.2.1 *Augmented Lagrangian method*

First we discuss the augmented Lagrangian method for the following constrained minimization problem

$$\min_{x \in \mathcal{C}} F(x) \quad \text{subject to } E(x) = 0, \; G(x) \leq 0. \tag{5.8}$$

The inequality constraint $G(x) \leq 0$ can be fairly general. For example, a polygonal constraint on x can be written as

$$G(x) = Gx - c \leq 0$$

for some $G \in \mathcal{L}(X, Z)$.

First we consider the case of equality constraint $E(x) = 0$. Then the augmented Lagrangian functional $L_c(x, \lambda)$ reads

$$L_c(x, \lambda) = F(x) + (\lambda, E(x)) + \frac{c}{2}\|E(x)\|^2,$$

where the constant $c > 0$ is arbitrary.

The augmented Lagrangian method is an iterative method, and it consists of the update step

$$\begin{aligned} x_{n+1} &= \text{argmin}_{x \in \mathcal{C}} L_c(x, \lambda_n), \\ \lambda_{n+1} &= \lambda_n + c\, E(x_n). \end{aligned}$$

The augmented Lagrangian method is a hybridization of the multiplier method ($c = 0$) and the penalty method ($\lambda = 0$). It can be shown [163] that the method is locally convergent provided that the Hessian $L_c''(x, \lambda)$ (with respect to x) is uniformly positive near the solution pair $(\bar{x}, \bar{\lambda})$. We observe that there holds

$$L_c''(\bar{x}, \bar{\lambda}) = F''(\bar{x}) + (\bar{\lambda}, E''(\bar{x})) + c\, E'(\bar{x})^* E'(\bar{x}).$$

Therefore, the augmented Lagrangian method has an enlarged convergent basin. In contrast to the penalty method, it is not necessary to let $c \to \infty$ in the augmented Lagrangian functional $L_c(x, \lambda)$ since the Lagrange multiplier update speeds up the convergence.

Next for the case of inequality constraint $G(x) \leq 0$, we consider the equivalent formulation:

$$\min_{(x,z) \in \mathcal{C} \times Z} F(x) + \frac{c}{2}\|G(x) - z\|^2 + (\mu, G(x) - z)$$

subject to

$$G(x) = z \quad \text{and} \quad z \leq 0.$$

Minimizing this objective functional with respect to z over the constraint set $\{z \leq 0\}$ yields that $z^* = \min(0, \frac{\mu + c\, G(x)}{c})$ attains the minimum. Consequently, we obtain

$$\min_{x \in \mathcal{C}} F(x) + \frac{1}{2c}(\| \max(0, \mu + c\, G(x))\|^2 - \|\mu\|^2).$$

Based on these discussions, for the general case (5.8), given (λ, μ) and $c > 0$, we define the following augmented Lagrangian functional

$$\begin{aligned} L_c(x, \lambda, \mu) = F(x) &+ (\lambda, E(x)) + \tfrac{c}{2}\|E(x)\|^2 \\ &+ \frac{1}{2c}(\| \max(0, \mu + c\, G(x))\|^2 - \|\mu\|^2). \end{aligned}$$

Algorithm 5.1 First-order augmented Lagrangian method

1: Initialize (λ^0, μ^0) and set $n = 0$
2: Let x_n be a solution to
$$\min_{x \in \mathcal{C}} L_c(x, \lambda_n, \mu_n).$$
3: Update the Lagrange multipliers
$$\lambda_{n+1} = \lambda_n + c\,E(x_n),$$
$$\mu_{n+1} = \max(0, \mu_n + c\,G(x_n)).$$
4: Stop or set $n = n + 1$ and return to step 1.

With the augmented Lagrangian functional $L_c(x, \lambda, \mu)$ at hand, we can now describe a first-order augmented Lagrangian method, cf. Algorithm 5.1. In essence, it is a sequential minimization of $L_c(x, \lambda, \mu)$ over $x \in \mathcal{C}$.

We have following remark concerning Algorithm 5.1.

Remark 5.2. Let the value function Φ be defined by $\Phi(\lambda, \mu) = \min_{x \in \mathcal{C}} L_c(x, \lambda, \mu)$. Then, one can show that the gradients Φ_λ and Φ_μ satisfy
$$\Phi_\lambda = E(x) \quad \text{and} \quad \Phi_\mu = \max(0, \mu + c\,G(x)).$$
Therefore, the Lagrange multiplier update in Algorithm 5.1 is a gradient ascent method for maximizing $\Phi(\lambda, \mu)$.

Now we examine the constraints more closely. First, for the equality constraint case, i.e., $E(x) = 0$, we have
$$L_c(x, \lambda) = F(x) + (\lambda, E(x)) + \frac{c}{2}\|E(x)\|^2,$$
and
$$L_c'(x, \lambda) = L_0'(x, \lambda + c\,E(x)).$$
The necessary optimality for the saddle point problem
$$\max_{\lambda} \min_{x} \; L_c(x, \lambda)$$
is given by
$$L_c'(x, \lambda) = L_0'(x, \lambda + c\,E(x)) = 0 \quad \text{and} \quad E(x) = 0.$$
A natural idea to achieve faster convergence is to apply the Newton method to the system. Hence, we obtain the Newton updates $\delta x = x^+ - x$ and $\delta\lambda = \lambda^+ - \lambda$ from
$$\begin{pmatrix} L_0''(x, \lambda + c\,E(x)) + cE'(X)^*E'(x) & E'(x) \\ E'(x) & 0 \end{pmatrix} \begin{pmatrix} \delta x \\ \delta\lambda \end{pmatrix} = -\begin{pmatrix} L_0'(x, \lambda + c\,E(x)) \\ E(x) \end{pmatrix}.$$

Next we turn to the case of inequality constraint $G(x) \leq 0$ (e.g., $Gx - \tilde{c} \leq 0$ for some $G \in \mathcal{L}(X, Z)$). Then the augmented Lagrangian functional $L_c(x, \mu)$ is given by

$$L_c(x, \mu) = F(x) + \frac{1}{2c}(\| \max(0, \mu + c\, G(x))\|^2 - \|\mu\|^2),$$

and the necessary optimality condition is given by

$$\begin{aligned} F'(x) + G'(x)^* \mu &= 0, \\ \mu &= \max(0, \mu + c\, G(x)), \end{aligned}$$

where the constant $c > 0$ is arbitrary. Based on the complementarity condition

$$\mu = \max(0, \mu + c\, G(x)),$$

we can derive a primal-dual active set method, cf. Algorithm 5.2, for the special case $G(x) = Gx - \tilde{c}$. In the algorithm, $G_{\mathcal{A}} = \{G_k\},\ k \in \mathcal{A}$. For a general nonlinear operator G, we take

$$G'(x)(x^+ - x) + G(x) = 0 \text{ on } \mathcal{A} = \{k : (\mu + c\, G(x))_k > 0\},$$

where we have linearized the nonlinear map $G(x)$ at the current iterate x.

Algorithm 5.2 Primal-dual active set method

1: Define the active index and inactive index by

$$\begin{aligned} \mathcal{A} &= \{k : (\mu + c\,(Gx - \tilde{c}))_k > 0\}, \\ \mathcal{I} &= \{j : (\mu + c\,(Gx - \tilde{c}))_j \leq 0\}. \end{aligned}$$

2: Let $\mu^+ = 0$ on $\mathcal{I}$ and $Gx^+ = \tilde{c}$ on $\mathcal{A}$.

3: Solve for (x^+, μ^+)

$$F''(x)(x^+ - x) + F'(x) + G_{\mathcal{A}}^* \mu^+ = 0, \quad G_{\mathcal{A}} x^+ - \tilde{c} = 0$$

4: Stop or set $n = n + 1$ and return to step 1,

The primal dual active set method is a semismooth Newton method. That is, if we define a generalized (Newton) derivative of $s \to \max(0, s)$ by 0 on $(-\infty, 0]$ and 1 on $(0, \infty)$. Note that $s \to \max(0, s)$ is not differentiable at $s = 0$ and we define the derivative at $s = 0$ as the limit of the derivatives from $s < 0$. So, we select the generalized derivative of $\max(0, s)$ as $0,\ s \leq 0$ and $1,\ s > 0$. Hence the generalized Newton update is given by

$$\begin{aligned} \mu^+ - \mu + 0 = 0, &\quad \text{if } \mu + c\,(Gx - \tilde{c}) \leq 0, \\ G(x^+ - x) + Gx - \tilde{c} = Gx^+ - \tilde{c} = 0, &\quad \text{if } \mu + c\,(Gx - \tilde{c}) > 0, \end{aligned}$$

which results in the following active set strategy

$$\mu_j^+ = 0, \ \ j \in \mathcal{I} \quad \text{and} \quad (Gx^+ - \tilde{c})_k = 0, \ \ k \in \mathcal{A}.$$

Later in Section 5.2.5, we will introduce a class of semismooth functions and the semismooth Newton method in function spaces. In particular, the pointwise (coordinate) operation $s \to \max(0, s)$ defines a semismooth function from $L^p(\Omega) \to L^q(\Omega)$, $p > q > 1$.

5.2.2 *Lagrange multiplier theory*

Now we present the Lagrange multiplier theory for nonsmmoth optimization problems of the form

$$\min_{x \in \mathcal{C}} f(x) + \alpha\varphi(\Lambda x). \tag{5.9}$$

The nonsmoothness is represented by the convex functional φ. We describe the Lagrange multiplier theory to deal with the nonsmoothness of the functional φ. To this end, we first note that (5.9) is equivalent to

$$\begin{aligned} &\min \quad f(x) + \alpha\varphi(\Lambda x - u) \\ &\text{subject to } x \in \mathcal{C} \quad \text{and} \quad u = 0 \text{ in } H. \end{aligned}$$

Upon treating the equality constraint $u = 0$ by the augmented Lagrangian method, we arrive at the following minimization problem

$$\min_{x \in \mathcal{C}, u \in H} f(x) + \alpha \left\{ \varphi(\Lambda x - u) + (\lambda, \, u)_H + \frac{c}{2}\|u\|_H^2 \right\},$$

where $\lambda \in H$ is a Lagrange multiplier and $c > 0$ is a penalty parameter. Next we can rewrite the optimization problem as

$$\min_{x \in C} \ L_c(x, \lambda) = f(x) + \alpha\varphi_c(\Lambda x, \lambda), \tag{5.10}$$

where the functional φ_c is defined by

$$\varphi_c(u, \lambda) = \inf_{v \in H} \left\{ \varphi(u - v) + (\lambda, \, v)_H + \frac{c}{2}\|v\|_H^2 \right\}.$$

For u, $\lambda \in H$ and $c > 0$, the functional $\varphi_c(u, \lambda)$ is called the generalized Yoshida-Moreau approximation of φ. It can be shown that $u \to \varphi_c(u, \lambda)$ is continuously Fréchet differentiable with a Lipschitz continuous derivative. Then, the augmented Lagrangian functional [161–163] of (5.9) is given by

$$L_c(x, \lambda) = f(x) + \alpha\varphi_c(\Lambda x, \lambda).$$

Let φ^* be the convex conjugate of φ, i.e.,

$$\varphi(z) = \sup_{y \in H} \left\{ (y, z) - \varphi^*(y) \right\}.$$

Then the Yoshida-Moreau approximation φ_c can be characterized as

$$\begin{aligned}\varphi_c(u,\lambda) &= \inf_{z\in H} \{\varphi(z) + (\lambda, u-z)_H + \tfrac{c}{2}\|u-z\|_H^2\} \\ &= \inf_{z\in H} \left\{ \sup_{y\in H}\{(y,z) - \varphi^*(y)\} + (\lambda, u-z)_H + \tfrac{c}{2}\|u-z\|_H^2 \right\} \\ &= \sup_{y\in H} \left\{-\tfrac{1}{2c}\|y-\lambda\|^2 + (y,u) - \varphi^*(y)\right\}.\end{aligned} \tag{5.11}$$

Therefore, there holds

$$\begin{aligned}\varphi_c'(u,\lambda) = y_c(u,\lambda) &= \mathrm{argmax}_{y\in H}\{-\tfrac{1}{2c}\|y-\lambda\|^2 + (y,u) - \varphi^*(y)\} \\ &= \lambda + c\, v_c(u,\lambda),\end{aligned}$$

where

$$v_c(u,\lambda) = \mathrm{armin}_{v\in H}\{\varphi(u-v) + (\lambda, v) + \frac{c}{2}\|v\|^2\}.$$

Let $x_c \in \mathcal{C}$ be the solution to (5.10). Then it satisfies

$$\begin{aligned}&\langle f'(x_c) + \Lambda^*\lambda_c, x - x_c\rangle_{X^*,X} \geq 0, \quad \text{for all } x \in \mathcal{C} \\ &\lambda_c = \varphi_c'(\Lambda x_c, \lambda).\end{aligned}$$

Under appropriate conditions [161–163], the pair $(x_c, \lambda_c) \in \mathcal{C} \times H$ has a (strong-weak) cluster point $(\bar{x}, \bar{\lambda})$ as $c \to \infty$ such that $\bar{x} \in \mathcal{C}$ is the minimizer of (5.9) and that $\bar{\lambda} \in H$ is a Lagrange multiplier in the sense that

$$\langle f'(\bar{x}) + \Lambda^*\bar{\lambda}, x - \bar{x}\rangle_{X^*,X} \geq 0, \quad \text{for all } x \in \mathcal{C}, \tag{5.12}$$

with the complementarity condition

$$\bar{\lambda} = \varphi_c'(\Lambda\bar{x}, \bar{\lambda}), \quad \text{for each } c > 0. \tag{5.13}$$

System (5.12)–(5.13) is a coupled nonlinear system for the primal-dual variable $(\bar{x}, \bar{\lambda})$. Here the distinct feature is that the frequently employed differential inclusion $\bar{\lambda} \in \partial\varphi(\Lambda\bar{x})$ is replaced by the equivalent nonlinear equation (5.13). With the help of system (5.12)-(5.13), we can give a first-order augmented Lagrangian method for (5.9), cf. Algorithm 5.3.

Algorithm 5.3 First-order augmented Lagrangian method

1: Select λ^0 and set $n = 0$
2: Let $x_n = \mathrm{argmin}_{x\in\mathcal{C}} L_c(x, \lambda_n)$
3: Update the Lagrange multiplier by $\lambda_{n+1} = \varphi_c(\Lambda x_n, \lambda_n)$.
4: Stop or set $n = n + 1$ and return to step 1.

In many applications, the convex conjugate functional φ^* of φ is given by an indicator function, i.e.,

$$\varphi^*(v) = I_{K^*}(v),$$

where K^* is a closed convex set in H and I_S denotes the indicator function of a set S. It follows from (5.11) that

$$\varphi_c(u,\lambda) = \sup_{y\in K^*} \{-\tfrac{1}{2c}\|y-\lambda\|^2 + (y,u)\}$$

and consequently (5.13) is equivalent to a projection

$$\bar\lambda = \mathrm{Proj}_{K^*}(\bar\lambda + c\,\Lambda\bar x), \tag{5.14}$$

which forms the basis of the augmented Lagrangian method.

Example 5.6. This example is concerned with inequality constraint. Let $H = L^2(\Omega)$ and $\varphi = I_{\mathcal{C}}$ with $\mathcal{C} = \{z \in L^2(\Omega) : z \le 0 \text{ a.e.}\}$, i.e.,

$$\min_{x\in\{x:\,\Lambda x-\tilde c\le 0\}} f(x).$$

In this case

$$\varphi^*(v) = I_{K^*}(v) \text{ with } K^* = -\mathcal{C},$$

and the complementarity relation (5.14) implies that

$$\begin{aligned} &f'(\bar x) + \Lambda^*\bar\lambda = 0\\ &\bar\lambda = \max(0, \bar\lambda + c\,(\Lambda\bar x - \tilde c)). \end{aligned}$$

Example 5.7. This example is concerned with $L^1(\Omega)$ optimization. Let $H = L^2(\Omega)$ and $\varphi(v) = \int_\Omega |v|\,d\omega$, i.e.,

$$\min\ f(x) + \alpha\int_\Omega |\Lambda x|_2\,d\omega.$$

In this case

$$\varphi^*(v) = I_{K^*}(v) \text{ with } K^* = \{z \in L^2(\Omega) : |z|_2 \le 1, \text{ a.e.}\},$$

and the complementarity relation (5.14) implies that

$$\begin{cases} f'(\bar x) + \alpha\Lambda^*\bar\lambda = 0,\\ \bar\lambda = \dfrac{\bar\lambda + c\,\Lambda\bar x}{\max(1, |\bar\lambda + c\,\Lambda\bar x|_2)} \end{cases} \quad \text{a.e..}$$

5.2.3 *Exact penalty method*

Now we consider the exact penalty method for the following inequality constraint minimization

$$\min\ F(x), \quad \text{subject to } Gx \le c,$$

where $G \in \mathcal{L}(X, L^2(\Omega))$. As we have seen earlier in Section 5.2.1, if x^* is a minimizer of the problem, the necessary optimality condition is given by

$$\begin{aligned} &F'(x^*) + G^*\mu = 0, \\ &\mu = \max(0, \mu + Gx^* - c), \quad \text{a.e. in } \Omega. \end{aligned}$$

For any fixed $\alpha > 0$, the penalty method is defined by

$$\min\ F(x) + \alpha\psi(Gx - c) \tag{5.15}$$

where the penalty functional ψ is defined by

$$\psi(y) = \int_\Omega \max(0, y)\, d\omega.$$

The necessary optimality condition reads

$$-F'(x) \in \alpha G^* \partial\psi(Gx - c),$$

where the subdifferential $\partial\psi$ is given by

$$\partial\psi(s) = \begin{cases} 0, & s < 0, \\ [0,1], & s = 0, \\ 1, & s > 0. \end{cases}$$

If the Lagrange multiplier μ satisfies

$$\sup_{\omega\in\Omega} |\mu(\omega)| \le \alpha, \tag{5.16}$$

then $\mu \in \alpha\partial\psi(Gx^* - c)$ and thus x^* satisfies the necessary optimality condition of (5.15). Moreover, if (5.15) has a unique minimizer x in a neighborhood of x^*, then $x = x^*$.

Due to the singularity and the non-uniqueness of the subdifferential of $\partial\psi$, the direct treatment of the condition (5.15) may not be convenient for numerical computation. Hence we define a regularized function $\max_\epsilon(0, s)$, for $\epsilon > 0$, of $\max(0, s)$ defined by

$$\max{}_\epsilon(0, s) = \begin{cases} \frac{\epsilon}{2}, & s \le 0, \\ \frac{1}{2\epsilon}|s|^2 + \frac{\epsilon}{2}, & 0 \le s \le \epsilon, \\ s, & s \ge \epsilon, \end{cases}$$

and consider the regularized problem of (5.15)

$$\min\ J_\epsilon(x) = F(x) + \alpha\psi_\epsilon(Gx - c), \tag{5.17}$$

where the regularized penalty functional ψ_ϵ is given by $\psi_\epsilon(y) = \int_\Omega \max_\epsilon(0, y)\, d\omega$.

We have the following consistency result for the solution x_ϵ to the regularized problem (5.17).

Theorem 5.4. *Let F be weakly lower semi-continuous. Then for any $\alpha > 0$, any weak cluster point of the solutions $\{x_\epsilon\}_{\epsilon>0}$ of the regularized problem (5.17) is a solution to the nonsmooth problem (5.15) as $\epsilon \to 0$.*

Proof. First, we note that $0 \le \max_\epsilon(s) - \max(0, s) \le \frac{\epsilon}{2}$ holds for all $s \in \mathbb{R}$. Let x be a solution to (5.15) and x_ϵ be a solution of the regularized problem (5.17). Then by the optimality of x and x_ϵ, we have

$$\begin{aligned} F(x_\epsilon) + \alpha\psi_\epsilon(x_\epsilon) &\le F(x) + \alpha\psi_\epsilon(x), \\ F(x) + \alpha\psi(x) &\le F(x_\epsilon) + \alpha\psi(x_\epsilon). \end{aligned}$$

Adding these two inequalities yields

$$\psi_\epsilon(x_\epsilon) - \psi(x_\epsilon) \le \psi_\epsilon(x) - \psi(x).$$

Further, we have

$$\begin{aligned} \psi_\epsilon(x_\epsilon) - \psi(\bar{x}) &= \psi_\epsilon(x_\epsilon) - \psi(x_\epsilon) + \psi(x_\epsilon) - \psi(\bar{x}) \\ &\le \psi_\epsilon(x) - \psi(x) + \psi(x_\epsilon) - \psi(\bar{x}). \end{aligned}$$

Consequently, the weak lower semicontinuity of ψ and the weak convergence of the subsequence $\{x_\epsilon\}$ yield

$$\lim_{\epsilon \to 0^+} (\psi_\epsilon(x) - \psi(x) + \psi(x_\epsilon) - \psi(\bar{x})) \le 0.$$

This together with the weakly lower semicontinuity of F implies that for any cluster point $\bar{x}$ of $\{x_\epsilon\}_{\epsilon>0}$, there holds

$$F(\bar{x}) + \alpha\psi(\bar{x}) \le F(x) + \alpha\psi(x),$$

i.e., $\bar{x}$ is a minimizer to (5.15). This concludes the proof of the theorem. □

As an immediate consequence of the theorem, we have:

Corollary 5.2. *If (5.15) has a unique minimizer in a neighborhood of x^* and (5.16) holds, then the sequence $\{x_\epsilon\}_{\epsilon>0}$ converges to the solution x^* of (5.15) as $\epsilon \to 0$.*

The necessary optimality condition for (5.17) is given by

$$F'(x) + \alpha G^* \psi_\epsilon'(Gx - c) = 0. \tag{5.18}$$

The optimality condition (5.18) bypasses the nonuniqueness issue of the subdifferential through regularization, however, it is still nonlinear. To design a numerical method, we first observe that the derivative $\max_\epsilon'(s)$ has an alternative representation, i.e.,

$$\max_\epsilon'(s) = \chi_\epsilon(s)s, \quad \chi_\epsilon(s) = \begin{cases} 0, & s \leq 0, \\ \frac{1}{\max(\epsilon, s)}, & s > 0. \end{cases}$$

This suggests the following semi-implicit fixed point iteration:

$$\begin{aligned} &P(x^{k+1} - x^k) + F'(x^k) + \alpha G^* \chi_k (Gx^{k+1} - c) = 0, \\ &\chi_k = \chi_\epsilon(Gx^k - c), \end{aligned} \tag{5.19}$$

where χ_k is a diagonal operator (matrix), the operator P is positive and self-adjoint and serves as a preconditioner for F'.

Upon letting $d^k = x^{k+1} - x^k$, equation (5.19) for x^{k+1} can be recast into the following equivalent equation for d^k

$$Pd^k + F'(x^k) + \alpha G^* \chi_k (Gx^k - c + Gd^k) = 0,$$

which gives rise to

$$(P + \alpha G^* \chi_k G) d^k = -F_\epsilon'(x^k), \tag{5.20}$$

where the right hand side $F_\epsilon'(x^k)$ is defined by

$$F_\epsilon'(x^k) = F'(x^k) + \alpha G^* \chi_k (Gx^k - c).$$

The direction d^k is a decent direction for $J_\epsilon(x)$ at x^k. Indeed, there holds

$$\begin{aligned} (d_k, J_\epsilon'(x^k)) &= -((P + \alpha G^* \chi_k G) d_k, d^k) \\ &= -(Pd^k, d^k) - \alpha(\chi_k G d^k, G d^k) < 0, \end{aligned}$$

since the operator P is strictly positive definite. So the iteration (5.19) can be seen as a preconditioned gradient decent method; see Algorithm 5.4 for the complete procedure.

Let us make some remarks on Algorithm 5.4. In many applications, F' and G are sparse block diagonal, and the system (5.20) for the direction d^k then becomes a linear system with a sparse symmetric positive-definite matrix; and can be efficiently solved, e.g., by the Cholesky decomposition method. The update (5.19) represents only one of many possible choices. For example, if $F(x) = \frac{1}{2}(x, Ax) - (b, x)$ (with A being symmetric positive

Algorithm 5.4 Algorithm (Fixed point iteration)

1: Set parameters: α, ϵ, P, tol
2: Compute the direction by $(P+\alpha G^*\chi_k G)d^k = -F'_\epsilon(x^k)$
3: Update $x^{k+1} = x^k + d^k$.
4: If $|J'_\epsilon(x^k)| < tol$, then stop. Otherwise repeat Steps 1-3.

definite), then we have $F'(x) = Ax - b$, and we may use the alternative update

$$P(x^{k+1}-x^k) + Ax^{k+1} - b + \alpha G^*\chi_k(Gx^{k+1}-c) = 0,$$

provided that it does not cost much to perform this fully implicit step. In practice, we observe that Algorithm 5.4 is globally convergent practically and the following results justify the fact.

Lemma 5.2. *Let* $R(x,\hat{x}) = -(F(x)-F(\widehat{x})-F'(\widehat{x})(x-\widehat{x}))+(P(x-\widehat{x}),x-\widehat{x})$. *Then, we have*

$$R(x^{k+1},x^k) + F(x^{k+1}) - F(x^k) + \frac{\alpha}{2}(\chi_k Gd^k, Gd^k) + \frac{\alpha}{2}(\chi_k, |Gx^{k+1}-c|^2 - |Gx^k - c|^2) = 0.$$

Proof. Upon taking inner product between (5.19) and $d^k = x^{k+1} - x^k$, we arrive at

$$(Pd^k, d^k) - (F(x^{k+1}) - F(x^k) - F'(x^k)d^k) + F(x^{k+1}) - F(x^k) + E_k = 0,$$

where the term E_k is defined by

$$E_k = \alpha(\chi_k(Gx^{k+1}-c), Gd^k) = \frac{\alpha}{2}(\chi_k Gd^k, Gd^k) + \frac{\alpha}{2}(\chi_k, |Gx^{k+1}-c|^2 - |Gx^k-c|^2),$$

where the last line follows from the elementary identity $2(a,a-b) = \|a-b\|^2 + \|a\|^2 - \|b\|^2$. □

Theorem 5.5. *Assume that* $R(x,\widehat{x}) \geq \omega\|x-\widehat{x}\|^2$ *for some* $\omega > 0$ *and for all* x *and* $\widehat{x}$, *and that all inactive indices remain inactive. Then*

$$\omega\|x^{k+1}-x^k\|^2 + J_\epsilon(x^{k+1}) - J_\epsilon(x^k) \leq 0$$

and the sequence $\{x^k\}$ *is globally convergent.*

Proof. By the concavity of the function $\psi_\epsilon(\sqrt{t})$ and the identity $\psi_\epsilon(\sqrt{t}) = \frac{1}{2\max(\sqrt{t},\epsilon)}$ for $t \geq 0$, we deduce that

$$\psi_\epsilon(\sqrt{s}) - \psi_\epsilon(\sqrt{t}) \leq \frac{s-t}{2\max(\epsilon, \sqrt{t})}, \quad \forall s,t \geq 0.$$

Consequently, by summing up the inequality componentwise we deduce

$$(\chi_k, |(Gx^{k+1} - c)|^2 - |(Gx^k - c)|^2) \geq 2(\psi_\epsilon(Gx^{k+1} - c) - \psi_\epsilon(Gx^k - c)).$$

Thus, we arrive at

$$\begin{aligned} F(x^{k+1}) + \alpha\psi_\epsilon(Gx^{k+1} - c) + \frac{\alpha}{2}(\chi_k Gd^k, Gd^k) \\ + R(x^{k+1}, x^k) \leq F(x^k) + \alpha\psi_\epsilon(Gx^k - c). \end{aligned}$$

Now it follows from the assumption $R(x, \widehat{x}) \geq \omega\|x - \widehat{x}\|^2$ for some $\omega > 0$ that the sequence $\{J_\epsilon(x^k)\}$ is monotonically decreasing and upon summing up over the index k, we arrive at

$$\sum_k \|x^{k+1} - x^k\|^2 < \infty.$$

This completes the proof of the theorem. □

Remark 5.3. The role of the operator P as a preconditioner is reflected in the condition $R(x, \widehat{x}) \geq \omega\|x - \widehat{x}\|^2$ for some $\omega > 0$. In the case $F(x) = \frac{1}{2}(x, Ax) - (x, b)$, A being symmetric positive definite, $R(x, \widehat{x})$ is given by

$$R(x, \widehat{x}) = (P(\widehat{x} - x), \widehat{x} - x) - \tfrac{1}{2}(A(x - \widehat{x}), x - \widehat{x}).$$

In particular, with the choice $P = \frac{1}{2}A + I$, the condition holds.

5.2.4 *Gauss-Newton method*

Consider the regularized nonlinear least L^1 fitting problem:

$$\min \|F(x)\|_{L^1(\Omega)} + \alpha\|x\|_{L^1(\Omega)},$$

subject to the pointwise constraint $|x| \leq 1$ a.e.. If the operator $F : L^2(\Omega) \to L^1(\Omega)$ is weakly continuous, then there exists at least one minimizer to the problem. The necessary optimality is given by

$$\begin{aligned} &(F'(x^*))^*\mu + p = 0, \quad x^* \in \Psi(p), \\ &\mu = \frac{\mu + cF(x^*)}{|\mu + cF(x^*)|}, \text{ a.e.} \end{aligned}$$

where the monotone graph Ψ is defined by (5.6). The Gauss-Newton method is an iterative method, i.e., given x_k the update x_{k+1} minimizes

$$\min \|F'(x_k)(x - x_k) + F(x_k)\|_{L^1(\Omega)} + \alpha\|x\|_{L^1(\Omega)}$$

subject to the pointwise constraint $|x| \leq 1$, a.e., which reads

$$(F'(x_k))^*\mu + p_{k+1} = 0, \quad x_{k+1} \in \Psi(p_{k+1}),$$
$$\mu = \frac{\mu + c\,(F'(x^k)(x^{k+1} - x^k) + F(x^k))}{|\mu + c(F'(x^k)(x^{k+1} - x^k) + F(x^k))|} \quad \text{a.e.}$$

5.2.5 *Semismooth Newton Method*

Now we present the semismooth Newton method for the nonsmooth necessary optimality condition. Consider the nonlinear equation

$$F(y) = 0$$

in a Banach space X. The generalized Newton update is given by

$$y^{k+1} = y^k - V_k^{-1}F(y^k), \tag{5.21}$$

where V_k is a generalized derivative of F at y^k. In the finite dimensional space, for a locally Lipschitz continuous function F, let D_F denote the set of points at which F is differentiable. Rademacher's theorem states that every locally Lipschitz continuous function in a finite dimensional space is differentiable almost everywhere. Hence, D_F is dense. Now for any $x \in X = \mathbb{R}^n$, we define $\partial_B F(x)$ by

$$\partial_B F(x) = \left\{ J : J = \lim_{x_i \to x,\ x_i \in D_F} \nabla F(x_i) \right\}.$$

Thus, we can take $V_k \in \partial_B F(y^k)$ in the generalized Newton method.

In infinite dimensional spaces, notions of generalized derivatives for functions which are not C^1 continuous cannot rely on Rademacher's theorem. Instead, we shall mainly utilize a concept of generalized derivative that is sufficient to guarantee the superlinear convergence of Newton's method [163]. This notion of differentiability is called Newton derivative. We refer to [163, 137, 61, 296] for further discussions of the notions and topics.

Definition 5.1 (Newton differentiability). *Let X and Z be real Banach spaces and $D \subset X$ be an open set.*

(1) $F\colon D \subset X \to Z$ *is called Newton differentiable at* x*, if there exists an open neighborhood* $N(x) \subset D$ *and mappings* $G\colon N(x) \to \mathcal{L}(X,Z)$ *such that*

$$\lim_{\|h\|_X \to 0} \frac{\|F(x+h) - F(x) - G(x+h)h\|_Z}{\|h\|_X} = 0.$$

The family $\{G(y) : y \in N(x)\}$ *is called a* N*-derivative of* F *at* x*.*

(2) F *is called semismooth at* x*, if it is Newton differentiable at* x *and*

$$\lim_{t \to 0^+} G(x + t\,h)h \quad \textit{exists uniformly in } \|h\|_X = 1.$$

Semismoothness was originally introduced in [235] for scalar functions. In finite-dimensional spaces, convex functions and real-valued C^1 functions are examples of semismooth functions [254, 255].

For example, if $F(y)(s) = \psi(y(s))$ point-wise, then $G(y)(s) \in \partial_B \psi(y(s))$ is an N-derivative in $L^p(\Omega) \to L^q(\Omega)$ under appropriate conditions [163]. We often use $\psi(s) = |s|$ and $\psi(s) = \max(0,s)$ in the necessary optimality conditions.

Suppose $F(y^*) = 0$. Then, $y^* = y^k + (y^* - y^k)$ and

$$\begin{aligned}\|y^{k+1} - y^*\| &= \|V_k^{-1}(F(y^*) - F(y^k) - V_k(y^* - y^k)\| \\ &\le \|V_k^{-1}\| o(\|y^k - y^*\|).\end{aligned}$$

Thus, the semismooth Newton method is q-superlinear convergent provided that the Jacobian sequence V_k is uniformly invertible as $y^k \to y^*$. That is, if one can select a sequence of quasi-Jacobian V_k that is consistent, i.e., $|V_k - G(y^*)|$ as $y^k \to y^*$, and uniformly invertible, the generalized Newton method (5.21) is still q-superlinear convergent.

In the rest of this part, we focus on the L^1-type optimization, i.e.,

$$\varphi(v) = \int_\Omega |v|_2 d\omega.$$

The complementarity condition is given by

$$\lambda \max(1 + \epsilon, |\lambda + c\,v|) = \lambda + c\,v, \tag{5.22}$$

for some small $\epsilon \ge 0$. It is often convenient (but not essential) to use a very small $\epsilon > 0$ to avoid the singularity for the implementation of algorithms. To this end, we let

$$\varphi_\epsilon(s) = \begin{cases} \dfrac{s^2}{2\epsilon} + \dfrac{\epsilon}{2}, & \text{if } |s| \le \epsilon, \\ |s|, & \text{if } |s| \ge \epsilon. \end{cases}$$

Then, equation (5.22) corresponds to the complementarity condition for a regularized L^1-optimization:

$$\min\ J_\epsilon(y) = F(y) + \alpha\varphi_\epsilon(\Lambda y).$$

The semismooth Newton update is given by

$$F'(y^+) + \alpha\Lambda^*\lambda^+ = 0,$$

$$\begin{cases} \lambda^+ = \frac{v^+}{\epsilon}, & \text{if } |\lambda + cv| \leq 1+\epsilon, \\ |\lambda + c\,v|\,\lambda^+ + \left(\lambda\left(\frac{\lambda+c\,v}{|\lambda+c\,v|}\right)^t\right)(\lambda^+ + c\,v^+) & \\ \qquad\qquad = \lambda^+ + c\,v^+ + |\lambda + c\,v|\,\lambda, & \text{if } |\lambda + c\,v| > 1+\epsilon. \end{cases}$$

There is no guarantee that the system is solvable for λ^+ and is stable. In order to obtain a compact yet unconditionally stable formula we use the following damped and regularized algorithm with $\beta \leq 1$:

$$\begin{cases} \lambda^+ = \frac{v}{\epsilon}, & \text{if } |\lambda + cv| \leq 1+\epsilon, \\ |\lambda + cv|\lambda^+ - \beta\left(\frac{\lambda}{\max(1,|\lambda|)}\left(\frac{\lambda+cv}{|\lambda+cv|}\right)^t\right)(\lambda^+ + cv^+) & \\ \qquad\qquad = \lambda^+ + cv^+ + \beta|\lambda + c\,v|\frac{\lambda}{\max(1,|\lambda|)}, & \text{if } |\lambda + cv| > 1+\epsilon. \end{cases} \tag{5.23}$$

Here, the purpose of the regularization $\frac{\lambda}{\max(1,|\lambda|)}$ is to constrain the dual variable λ into the unit ball. The damping factor β is selected to achieve the numerical stability. Let

$$d = |\lambda + cv|, \quad \eta = d - 1, \quad a = \frac{\lambda}{\max(1,|\lambda|)}, \quad b = \frac{\lambda + cv}{|\lambda + cv|}, \quad F = ab^t.$$

Then, (5.23) is equivalent to

$$\lambda^+ = (\eta I + \beta F)^{-1}((I - \beta F)(cv^+) + \beta da),$$

where by Sherman–Morrison formula

$$(\eta I + \beta F)^{-1} = \frac{1}{\eta}\left(I - \frac{\beta}{\eta + \beta\,a\cdot b}F\right).$$

Then,

$$(\eta I + \beta F)^{-1}\beta da = \frac{\beta d}{\eta + \beta a \cdot b}a.$$

In view of the relation $F^2 = (a \cdot b)F$,

$$(\eta I + \beta F)^{-1}(I - \beta F) = \frac{1}{\eta}\left(I - \frac{\beta d}{\eta + \beta a \cdot b}F\right).$$

In order to achieve the stability, we set

$$\frac{\beta d}{\eta + \beta a \cdot b} = 1, \ \ i.e., \ \ \beta = \frac{d-1}{d - a \cdot b} \leq 1.$$

Consequently, we obtain a compact Newton step

$$\lambda^+ = \frac{1}{d-1}(I - F)(cv^+) + \frac{\lambda}{\max(1, |\lambda|)}, \tag{5.24}$$

which results in a primal-dual active set method for L^1 regularization, cf. Algorithm 5.5. Our experiences indicate that the algorithm is unconditionally stable and is rapidly convergent [159].

Algorithm 5.5 Primal-dual active set method (L^1-optimization)

1: Initialize: $\lambda^0 = 0$ and solve $F'(y^0) = 0$ for y^0. Set $k = 0$.
2: Set inactive set $\mathcal{I}_k$ and active set $\mathcal{A}_k$ by

$$\mathcal{I}_k = \{|\lambda^k + c\,\Lambda y^k| > 1 + \epsilon\}, \quad \text{and} \quad \mathcal{A}_k = \{|\lambda^k + c\,\Lambda y^k| \leq 1 + \epsilon\}.$$

3: Solve for $(y^{k+1}, \lambda^{k+1}) \in X \times H$:

$$\begin{cases} F'(y^{k+1}) + \alpha \Lambda^* \lambda^{k+1} = 0 \\ \lambda^{k+1} = \dfrac{1}{d^k - 1}(I - F^k)(c\,\Lambda y^{k+1}) + \dfrac{\lambda^k}{\max(1, |\lambda^k|)} \ \text{in } \mathcal{A}_k \quad \text{and} \\ \Lambda y^{k+1} = \epsilon\, \lambda^{k+1} \ \text{in } \mathcal{I}_k. \end{cases}$$

4: Convergent or set $k = k + 1$ and return to Steps 2 and 3.

We have the following two remarks concerning Algorithm 5.5.

Remark 5.4.

(i) If $a \cdot b \to 1^+$, then $\beta \to 1$. Suppose that (λ, v) is a fixed point of the iteration (5.24). Then

$$\left(1 - \frac{1}{\max(1, |\lambda|)}\left(1 - \frac{1}{d-1}\frac{(\lambda + c\,v)\cdot(cv)}{|\lambda + c\,v|}\right)\right)\lambda = \frac{1}{d-1}cv,$$

and thus the angle between λ and $\lambda + c\,v$ is zero. Consequently,

$$1 - \frac{1}{\max(1, |\lambda|)} = \frac{c}{|\lambda| + c\,|v| - 1}\left(\frac{|v|}{|\lambda|} - \frac{|v|}{\max(1, |\lambda|)}\right),$$

which implies $|\lambda| = 1$. It follows that

$$\lambda + c\,v = \frac{\lambda + c\,v}{\max(|\lambda + c\,v|, 1 + \epsilon)}.$$

That is, if the algorithm converges and $a \cdot b \to 1$, then $|\lambda| \to 1$ and the algorithm is consistent.

(ii) Consider the substitution iterate:

$$\begin{cases} F'(y^+) + \alpha\Lambda^*\lambda^+ = 0, \\ \lambda^+ = \dfrac{1}{\max(\epsilon, |v|)} v^+, \quad v = \Lambda y. \end{cases} \tag{5.25}$$

Note that

$$\begin{aligned}\left(\frac{v^+}{|v|}, v^+ - v\right) &= \left(\frac{|v^+|^2 - |v|^2 + |v^+ - v|^2}{2|v|}\right) \\ &= \left(|v^+| - |v| - \frac{(|v^+| - |v|)^2 + |v^+ - v|^2}{2|v|}\right).\end{aligned}$$

In case a quadratic functional F, i.e., $F(y) = \frac{1}{2}\|Ay - b\|^2$, we have

$$J_\epsilon(v^+) \leq J_\epsilon(v),$$

with $J_\epsilon(v) = \int \psi_\epsilon(|v|)d\omega$ and $\psi_\epsilon(t)$

$$\psi_\epsilon(t) = \begin{cases} \frac{1}{2\epsilon}t^2 + \frac{\epsilon}{2}, & 0 \leq t \leq \epsilon, \\ t, & t \geq \epsilon. \end{cases}$$

Further, the equality holds only if $v^+ = v$. This fact can be used to prove the iterative method (5.25) is globally convergent [163, 165]. It also suggests the use of a hybrid method: for $0 < \mu < 1$

$$\lambda^+ = \frac{\mu}{\max(\epsilon, |v|)} v^+ + (1 - \mu)\left(\frac{1}{d-1}(I - F)v^+ + \frac{\lambda}{\max(1, |\lambda|)}\right),$$

in order to enhance the global convergence property without losing the fast convergence of the Newton method.

5.3 ℓ^p sparsity optimization

In this section we consider ℓ^p minimization of the form

$$\tfrac{1}{2}\|Ax - b\|^2 + \alpha\|x\|_{\ell^p}^p,$$

where $A \in \mathcal{L}(\ell^2)$, $b \in \ell^2$, and $p \in [0, 1)$. We shall discuss the cases $p = 0$ and $p \in (0, 1)$, separately, and derive numerical schemes based on regularization, following [165].

5.3.1 ℓ^0 optimization

First we consider the ℓ^0 minimization

$$\tfrac{1}{2}\|Ax-b\|^2+\alpha N_0(x) \tag{5.26}$$

where $N_0(x)$ denotes the number of nonzero elements of x, i.e.,

$$N_0(x)=\sum |x_k|^0,$$

where $|s|^0=1,\ s\neq 0$ and $|0|^0=0$. It can be shown that N_0 is a complete metric.

Theorem 5.6. *Problem* (5.26) *has a solution* $\bar{x}\in\ell^0$.

Proof. Let $N(s)=0$ if $s=0$, and otherwise $N(s)=1$. Let $\{x^n\}$ be a minimizing sequence of (5.26). Since $c'=\ell^1$, there exists a subsequence of $\{N(x^n)\}$ converging in ℓ^1 weakly star to $N(\bar{x})$ and x^n converges to $\bar{x}$ in ℓ^2. Thus, $\lim_{n\to\infty} N_0(x^n)=\langle N(x^n),1\rangle\to\langle N(\bar{x}),1\rangle=N_0(\bar{x})$. Thus, we have

$$\tfrac{1}{2}\|A\bar{x}-b\|^2+\alpha N_0(\bar{x})\le\inf_{x\in\ell_0}\tfrac{1}{2}\|Ax-b\|^2+\alpha N_0(x),$$

and $\bar{x}$ is a minimizer. □

Theorem 5.7. *Let $\bar{x}$ be a minimizer to Problem* (5.26). *Then it satisfies the following necessary optimality condition*

$$\begin{cases}\bar{x}_i=0, & \text{if } |(A_i,f_i)|<\sqrt{2\alpha}|A_i|,\\ (A_i,A\bar{x}-b)=0 & \text{if } |(A_i,f_i)|>\sqrt{2\alpha}|A_i|,\end{cases} \tag{5.27}$$

where $A_i=Ae_i$, $f_i=b-A(\bar{x}-\bar{x}_ie_i)$. For the second case of (5.27), *$|(A_i,f_i)|>\sqrt{2\alpha}|A_i|$ is equivalent to $|\bar{x}_i|>\frac{\sqrt{2\alpha}}{|A_i|}$.*

Proof. By the minimizing property of $\bar{x}\in\ell^0$, we deduce that $\bar{x}_i\in\mathbb{R}$ minimizes

$$G(x_i)=\tfrac{1}{2}|A_ix_i-f_i|^2+\alpha|x_i|^0.$$

If $x_i=z\neq 0$ is a minimizer of $G(x_i)$, then

$$|A_i|^2z-(A_i,f_i)=0,$$

$$G(z)=-\frac{(A_i,f_i)^2}{2|A_i|^2}+\alpha+G(0).$$

First, if $G(z)<G(0)$, i.e., $|(A_i,f_i)|>\sqrt{2\alpha}|A_i|$, then $\bar{x}_i=z$ is the unique minimizer and $|\bar{x}_i|>\frac{\sqrt{2\alpha}}{|A_i|}$. Second, if $|(A_i,f_i)|<\sqrt{2\alpha}|A_i|$, then $\bar{x}_i=0$ is the unique minimizer of $G(x_i)$. Last, if $|(A_i,f_i)|=\sqrt{2\alpha}|A_i|$, then there are two minimizers, i.e., 0 and z. This completes the proof of the theorem. □

It follows from the proof of Theorem 5.7 that a minimizer to (5.26) is not necessarily unique. In general, for a nonlinear functional $F(x)$, we can define

$$G_i(x) = F(x) - F(\tilde{x}_i), \quad \text{where } \tilde{x}_k = x_k,\ k \neq i \text{ and } \tilde{x}_i = 0.$$

This definition in the case of (5.26) recovers $G_i(x) = \frac{1}{2}(|A_i x_i - f_i|^2 - |f_i|^2)$. Further, we have the following necessary optimality condition.

Corollary 5.3. *Let $\bar{x} \in \ell^0$ is a minimizer of*

$$\min\ F(x) + \alpha N_0(x).$$

Then it satisfies the following necessary optimality condition

$$F'(\bar{x}) + \bar{\lambda} = 0, \quad \bar{\lambda}_i \bar{x}_i = 0 \textit{ for all } i.$$

where $\bar{x}$ and $\bar{\lambda}$ satisfy the complementarity condition

$$\begin{cases} \bar{x}_i = 0, & \textit{if } G_i(\bar{x}) \geq \alpha, \\ \bar{\lambda}_i = 0, & \textit{if } G_i(\bar{x}) < \alpha. \end{cases}$$

5.3.2 ℓ^p $(0 < p < 1)$-*optimization*

Now we turn to ℓ^p, $0 < p < 1$, minimization of the form

$$\tfrac{1}{2}\|Ax - b\|^2 + \alpha N_p(x) \tag{5.28}$$

where $N_p(x) = \sum |x_k|^p$. $N_p(x)$ is a complete metric and weakly sequentially lower continuous for $p > 0$ [165], and $N_p(x) \to N_0(x)$ as $p \to 0^+$. Any convergent subsequence of minimizers x_p to (5.28) converges to a minimizer of (5.26) as $p \to 0^+$. The goal of this part is to derive a globally convergent method based on a regularized formulation.

In order to overcome the singularity near $s = 0$ for $(|s|^p)' = \frac{ps}{|s|^{2-p}}$, for $\epsilon > 0$, we consider a regularized problem

$$J_\epsilon(x) = \tfrac{1}{2}\|Ax - b\|^2 + \alpha \Psi_\epsilon(|x|^2), \tag{5.29}$$

where the regularized functional $\Psi_\epsilon(t)$, $t \geq 0$, is defined by

$$\Psi_\epsilon(t) = \begin{cases} \dfrac{p}{2}\dfrac{t}{\epsilon^{2-p}} + (1 - \dfrac{p}{2})\,\epsilon^p, & t \leq \epsilon^2, \\ t^{\frac{p}{2}}, & t \geq \epsilon^2. \end{cases}$$

For any $\epsilon > 0$, we consider the following iterative algorithm for the solution of the regularized problem (5.29):

$$A^* A x^{k+1} + \frac{\alpha p}{\max(\epsilon^{2-p}, |x^k|^{2-p})}\, x^{k+1} = A^* b, \tag{5.30}$$

where the second term on the left hand side is short for the vector component $\frac{\alpha p}{\max(\epsilon^{2-p},|x_i^k|^{2-p})} x_i^{k+1}$. Taking inner product between this identity and $x^{k+1} - x^k$, we arrive at

$$\|Ax^{k+1}\|^2 - \|Ax^k\|^2 + \|A(x^{k+1} - x^k)\|^2 - 2(A^*b, x^{k+1} - x^k) \\ + \sum_i \frac{\alpha p}{\max(\epsilon^{2-p}, |x_i^k|^{2-p})}(|x_i^{k+1}|^2 - |x_i^k|^2 + |x_i^{k+1} - x_i^k|^2) = 0.$$

Now from the relation

$$\frac{1}{\max(\epsilon^{2-p}, |x_i^k|^{2-p})} \frac{p}{2} (|x_i^{k+1}|^2 - |x_i^k|^2) = \Psi_\epsilon'(|x_i^k|^2)(|x_i^{k+1}|^2 - |x_i^k|^2),$$

and the concavity of the map $t \to \Psi_\epsilon(t)$, we deduce

$$\Psi_\epsilon(|x_i^{k+1}|^2) - \Psi_\epsilon(|x_i^k|^2) - \frac{1}{\max(\epsilon^{2-p}, |x_i^k|^{2-p})} \frac{p}{2} (|x_i^{k+1}|^2 - |x_i^k|^2) \leq 0,$$

and thus

$$J_\epsilon(x^{k+1}) + \tfrac{1}{2}\|A(x^{k+1} - x^k)\|^2 \\ + \sum_i \frac{\alpha p}{\max(\epsilon^{2-p}, |x_i^k|^{2-p})} \tfrac{1}{2}|x_i^{k+1} - x_i^k|^2 \leq J_\epsilon(x^k). \tag{5.31}$$

We have the following convergence result:

Theorem 5.8. *For $\epsilon > 0$, let $\{x^k\}$ be generated by the fixed-point iteration (5.30). Then the sequence $\{J_\epsilon(x^k)\}$ is monotonically nonincreasing, and $\{x_k\}$ converges to the minimizer of J_ϵ defined in (5.29).*

Proof. It follows from (5.31) that the sequence $\{x^k\}$ is bounded in ℓ^2 and thus in ℓ^∞. Consequently, from (5.31) there exists $\kappa > 0$ such that

$$J_\epsilon(x^{k+1}) + \tfrac{1}{2}\|A(x^{k+1} - x^k)\|^2 + \kappa\|x^{k+1} - x^k\|^2 \leq J_\epsilon(x^k).$$

This implies the first assertion. Further, it yields

$$\sum_{k=0}^{\infty} \|x^{k+1} - x^k\|^2 < \infty.$$

Now by the uniform boundedness of $\{x^k\}$ in ℓ^2, there exists a subsequence of $\{x^k\}$, also denoted by $\{x^k\}$, and $x^* \in \ell^p$ such that $x^k \to x^*$ weakly. By the above inequality, $\lim_{k\to\infty} x_i^k = x_i^*$. Now testing (5.30) componentwise with e_i gives

$$A^*Ax^* + \frac{\alpha p}{\max(\epsilon^{2-p}, |x^*|^{2-p})} x^* = A^*b,$$

i.e., x^* is a minimizer to J_ϵ. □

The iterative method (5.30) can be used as a globalization step for the semismooth Newton method discussed below.

5.3.3 *Primal-dual active set method*

In this part we develop the augmented Lagrangian formulation and the primal-dual active set strategy for the ℓ^0 optimization (5.26). Let P be a nonnegative self-adjoint operator P and $\Lambda_k = \|A_k\|_2^2 + P_{kk}$. We consider the following augmented Lagrangian functional

$$L(x, v, \lambda) = \tfrac{1}{2}\|Ax - b\|^2 + \tfrac{1}{2}(Px, x) + \alpha \sum_k |v_k|^0 + \sum_k \left(\tfrac{\Lambda_k}{2} |x_k - v_k|^2 + (\lambda_k, x_k - v_k) \right).$$

The purpose of the introducing the regularization functional (x, Px) is to regularize the problem, in case that the operator A is nearly singular.

For a given (x, λ), the functional $L(x, v, \lambda)$ is minimized at

$$v = \Phi(x, \lambda) = \begin{cases} \frac{\lambda_k + \Lambda_k x_k}{\Lambda_k} & \text{if } |\lambda_k + \Lambda_k x_k|^2 > 2\Lambda_k \alpha, \\ 0 & \text{otherwise.} \end{cases}$$

Meanwhile, for a given (v, λ), the functional $L(x, v, \lambda)$ is minimized at x that satisfies

$$A^*(Ax - b) + Px + \Lambda(x - v) + \lambda = 0,$$

where Λ is a diagonal operator with entries Λ_k.

Thus, the augmented Lagrangian method [163] uses the update:

$$\begin{cases} A^*(Ax^{n+1} - b) + Px^{n+1} + \Lambda(x^{n+1} - v^n) + \lambda^n = 0, \\ v^{n+1} = \Phi(x^{n+1}, \lambda^n), \\ \lambda^{n+1} = \lambda^n + \Lambda(x^{n+1} - v^{n+1}). \end{cases}$$

If the method converges, i.e., x^n, $v^n \to x$ and $\lambda^n \to \lambda$, then

$$\begin{cases} A^*(Ax - b) + Px + \lambda = 0, \\ \lambda_k = 0, \quad \text{if} \quad |x_k|^2 > \dfrac{2\alpha}{\Lambda_k}, \\ x_k = 0, \quad \text{if} \quad |\lambda_k|^2 \le 2\alpha\Lambda_k. \end{cases}$$

That is, the limit (x, λ) satisfies the necessary optimality condition (5.27).

Motivated by the augmented Lagrangian formulation we propose the following primal-dual active set method, cf. Algorithm 5.6.

Note that

$$\begin{cases} \lambda_k = 0 & \text{if } k \in \{k : |\lambda_k + \Lambda_k x_k|^2 > 2\alpha\Lambda_k\}, \\ x_j = 0 & \text{if } j \in \{j : |\lambda_j + \Lambda_j x_j|^2 \le 2\alpha\Lambda_j\}, \end{cases}$$

Algorithm 5.6 Primal-dual active set method (sparsity optimization)

1: Initialize: $\lambda^0 = 0$ and x^0 is determined by $A^*(Ax^0 - b) + Px^0 = 0$. Set $n = 0$

2: Solve for (x^{n+1}, λ^{n+1});

$$A^*(Ax^{n+1} - b) + Px^{n+1} + \lambda^{n+1} = 0,$$

where

$$\begin{cases} \lambda_k^{n+1} = 0, & \text{if } k \in \{k : |\lambda_k^n + \Lambda_k x_k^n|^2 > 2\alpha\Lambda_k\} \\ x_j^{n+1} = 0, & \text{if } j \in \{j : |\lambda_j^n + \Lambda_j x_j^n|^2 \le 2\alpha\Lambda_j\} \end{cases}$$

3: Convergent or set $k = k + 1$ and Return to Step 2.

provides a complementarity condition for (5.27). Thus, if the active set method converges, then the converged (x, λ) satisfies the necessary optimality (5.27). Numerically, the active set method converges globally.

We have the following remark on Algorithm 5.6.

Remark 5.5.

(i) It follows from the complementarity condition $(x^{n+1}, \lambda^{n+1}) = 0$ that
$$((A^*A + P)x^{n+1}, x^{n+1}) \le (Ax^{n+1}, f),$$
and thus the sequence $\|Ax^{n+1}\|^2 + (Px^{n+1}, x^{n+1})$ is uniformly bounded.

(ii) Numerically, Algorithm 5.6 converges globally. The following estimate is a key to establish its global convergence. Note that
$$\begin{aligned} 0 &= (A^*(Ax^{n+1} - b) + Px^{n+1} + \lambda^{n+1}, x^{n+1} - x^n) \\ &= \tfrac{1}{2}(\|Ax^{n+1} - b\|^2 + (x^{n+1}, Px^{n+1})) - \tfrac{1}{2}(\|Ax^n - b\|^2 + (x_n, Px^n)) \\ &\quad + \tfrac{1}{2}\|A(x^{n+1} - x^n)\|^2 + (x^{n+1} - x^n, P(x^{n+1} - x^n)) - (\lambda^{n+1}, x^n). \end{aligned}$$
The term (λ^{n+1}, x^n) is related to the switching between the active and inactive sets.

(iii) The global convergence of Algorithm 5.6 has been analyzed in [165] for the case when $Q = A^*A + P$ is an M-matrix and $f = A^*b > 0$, and under the following uniqueness assumption.

Let (x, λ) be a solution to (5.27), and Λ_k is the kth diagonal entry of $Q = A^*A + P$. The necessary optimality condition (5.27) can be equivalently written as

$$\begin{cases} A^*(Ax - b) + Px + \lambda = 0, & \lambda_k x_k = 0 \text{ for all } k, \\ \lambda_k = 0, \quad \text{if} \quad k \in \mathcal{I}, \\ x_j = 0, \quad \text{if} \quad j \in \mathcal{A}, \end{cases} \tag{5.32}$$

where the sets $\mathcal{I}$ and $\mathcal{A}$ are respectively given by

$$\mathcal{I} = \{k : |\lambda_k + \Lambda_k x_k|^2 > 2\alpha\Lambda_k\},$$
$$\mathcal{A} = \{k : |\lambda_k + \Lambda_k x_k|^2 \le 2\alpha\Lambda_k\}.$$

We have the following uniqueness under a diagonal dominance assumption on $Q = P + A^*A$.

Theorem 5.9. *We assume the strict complementarity, i.e., there exists a* $\delta > 0$ *such that*

$$\min_{\mathcal{I}} \|\Lambda^{-\frac{1}{2}}(\Lambda x + \lambda)\| - \max_{\mathcal{A}} \|\Lambda^{-\frac{1}{2}}(\Lambda x + \lambda)\| \ge \delta\sqrt{2\alpha},$$

and Q *is diagonally dominant, i.e., there exists* $0 \le \rho < 1$ *such that*

$$\|\sqrt{\Lambda}^{-\frac{1}{2}}(Q - \Lambda)\sqrt{\Lambda}^{-\frac{1}{2}}\|_\infty \le \rho.$$

If $\delta > \frac{2\rho}{1-\rho}$, *then* (5.32) *has a unique solution.*

Proof. Assume that there exist two pairs (x, λ) and $(\widehat{x}, \widehat{\lambda})$ satisfying the necessary optimality (5.32). Then we have

$$Q(x - \widehat{x}) + \lambda - \widehat{\lambda} = 0,$$
$$\Lambda^{-\frac{1}{2}}(\Lambda x + \lambda - (\Lambda\widehat{x} + \widehat{\lambda})) = \Lambda^{-\frac{1}{2}}(\Lambda - Q)\Lambda^{-\frac{1}{2}}\Lambda^{\frac{1}{2}}(x - \widehat{x}).$$

Now we let $S = \{k : \ x_k = \widehat{x}_k = 0\}^c$. By the triangle inequality,

$$|\Lambda_k^{-\frac{1}{2}}(\lambda_k - \Lambda_k x_k)| - |\Lambda_k^{-\frac{1}{2}}(\widehat{\lambda}_k - \Lambda_k\widehat{x}_k)| \le |\Lambda_k^{-\frac{1}{2}}(\lambda_k - \widehat{\lambda}_k + \Lambda_k(x_k - \widehat{x}_k))|.$$

The diagonal dominance assumption implies

$$\begin{aligned}\|\Lambda^{-\frac{1}{2}}(\lambda - \widehat{\lambda} + \Lambda(x - \widehat{x}))\| &= \|\Lambda^{-\frac{1}{2}}(\Lambda - Q)\Lambda^{-\frac{1}{2}}\Lambda^{\frac{1}{2}}(x - \widehat{x})\| \\ &\le \rho\|\Lambda^{\frac{1}{2}}(x - \widehat{x})\| = \rho\|\left(\Lambda^{\frac{1}{2}}(x - \widehat{x})\right)_S\|.\end{aligned}$$

In particular, this yields

$$\begin{aligned}&\min_{\mathcal{I}} \|\Lambda^{-\frac{1}{2}}(\Lambda x + \lambda)\| - \max_{\mathcal{A}} \|\Lambda^{-\frac{1}{2}}(\Lambda x + \lambda)\| \\ &\le \|\Lambda^{-\frac{1}{2}}(\Lambda x + \lambda)\| - \|\Lambda^{-\frac{1}{2}}(\Lambda\widehat{x} + \widehat{\lambda})\| \\ &\le \rho\|\left(\Lambda^{\frac{1}{2}}(x - \widehat{x})\right)_S\|.\end{aligned}$$

A second application of the triangle inequality yields

$$\|\left(\Lambda^{\frac{1}{2}}(x - \widehat{x})\right)_S\| \le \|\Lambda^{-\frac{1}{2}}(\lambda - \widehat{\lambda})\| + \rho\|\left(\Lambda^{\frac{1}{2}}(x - \widehat{x})\right)_S\|.$$

Now by the complementarity condition, we have

$$|(\Lambda^{-\frac{1}{2}}\lambda)_k| \le \sqrt{2\alpha} \ \text{ if } x_k = 0, \qquad |(\Lambda^{-\frac{1}{2}}\widehat{\lambda})_k| \le \sqrt{2\alpha} \ \text{ if } \widehat{x}_k = 0.$$

Consequently,

$$\|\Lambda^{\frac{1}{2}}(x - \widehat{x})\|_S \le \frac{2}{1 - \rho}\sqrt{2\alpha}.$$

Combining the preceding estimates give

$$\min_{\mathcal{I}} \|\Lambda^{-\frac{1}{2}}(\Lambda x + \lambda)\| - \max_{\mathcal{A}} \|\Lambda^{-\frac{1}{2}}(\Lambda x + \lambda)\| \le \frac{2\rho}{1 - \rho}\sqrt{2\alpha}.$$

It thus follows that $\delta \le \frac{2\rho}{1-\rho}$, which contradicts the assumption $\delta > \frac{2\rho}{1-\rho}$. □

5.4 Nonsmooth nonconvex optimization

In this section we consider a class of nonsmooth non-convex constrained optimization

$$\min_{x \in \mathcal{X}} F(x) + \psi(x),$$

where the constraint $\mathcal{C}$ reads

$$\mathcal{C} = \{x(\omega) \in U \text{ a.e.}\},$$

with U being a closed convex set in $\mathbb{R}^m$. In particular, we discuss $L^0(\Omega)$ optimization and $L^p(\Omega)$, $0 < p < 1$, optimization in detail, following [164]. Throughout, we assume that the functional $\psi(x)$ admits the following integral representation

$$\psi(x) = \int_\Omega h(x(\omega))\, d\omega,$$

where h is a lower semi-continuous, but nonconvex and nonsmooth, regularization functional on $\mathbb{R}^m$. Here X is a Banach space and continuously embeds into $H = (L^2(\Omega))^m$ and let $F : X \to \mathbb{R}^+$ be C^1 continuous.

For an integrable function x^*, $\delta > 0$, $s \in \Omega$ and $u \in U$, we define another measurable function x, known as the needle perturbation of (the optimal solution) x^*, by

$$x = \begin{cases} x^*, & \text{on } \Omega \setminus B(s,\delta), \\ u, & \text{on } B(s,\delta), \end{cases}$$

where the set $B(s,\delta) = \{\omega : |\omega - s| \leq \delta\}$. Further, we assume that the functional F satisfies

$$|F(x) - F(x^*) - (F'(x^*), x - x^*)| \sim o(\text{vol}(B(s,\delta))). \tag{5.33}$$

Further, we introduce the Hamiltonian $\mathcal{H}$ of h defined by

$$\mathcal{H}(x,\lambda) = (\lambda, x) - h(x), \quad x,\ \lambda \in \mathbb{R}^m.$$

Then we have the following pointwise necessary optimality condition.

Theorem 5.10. *Let condition* (5.33) *hold, and x^* be an integrable function attaining the minimum. Then for a.e. $\omega \in \Omega$, there holds*

$$\begin{aligned} &F'(x^*) + \lambda = 0, \\ &\mathcal{H}(u, \lambda(\omega)) \leq \mathcal{H}(x^*(\omega), \lambda(\omega)) \quad \textit{for all } u \in U. \end{aligned}$$

Proof. By the minimizing property of x^*, we observe that

$$\begin{aligned}0 &\geq F(x) + \psi(x) - (F(x^*) + \psi(x^*)) \\ &= F(x) - F(x^*) - (F'(x^*), x - x^*) \\ &\quad + \int_{B(s,\delta)} (F'(x^*), u - x^*) + h(u) - h(x^*))\, d\omega.\end{aligned}$$

By the lower semicontinuity of h and condition (5.33) that at a Lebesgue point $s = \omega$ of x^* the desired necessary optimality holds. □

For $\lambda \in \mathbb{R}^m$, we define a set-valued function by

$$\Phi(\lambda) = \text{argmax}_{x \in U}\{(\lambda, x) - h(x)\}.$$

Then the necessary optimality condition in Theorem 5.10 reads

$$x^*(\omega) \in \Phi(-F'(x^*(\omega))). \tag{5.34}$$

5.4.1 *Biconjugate function and relaxation*

The graph of Φ is monotone, i.e.,

$$(\lambda_1 - \lambda_2, y_1 - y_2) \geq 0 \text{ for all } y_1 \in \Phi(\lambda_1),\ y_2 \in \Phi(\lambda_2).$$

but it not necessarily maximal monotone. Thus, it is not easy to show the existence of solutions to (5.34). Let h^* be the conjugate function of function h defined by

$$h^*(\lambda) = \max_x \mathcal{H}(x, \lambda).$$

Then, h^* is necessarily convex. Further, note that if $u \in \Phi(\lambda)$, then for any $\widehat{\lambda} \in \mathbb{R}^m$,

$$h^*(\widehat{\lambda}) \geq (\widehat{\lambda}, y) - h(u) = (\widehat{\lambda} - \lambda, y) + h^*(u),$$

and thus $u \in \partial h^*(\lambda)$. By the convexity of h^*, ∂h^* is maximal monotone and thus ∂h^* is the maximal monotone extension of Φ.

Let h^{**} be the biconjugate function of h, i.e.,

$$h^{**}(x) = \sup_\lambda \left\{ (x, \lambda) - \sup_y \mathcal{H}(y, \lambda) \right\},$$

whose epigraph $epi(h^{**})$ is the convex envelope of the epigraph of h, i.e.,

$$epi(h^{**}) = \{\lambda_1\, y_1 + (1 - \lambda)\, y_2 \text{ for all } y_1,\ y_2 \in epi(h) \text{ and } \lambda \in [0, 1]\}.$$

Then, h^{**} is necessarily convex and is the convexification of h and

$$\lambda \in \partial h^{**}(u) \text{ if and only if } u \in \partial h^*(\lambda)$$

and

$$h^*(\lambda) + h^{**}(u) = (\lambda, u).$$

Now we consider the following relaxed problem:

$$\min \quad F(x) + \psi^{**}(x), \tag{5.35}$$

where the functional $\psi^{**}(x)$ is defined by

$$\psi^{**}(x) = \int_\Omega h^{**}(x(\omega))\,d\omega.$$

By the convexity of the function h^{**}, the functional $\psi^{**}(x)$ is weakly lower semi-continuous, and thus there exists at least one minimizer x to the relaxed problem (5.35). Further, it follows from Theorem 5.10 that the necessary optimality condition of (5.35) reads

$$-F'(x)(\omega) \in \partial h^{**}(x(\omega)),$$

or equivalently

$$x \in \partial h^*(-F'(x)).$$

If the strict complementarity holds, i.e., $x(\omega)$ satisfies (5.34) a.e., then x also minimizes the original cost functional.

Example 5.8. In this example, we apply the analysis to the case of volume control by $L^0(\Omega)$, i.e., $h : \mathbb{R} \to \mathbb{R}$ is given by

$$h(u) = \tfrac{\alpha}{2}|u|^2 + \beta|u|_0,$$

where the function $|\cdot|_0$ is defined by

$$|u|_0 = \begin{cases} 0, & \text{if } u = 0, \\ 1, & \text{if } u \neq 0. \end{cases}$$

Hence

$$\int_\Omega |u(\omega)|_0\,d\omega = \text{meas}(\{u(\omega) \neq 0\}).$$

In this case Theorem 5.10 yields

$$\begin{aligned} \Phi(\lambda) &:= \arg\max_{u\in\mathbb{R}} \ (\lambda u - h(u)) \\ &= \begin{cases} \frac{\lambda}{\alpha}, & \text{for } |\lambda| \geq \sqrt{2\alpha\beta}, \\ 0, & \text{for } |\lambda| < \sqrt{2\alpha\beta}, \end{cases} \end{aligned}$$

and the conjugate function h^* of h is given by

$$h^*(\lambda) = \lambda\,\Phi(\lambda) - h(\Phi(\lambda))$$
$$= \begin{cases} \frac{1}{2\alpha}|\lambda|^2 - \beta, & \text{for } |\lambda| \geq \sqrt{2\alpha\beta}, \\ 0, & \text{for } |\lambda| < \sqrt{2\alpha\beta}. \end{cases}$$

The biconjugate function $h^{**}(x)$ is given by

$$h^{**}(x) = \begin{cases} \frac{\alpha}{2}|x|^2 + \beta, & |x| > \sqrt{\frac{2\beta}{\alpha}}, \\ \sqrt{2\alpha\beta}|x|, & |x| \leq \sqrt{\frac{2\beta}{\alpha}}. \end{cases}$$

Clearly, the function $\Phi : \mathbb{R} \to \mathbb{R}$ is monotone, but not maximal monotone. The maximal monotone extension $\tilde{\Phi} = \partial h^*$ of Φ is given by

$$\tilde{\Phi}(\lambda) \in \begin{cases} \frac{\lambda}{\alpha}, & \text{for } |\lambda| > \sqrt{2\alpha\beta}, \\ [0, \frac{\lambda}{\alpha}], & \text{for } |\lambda| = \sqrt{2\alpha\beta}, \\ 0, & \text{for } |\lambda| < \sqrt{2\alpha\beta}. \end{cases}$$

If the functional $F(x)$ is strictly convex, i.e., $F(x) - F(x^*) - F'(x^*)(x - x^*) > 0$ for $|x - x^*| > 0$, then the choice $x^*(\omega) = 0$, $\frac{\lambda}{\alpha}$ at $|\lambda| = \sqrt{2\alpha\beta}$ is a measurable selection from ∂h^* and it is optimal, i.e., the original optimization problem has a solution.

Example 5.9. In this example, we consider the $L^p(\Omega)$ optimization, $0 \leq p < 1$, i.e., $h(u) = \frac{\alpha}{2}|u|^2 + \beta\,|u|^p$ and

$$\min_{u \in \mathbb{R}} \; -(\lambda, u) + \frac{\alpha}{2}|u|^2 + \beta\,|u|^p. \tag{5.36}$$

For $0 < p < 1$, if $u > 0$ attains the minimum, then there holds

$$\alpha\,u - \lambda + \beta p u^{p-1} = 0 \quad \text{and} \quad \tfrac{\alpha}{2}u^2 - \lambda\,u + \beta u^p \leq 0,$$

where the first is the first-order optimality condition, and the second compares the function value with that at $u = 0$. Let $\bar{u}$ satisfy

$$\begin{aligned} &\alpha\,\bar{u}^{2-p} - \lambda\bar{u}^{1-p} + \beta p = 0, \\ &\tfrac{\alpha}{2}\,\bar{u}^{2-p} - \lambda\bar{u}^{1-p} + \beta = 0. \end{aligned} \tag{5.37}$$

That is, both $\bar{u}$ and 0 are minimum points to (5.36). Therefore,

$$\bar{u} = u_p = \left(\frac{2(1-p)\beta}{\alpha}\right)^{\frac{1}{2-p}}.$$

It follows from (5.37) that $u \geq \bar{u}$ if $q \geq \mu_p$, where

$$\mu_p = \alpha\bar{u} + \beta\bar{u}^{1-p} = (2(1-p))^{\frac{p-1}{2-p}}(2-p)\,(\alpha^{1-p}\beta)^{\frac{1}{2-p}}.$$

If $p = 0$, $\mu_p = \sqrt{2\alpha\beta}$ and if $p = 1$, $\mu_p = \beta$. Let $z = z(\lambda)$ be a maximum solution to

$$\alpha\, z + \beta p z^{1-p} = \lambda. \tag{5.38}$$

Note that $z(\lambda) = \frac{\lambda}{\alpha}$ asymptotically as $|\lambda| \to \infty$. Thus, we obtain the following complementarity condition

$$u = \Phi_p(\lambda) = \begin{cases} 0, & |\lambda| < \mu_p, \\ \{0, u_p\}, & |\lambda| = \mu_p, \\ z(\lambda), & |\lambda| > \mu_p, \end{cases}$$

for the minimizer u of (5.36). The conjugate function $h^*(\lambda)$ is given by

$$h^*(\lambda) = \begin{cases} 0, & |\lambda| \le \mu_p, \\ \frac{\alpha}{2}|z(\lambda)|^2 + (p-1)|z(\lambda)|^p, & |\lambda| > \mu_p, \end{cases}$$

and

$$\partial h^*(\lambda) = \begin{cases} 0, & |\lambda| < \mu_p, \\ [0, u_p], & |\lambda| = \mu_p, \\ z(\lambda), & |\lambda| > \mu_p. \end{cases}$$

On the $L^p(\Omega)$ optimization, we have the following remark.

Remark 5.6.

(i) For $|q| > \mu_p$, equation (5.38) is nonlinear in z. One may use successive linearization to approximately solve (5.38), i.e.,

$$z^+ = \frac{q}{\alpha + \beta p z^{p-2}}.$$

(ii) The sparsity condition, i.e., $|\lambda| \le \mu_p$, is determined by a function of α, β and p.

(iii) The graph Φ is not maximal and the maximized graph ∂h^* is set-valued at $|q| = \mu_p$, except $p = 1$.

5.4.2 *Semismooth Newton method*

Like in the case of $L^0(\Omega)$ optimization, $\partial h^{**}(x)$ is a set-valued function in general. Thus, we consider the Yoshida-Moreau approximation of h

$$h_c(x, \lambda) = \inf_z \left\{ h^{**}(z) + (\lambda, x - z)_H + \frac{c}{2}\|x - z\|^2 \right\}.$$

It follows from [163] that

$$\lambda \in \partial h^{**}(x) \text{ iff } \lambda \in h'_c(x,\lambda) \quad \text{for some } c>0,$$
$$h'_c(x,\lambda) = p_c(\lambda + cx),$$

where the function $p_c(\lambda)$ is given by

$$p_c(\lambda) = \operatorname{argmin}\left\{\tfrac{1}{2c}\|p-\lambda\|^2 + h^*(p)\right\}.$$

The necessary optimality condition is written as

$$-F'(x) = \lambda \quad \text{and} \quad \lambda = h'_c(x,\lambda). \tag{5.39}$$

In general the function $h'_c(x,\lambda)$ is Lipshitz continuous and it is a system of semismooth equations. Hence, one can apply the semismooth Newton method. We illustrate this with Examples 5.8 and 5.9.

Example 5.10. Here we derive the semismooth Newton update formulas for Examples 5.8 and 5.9. For Example 5.8, $h(x) = \frac{\alpha}{2}\|x\|^2 + \beta\|x\|_0$. Further, we have

$$h'_c(x,\lambda) = \begin{cases} \lambda, & \text{if } |\lambda + cx| \le \sqrt{2\alpha\beta}, \\ \sqrt{2\alpha\beta}, & \text{if } \sqrt{2\alpha\beta} \le \lambda + cx \le (1+\frac{c}{\alpha})\sqrt{2\alpha\beta}, \\ -\sqrt{2\alpha\beta}, & \text{if } \sqrt{2\alpha\beta} \le -(\lambda + cx) \le (1+\frac{c}{\alpha})\sqrt{2\alpha\beta}, \\ \frac{\lambda}{\alpha+c} & \text{if } |\lambda + cx| \ge (1+\frac{c}{\alpha})\sqrt{2\alpha\beta}. \end{cases}$$

Hence the semismooth Newton update is given by

$$-F'(x^{n+1}) = \lambda^{n+1},$$
$$\begin{cases} x^{n+1} = 0, & \text{if } |\lambda^n + cx^n| \le \sqrt{2\alpha\beta}, \\ \lambda^{n+1} = \sqrt{2\alpha\beta}, & \text{if } \sqrt{2\alpha\beta} < \lambda^n + cx^n < (1+\frac{c}{\alpha})\sqrt{2\alpha\beta}, \\ \lambda^{n+1} = -\sqrt{2\alpha\beta}, & \text{if } \sqrt{2\alpha\beta} < -(\lambda^n + cx^n) < (1+\frac{c}{\alpha})\sqrt{2\alpha\beta}, \\ x^{n+1} = \frac{\lambda^{n+1}}{\alpha}, & \text{if } |\lambda^n + cx^n| \ge (1+\frac{c}{\alpha})\sqrt{2\alpha\beta}. \end{cases}$$

Similarly, for the case of $L^p(\Omega)$ optimization from Example 5.9, $h(u) = \frac{\alpha}{2}|u|^2 + \beta|u|^p$, the complementarity equation becomes

$$\begin{cases} x = 0, & |\lambda + cu| \le \mu_p, \\ \lambda = \mu_p, & |\lambda + cu| = \mu_p + cu_p, \\ x = z(\lambda), & |\lambda + cu| > \mu_p + cu_p. \end{cases} \tag{5.40}$$

This relation can be exploited for designing a primal-dual active set algorithm for the $L^p(\Omega)$-optimization; see Algorithm 5.7 below for details.

5.4.3 *Constrained optimization*

Now we consider the constrained minimization problem of the form:

$$\min_{u\in\mathcal{U}} J(x,u) = \int_\Omega (\ell(x)+h(u))\,d\omega, \tag{5.41}$$

subject to the equality constraint

$$\tilde{E}(x,u) = Ex + f(x) + Bu = 0.$$

Here $\mathcal{U} = \{u \in L^2(\Omega)^m : u(\omega) \in U \text{ a.e. in } \Omega\}$, where U is a closed convex subset of $\mathbb{R}^m$. The space X is a closed subspace of $L^2(\Omega)^n$ and $E : X \times \mathcal{U} \to X^*$ with $E \in \mathcal{L}(X, X^*)$ being boundedly invertible, $B \in \mathcal{L}(L^2(\Omega)^m, L^2(\Omega)^n)$ and $f : L^2(\Omega)^n \to L^2(\Omega)^n$ are Lipschitz continuous. We assume that $\tilde{E}(x,u) = 0$ has a unique solution $x = x(u) \in X$, for any given $u \in \mathcal{U}$, that there exist a solution $(\bar{x},\bar{u})$ to Problem (5.41), and that the adjoint equation

$$(E + f_x(\bar{x}))^* p + \ell'(\bar{x}) = 0, \tag{5.42}$$

has a solution in X.

To derive a necessary condition for this class of (nonconvex) problems, we use a maximum principle approach. For an arbitrary $s \in \Omega$, we shall utilize needle perturbations of the optimal solution $\bar{u}$ defined by

$$v(\omega) = \begin{cases} u, & \text{on } B(s,\delta), \\ \bar{u}(\omega), & \text{otherwise,} \end{cases} \tag{5.43}$$

where $u \in U$ is a constant and $\delta > 0$ is sufficiently small so that the ball $B(s,\delta) := \{\omega : |\omega - s| < \delta\} \subset \Omega$.

The following additional properties on the optimal state $\bar{x}$ and each perturbed state $x(v)$ will be assumed.

Assumption 5.2. The optimal state $\bar{x}$ and $x(v)$, with v being a needle perturbation defined in (5.43), satisfy

(a) $\|x(v) - \bar{x}\|^2_{L^2(\Omega)} = o(\mathrm{meas}(B(s,\delta)))$;
(b) the linear form $\ell(\cdot,\cdot)$ satisfies

$$\int_\Omega (\ell(\cdot,x(v)) - \ell(\cdot,\bar{x}) - \ell_x(\cdot,\bar{x})(x(v)-\bar{x}))d\omega = O(\|x(v)-\bar{x}\|^2_{L^2(\Omega)});$$

(c) The function $f(\cdot,\cdot)$ satisfies

$$\langle f(\cdot,x(v)) - f(\cdot,\bar{x}) - f_x(\cdot,\bar{x})(x(v)-\bar{x}), p\rangle_{X^*,X} = O(\|x(v)-\bar{x}\|^2_{L^2(\Omega)}).$$

Theorem 5.11. *Let $(\bar{x}, \bar{u}) \in X \times \mathcal{U}$ be optimal for problem (5.41), $p \in X$ satisfy the adjoint equation (5.42) and Assumption 5.2 hold. Then the necessary optimality condition is that $\bar{u}(\omega)$ minimizes $h(u) + (p, Bu)$ pointwise a.e. in Ω.*

Proof. Let v be a needle perturbation and $x = x(v)$. Then by the optimality of $\bar{u}$ and Assumption 5.2(b), we have

$$\begin{aligned} 0 &\leq J(v) - J(\bar{u}) \\ &= \int_\Omega \big(\ell(\cdot, x(v)) - \ell(\cdot, x(\bar{u})) + h(v) - h(\bar{u})\big)\, d\omega \\ &= \int_\Omega (\ell_x(\cdot, \bar{x})(x - \bar{x}) + h(v) - h(\bar{u}))\, d\omega + O(\|x - \bar{x}\|^2_{L^2(\Omega)}). \end{aligned}$$

Now using the adjoint variable $p \in X$ defined in (5.42), we deduce

$$(\ell_x(\cdot, \bar{x})(x - \bar{x}), p) = -\langle (E + f_x(\cdot, \bar{x})(x - \bar{x})), p\rangle,$$

and consequently

$$\begin{aligned} 0 &= \langle E(x - \bar{x}) + f(\cdot, x) - f(\cdot, \bar{x}) + B(v - \bar{u}), p\rangle \\ &= \langle E(x - \bar{x}) + f_x(\cdot, \bar{x})(x - \bar{x}) + B(v - \bar{u}), p\rangle + O(\|x - \bar{x}\|^2_{L^2(\Omega)}). \end{aligned}$$

Thus, we obtain

$$\begin{aligned} 0 &\leq J(v) - J(\bar{u}) \\ &\leq \int_\Omega ((-B^*p, v - \bar{u}) + h(v) - h(\bar{u}))d\omega + O(\|x - \bar{x}\|^2_{L^2(\Omega)}). \end{aligned}$$

Now we restrict s to be a Lebesgue point of the mapping

$$\omega \to (-B^*p, v - \bar{u}) + h(v) - h(\bar{u})).$$

Let S denote the set of these Lebesgue points and note that $\text{meas}(\Omega \backslash S) = 0$. Upon dividing the inequality by $\text{meas}(B(s, \delta)) > 0$, letting $\delta \to 0$, and using Assumption 5.2(a) and lower semiconitinuity of the function h, we obtain the desired assertion at a Lebesgue point s of the mapping, and the desired claim follows directly. □

Consider the relaxed problem of (5.41):

$$\min_{u \in \mathcal{U}} \quad J(x, u) = \int_\Omega (\ell(x) + h^{**}(u))\, d\omega.$$

The pointwise necessary optimality is given by

$$\begin{cases} E\bar{x} + f'(\bar{x}) + B\bar{u} = 0, \\ (E + f'(\bar{x}))^* p + \ell'(\bar{x}) = 0, \\ \bar{u} \in \partial h^*(-B^*p), \end{cases}$$

or equivalently

$$\lambda = h_c'(u, \lambda) \quad \text{and} \quad \lambda = B^* p,$$

where h_c is the Yoshida-Moreau approximation of the biconjugate function h^{**}.

Based the complementarity condition (5.40) we have a primal dual active set method for the $L^p(\Omega)$ optimization, $0 < p < 1$, i.e., the case $h(u) = \frac{\alpha}{2}|u|^2 + \beta\,|u|^p$; cf. Algorithm 5.7.

Algorithm 5.7 Primal dual active set method for $L^p(\Omega)$ optimization.

1: Initialize $u^0 \in X$ and $\lambda^0 \in H$. Set $n = 0$.

2: Solve for $(y^{n+1}, u^{n+1}, p^{n+1})$

$$Ey^{n+1} + Bu^{n+1} = g, \quad E^* p^{n+1} + \ell'(y^{n+1}) = 0, \quad \lambda^{n+1} = B^* p^{n+1}$$

and

$$\begin{aligned}
u^{n+1} &= -\frac{\lambda^{n+1}}{\alpha + \beta p |u^n|^{p-2}} \text{ if } \omega \in \{|\lambda^n + c u^n| > (\mu_p + c u_p\} \\
\lambda^{n+1} &= \mu_p \text{ if } \omega \in \{\mu_p \le \lambda^n + c u^n \le \mu_p + c u_n\}. \\
\lambda^{n+1} &= -\mu_p \text{ if } \omega \in \{\mu_p \le -(\lambda^n + c u^n) \le \mu_p + c u_n\}. \\
u^{n+1} &= 0, \text{ if } \omega \in \{|\lambda^n + c u^n| \le \mu_p\}.
\end{aligned}$$

3: Stop, or set $n = n + 1$ and return to the second step.

Chapter 6

Direct Inversion Methods

In this chapter, we describe several direct inversion methods for extracting essential characteristics of inverse solutions, e.g., location and size of the inclusions, which might be sufficient for some applications. Usually, direct inversion methods are very fast and easy to implement, however, their reconstructions are not very accurate when compared with that obtained by other more expensive inversion techniques, like Tikhonov regularization. Nonetheless, the reconstructions by direct inversion methods can serve as good initial guesses for Tikhonov regularization, which may provide a more refined estimate but at the expense of (much) increased computational efforts. Meanwhile, a good initial guess can essentially reduce the computational efforts, e.g., by refining the region of interest only. The development of direct inversion methods is an important ongoing research topic. In this chapter, we describe a number of direct inversion methods, including the MUSIC algorithm, linear sampling method, direct sampling method, El Badia-HaDuong algorithm, numerical continuation method, and Gel'fand-Levitan-Marchenko transformation, for a number of representative (nonlinear) inverse problems, including inverse scattering, inverse source problem, analytic continuation, and inverse Sturm-Liouville problem. The main goal of describing these methods is to give a flavor of such techniques, instead of a comprehensive survey on all direct inversion techniques.

6.1 Inverse scattering methods

In this section, we briefly describe three direct inversion methods for inverse scattering, including the MUSIC algorithm, linear sampling method and direct sampling method. All three methods rely on constructing a certain easy-to-compute indicator function over the sampling domain. The differences between them lie in their motivation and the indicator function, and hence the requirements on the data and implementation.

6.1.1 *The MUSIC algorithm*

One popular direct inversion technique in engineering is *MUltiple SIgnal Classification* (MUSIC). Originally, it was developed in the signal processing community for estimating the frequency contents of a given signal. Then it was extended to inverse scattering problems. We first describe the method for signal processing, and then for inverse scattering problems.

MUSIC in signal processing In many signal processing problems, the objective is to estimate a set of constant parameters from noisy measurements of the received signal. It was first systematically formulated by R. O. Schmidt in 1977, but the final paper appeared only in 1986 [270]. He was the first to correctly exploit the measurement model for parameter estimation. The resulting algorithm was called MUSIC. Since then it has attracted considerable interest in signal processing, and now is a standard tool for frequency analysis. Mathematically, MUSIC is essentially a method for characterizing the range of an operator.

Let us first consider the self-adjoint case. Let A be a self-adjoint operator with eigenvalues $\lambda_1 \geq \lambda_2 \geq \ldots$ and the respective (orthonormal) eigenvectors $v_1, v_2, \ldots$. Now suppose that the eigenvalues $\lambda_{M+1}, \lambda_{M+2}, \ldots$ are all zero for some $M \in \mathbb{N}$. In other words, the vectors $v_{M+1}, v_{M+2}, \ldots$ span the null space of the operator A. In practice, due to the presence of measurement noise, the eigenvalues $\lambda_{M+1}, \lambda_{M+2}, \ldots$ generally do not vanish but level off at a small value determined by the noise level of the system. We denote by the noise subspace of A the subspace $\text{span}(\{v_j\}_{j\geq M+1})$, and let P_{ns} be the orthogonal projection onto the noise subspace, i.e.,

$$P_{\text{ns}} = \sum_{j>M} v_j \otimes \bar{v}_j.$$

The idea of MUSIC is as follows: because the operator A is self-adjoint, its noise subspace is orthogonal to the essential range. Therefore, a vector f is in the range of A if and only if its projection on the noise subspace is zero, i.e., $\|P_{\text{ns}} f\| = 0$. However, this happens if and only if $\frac{1}{\|P_{\text{ns}} f\|} = \infty$. This motivates the following definition of the indicator function

$$\Psi(f) = \frac{1}{\|P_{\text{ns}} f\|},$$

and then one plots the indicator function over the sampling domain. The peaks in the plot corresponds to the essential signal components.

For a general non-self-adjoint compact operator A, MUSIC can be accomplished with singular value decomposition, cf. Appendix A, instead of eigenvalue decomposition. Let the singular system of A be

$$A = \sum_j \sigma_j u_j \otimes \bar{v}_j,$$

where the (always nonnegative) singular values σ_j are ordered decreasingly, i.e., $\sigma_1 \geq \sigma_2 \geq \ldots$. The orthonormal vectors u_j and v_j are the left- and right-singular vectors corresponding to σ_j, respectively, and further there holds

$$AA^* u_j = \sigma_j^2 u_j \quad \text{and} \quad A^* A v_j = \sigma_j^2 v_j.$$

Now suppose that there is an $M \in \mathbb{N}$ such that $\sigma_{M+1}, \sigma_{M+2} \ldots$ all vanish. Then the null space of the operator A is spanned by $\{v_j\}_{j>M}$, and the range of the operator A is given by $\mathrm{span}\{u_j\}_{j=1}^M$. Accordingly, the noise subspace of A is the span of the vectors $\{u_j\}_{j\geq M+1}$, and the respective orthogonal projection onto the noise subspace is given by

$$P_{\mathrm{ns}} = \sum_{j>M} u_j \otimes \bar{u}_j.$$

This leads to an analogous definition of the indicator function. Practically speaking, the choice of M is essential for the performance of the algorithm, playing a role analogous to the regularization parameter in classical Tikhonov regularization, and its value should depend on the noise level. In particular, if the spectrum has a significant gap, then M may be set to the index of the onset of the gap. Alternatively, the projection P_{ns} can be computed as

$$P_{\mathrm{ns}} = I - \sum_{j=1}^M u_j \otimes \bar{u}_j.$$

This formulation is convenient if the dimension M of the signal subspace is small, which is the case for a limited amount of data.

Example 6.1. In this example, we reconstruct a signal f containing two sinusoidal components f_1 and f_2, i.e.,

$$f_1 = e^{-\mathrm{i}\frac{\pi}{10}x} \quad \text{and} \quad f_2 = e^{\mathrm{i}\frac{\pi}{6}x},$$

and we measure ten noisy versions of the random combinations of the two components: for $j = 1, 2, \ldots, 6$

$$g_j^\delta(x) = f_1\xi_1 + f_2\xi_2 + \zeta,$$

with $\xi_1, \xi_2 \sim N(\xi; 0, 1)$, and both the real and imaginary parts of $\zeta(x)$ follow $(\zeta; 0, \sigma^2)$, with $\sigma = 0.1$ being the noise level. The measurements were sampled at six points, i.e., $x = 0, \ldots, 5$. Hence the measurements form a matrix of size 6×10, and the left singular vectors are shown in Fig. 6.1(a) and (b), and the singular values are in Fig. 6.1(c). One observes that there are only two significant singular values, and the remaining four are much smaller in magnitude. The magnitudes of small singular values are of the order of the noise level. This indicates a choice $p = 2$ in the MUSIC indicator function $\Psi(\theta) \equiv \Psi(f_\theta)$, with $f_\theta = e^{-\mathrm{i}\theta x}$. The function $\Psi(\theta)$ is plotted in Fig. 6.1(d), where the dashed lines refer to the locations of the true frequency components. The locations of the recovery agrees perfectly with the true ones, indicating the accuracy of the algorithm.

MUSIC in inverse scattering The MUSIC algorithm has been extended to inverse scattering problems to estimate the locations of point like scatterers, first done by Devaney [81] and later extended to electromagnetic scattering by Ammari et al [4, 5] as well as extended obstacles; see also [200, 63] for comparisons with the factorization method, cf. Section 6.1.2. We describe the method in the context of wave propagation in free space. The mathematical model is the Helmholtz equation

$$\Delta u + k^2 u = 0.$$

Suppose there are n antennas located at the points $\{y_j\}_{j=1}^n$, transmitting spherically spreading waves. If the jth antenna is excited by an input e_j, the induced field u_j^i at the point x by the jth antenna is $u_j^i = G(x, y_j)e_j$, where $G(x, y)$ is the fundamental solution to the free-space Helmholtz equation.

We assume that the scatterers, located at $\{x_i\}_{i=1}^m$, are small, weak and well separated, so that the Born approximation, cf. Section 2.2.3, is valid. Thus if the induced field u^i is incident on the ith scatterer (at x_i), it produces at x the scattered field $G(x, x_i)\tau_i u^i(x_i)$, where τ_i is the strength of the ith scatterer. The scattered field from the whole cloud of scatterers is $\sum_i G(x, x_i)\tau_i u^i(x_i)$. Thus the total scattered field due to the field emanating from the jth antenna is given by $u_j^s(x) = \sum_i G(x, x_i)\tau_i G(x_i, y_j)e_j$, and the measured field at the lth antenna is

$$u_j^s(y_l) = \sum_i G(y_l, x_i)\tau_i G(x_i, y_j)e_j.$$

This gives rise to the multi-static response matrix $K = [k_{lj}] \in \mathbb{C}^{n\times n}$, with k_{lj} given by

$$k_{lj} = \sum_i G(y_l, x_i)\tau_i G(x_i, y_j).$$

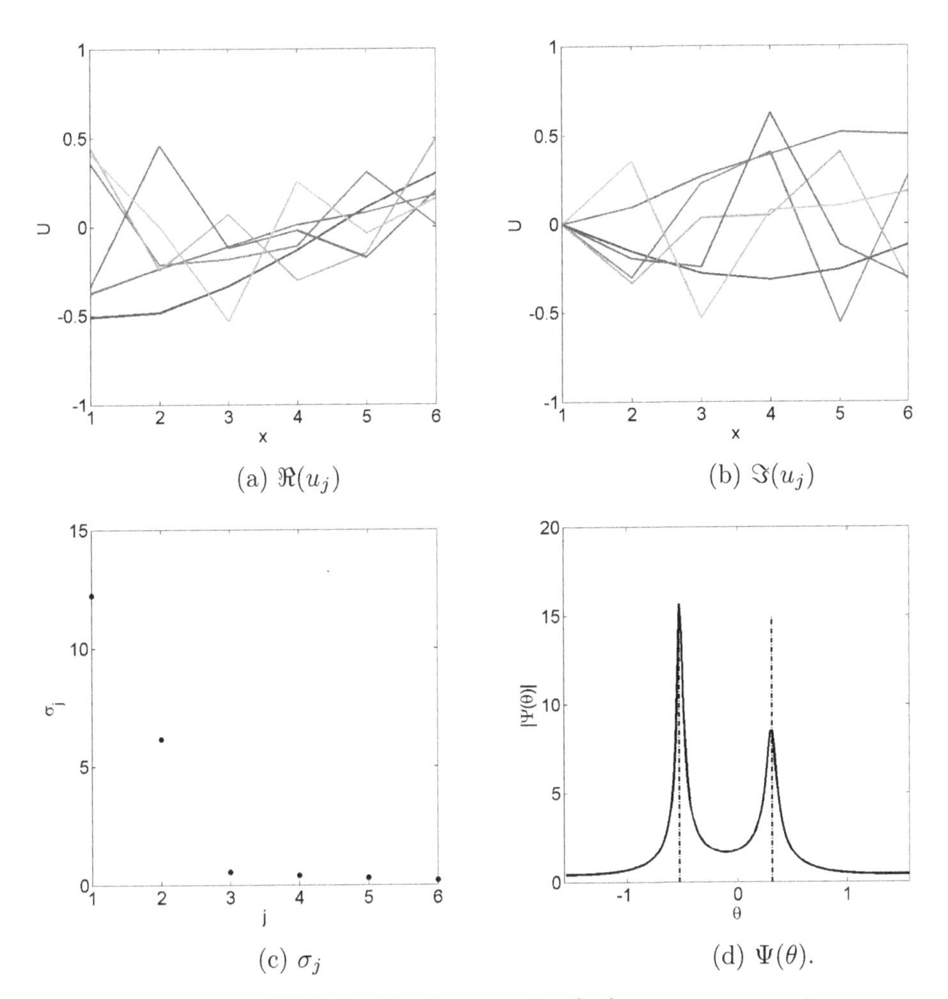

Fig. 6.1 MUSIC algorithm for recovering the frequency components.

The multi-static response matrix maps the vector of input amplitudes $(e_1, e_2, \ldots, e_n)^{\mathrm{t}}$ to the vector of measured amplitudes on the n antennas. By letting $g_i = (G(y_1, x_i), G(y_2, x_i), \ldots, G(y_n, x_i))^{\mathrm{t}}$, then the matrix K has the following rank-one representation

$$K = \sum_i \tau_i g_i g_i^{\mathrm{t}}. \tag{6.1}$$

In view of the symmetry of the fundamental solution $G(x, y)$, i.e., $G(x, y) = G(y, x)$, the matrix K is symmetric, but not self-adjoint. Further, the representation (6.1) indicates that the range of K is spanned by $\{g_i\}$, which can be used to compute the range, its orthogonal complement – the noise subspace, and the orthogonal projection $P_{\rm ns}$. Alternatively, we may form a self-adjoint matrix $A = K^*K$. Physically speaking, the matrix K^* is the time-reversed multi-static response matrix, hence K^*K performs a scattering experiment, time-reversing the received signal and using them as input for a second scattering experiment.

Devaney's insight is that the MUSIC algorithm in Section 6.1.1 can be used to determine the locations $\{x_i\}$ of the scatterers as follows. Given any point x_p from the sampling domain, we form the vector $g^p = (G(y_1, x_p), G(y_2, x_p), \ldots, G(y_n, x_p))^{\rm t}$. The point x_p coincides with the location of a scatterer, i.e., $x_p \in \{x_i\}_{i=1}^m$, if and only if

$$P_{\rm ns} g^p = 0.$$

Thus we can form an image of the scatterers by plotting, at each point x_p in the sampling domain, the quantity

$$\Psi(x_p) = \frac{1}{\|P_{\rm ns} g^p\|}.$$

The resulting plot will have large peaks at the locations of the scatterers, which provides an indicator of the presence of scatterers. The mathematical justification of the MUSIC algorithm has also received considerable interest. We refer interested readers to Chapter 4 of [201]. The above derivation is for point scatterers, and there have also been much attention to extend the algorithm to extended objects [146, 60].

Example 6.2. In this example we illustrate the MUSIC algorithm for recovering point scatterers. The true scatterers are located at $(0, -0.5)$, $(0, 0)$ and $(0, 0.5)$, cf. Fig. 6.2(a), with a wave number $k = 2\pi$ and wavelength $\lambda = 1$. The data consist of the scattered field at 30 points uniformly distributed on a circle of radius 5, and for 8 incident directions. The scattered field is contaminated with 20% noise for both real and imaginary parts. The singular values of the multi-static matrix is given in Fig. 6.2(b). There is no distinct gap in the singular value spectrum, indicating a numerical rank of eight. We note that for exact data, there are only three significant singular values, and the rest are almost zero, reminiscent of the fact that there are only three point scatterers, which actually leads to almost perfect focusing on the true scatterers. The reconstruction for the noisy data is shown in Fig. 6.2(c). The locations of the scatterers are well resolved.

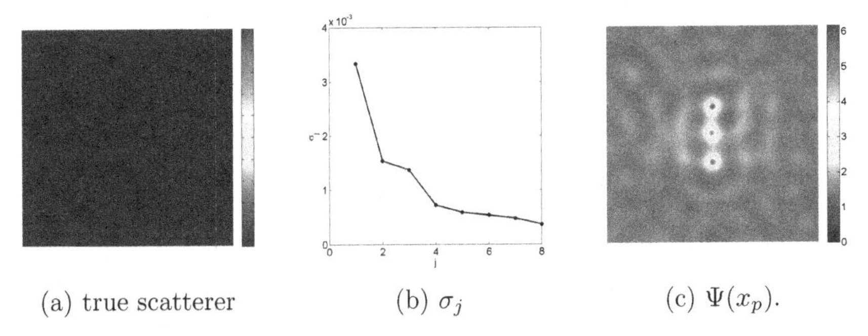

(a) true scatterer (b) σ_j (c) $\Psi(x_p)$.

Fig. 6.2 MUSIC algorithm for recovering point scatterers.

6.1.2 *Linear sampling method*

The linear sampling method [72] is one of the most prominent direct methods for inverse scattering. In this part, we briefly describe the method, and refer to complete technical developments to the monograph [51] and surveys [252, 74].

Consider a sound-soft scatterer D, which is assumed to the open complement of an unbounded domain of class C^2 in $\mathbb{R}^{\mathrm{d}}$ ($\mathrm{d} = 2, 3$). The obstacle may consist of a finite number of open connected subdomains $\{D_i\}$, i.e., $D = \cup D_i$. Given a plane-wave incident field $u^i = e^{\mathrm{i}kx\cdot d}$, with $d \in \mathbb{S}^{\mathrm{d}-1}$ and $k > 0$ being respectively the incident direction and the wave number, the presence of the obstacle D induces a scattered field u^s. Let $u(x) = u^i(x) + u^s(x)$ be the total field. Then the total field u satisfies, cf. Section 2.2.3

$$\begin{cases} \Delta u + k^2 u = 0 \quad \text{in } \mathbb{R}^{\mathrm{d}} \setminus \overline{D}, \\ u = 0 \quad \text{on } \partial D, \\ \lim_{r\to\infty} r^{(\mathrm{d}-1)/2}(\frac{\partial u^s}{\partial r} - \mathrm{i}ku^s) = 0, \end{cases} \tag{6.2}$$

where $r = |x|$ for any $x \in \mathbb{R}^{\mathrm{d}}$. It is known that the asymptotic behavior at infinity of the scattered field $u^s(x)$ satisfies

$$u^s(x) = \frac{e^{\mathrm{i}k|x|}}{|x|^{\frac{\mathrm{d}-1}{2}}}\left(u^\infty(\hat{x}) + \mathcal{O}\left(\frac{1}{|x|}\right)\right) \quad \text{as } |x| \to \infty$$

uniformly for all directions $\hat{x} = x/|x| \in \mathbb{S}^{\mathrm{d}-1}$. The analytic function $u^\infty = u^\infty(\hat{x}, d)$ is called the far-field pattern. Then the inverse obstacle scattering problem is to determine the boundary ∂D from the measurement of $u^\infty(\hat{x}, d)$ for all $\hat{x}, d \in \mathbb{S}^{\mathrm{d}-1}$.

Next we introduce the far-field operator $F : L^2(\mathbb{S}^{\mathrm{d}-1}) \mapsto L^2(\mathbb{S}^{\mathrm{d}-1})$ defined by

$$(Fg)(\hat{x}) = \int_{\mathbb{S}^{\mathrm{d}-1}} u^\infty(\hat{x}, d) g(d) ds(d), \quad \hat{x} \in \mathbb{S}^{\mathrm{d}-1}.$$

The linear sampling method, proposed by Colton and Kirsch in 1996 [72], solves the following far-field equation

$$(Fg)(\hat{x}) = G^\infty(\hat{x}, z), \quad \hat{x} \in \mathbb{S}^{\mathrm{d}-1},\ z \in \mathbb{R}^{\mathrm{d}}, \tag{6.3}$$

where $G^\infty(\hat{x}, z)$ is the far-field pattern of the radiating fundamental solution $G(x, z)$ (see Section 2.2.3), i.e.,

$$G^\infty(\hat{x}, z) = \gamma e^{-\mathrm{i}k\hat{x}\cdot z}, \tag{6.4}$$

with the constant $\gamma = \frac{1}{4\pi}$ in d $= 3$ and $\gamma = e^{\mathrm{i}\pi}/\sqrt{8\pi k}$ in d $= 2$.

The linear sampling method is motivated by the next result.

Theorem 6.1. *Assume that k^2 is not a Dirichlet eigenvalue for $-\Delta$ in D. Then the following two statements hold.*

(i) For $z \in D$ and a fixed $\epsilon > 0$, there exists a $g_\epsilon^z \in L^2(\mathbb{S}^{\mathrm{d}-1})$ such that

$$\|Fg_\epsilon^z - G^\infty(\cdot, z)\|_{L^2(\mathbb{S}^{\mathrm{d}-1})} < \epsilon \quad \text{and} \quad \lim_{z \to \partial D} \|g_\epsilon^z\|_{L^2(\mathbb{S}^{\mathrm{d}-1})} = \infty.$$

(ii) For $z \in \mathbb{R}^{\mathrm{d}} \backslash \overline{D}$ and any given $\epsilon > 0$, every $g_\epsilon^z \in L^2(\mathbb{S}^{\mathrm{d}-1})$ satisfying

$$\|Fg_\epsilon^z - G^\infty(\cdot, z)\|_{L^2(\mathbb{S}^{\mathrm{d}-1})} < \epsilon$$

satisfies

$$\lim_{\epsilon \to 0} \|g_\epsilon^z\|_{L^2(\mathbb{S}^{\mathrm{d}-1})} = \infty.$$

By Theorem 6.1, the (approximate) solution of the far-field equation (6.3) will have a large $L^2(\mathbb{S}^{\mathrm{d}-1})$-norm outside and close to the boundary ∂D. This suggests to obtain an image of the obstacle D by plotting the norm of the (approximate) solution g_ϵ^z to (6.3). This leads to the following elegant direct algorithm for the inverse obstacle scattering problem: given test points $z \in \mathbb{R}^{\mathrm{d}}$, solve approximately the linear far-field equation, compute the norm of the solution g_ϵ^z and plot the solution norms. The complete procedure is given in Algorithm 6.1. The distinct features of the method are manifold: with necessarily data available, it is easy to implement and the reconstruction is readily obtained.

Note that the far-field equation (6.3) involves an analytic kernel u^∞, and thus it is severely ill-posed. Hence, in practice, regularization techniques,

Algorithm 6.1 Linear sampling method

1: Select a sampling domain Ω with $D \subset \Omega$, and a set of sampling points;
2: Solve approximately (6.3) for each sampling point z;
3: Plot the solution norm.

e.g., Tikhonov regularization, are employed for the stable solution of equation (6.3). Numerically, it has been very successful, and is one of the most popular methods for inverse scattering, including many other scattering scenarios, e.g., sound-hard obstacles, medium scattering, crack scattering, electromagnetic and elastic scattering.

Some arguments were presented in [72] to justify the linear sampling method. However, it is not self evident why the method should work at all: The right hand side of equation (6.3) is almost never an element of the range of the operator F, and hence the classical regularization theory is not directly applicable. Thus, it is unclear why an approximate solution computed by regularization techniques will actually behave like the theoretical prediction in Theorem 6.1. Some preliminary results in this direction have been provided in [11, 12].

The mathematical gap in the linear sampling method motivated A. Kirsch [199] to develop the factorization method, again based on the singularity behavior of the fundamental solution. To this end, we introduce the following scattering problem: given the boundary data $f \in H^{1/2}(\partial D)$, find a solution $v \in H^1_{\mathrm{loc}}(\mathbb{R}^3 \setminus \overline{D})$ such that

$$\begin{cases} \Delta v + k^2 v = 0 \quad \text{in } \mathbb{R}^{\mathrm{d}} \setminus \overline{D}, \\ v = f \quad \text{on } \partial D, \\ \lim\limits_{r\to\infty} r(\frac{\partial v}{\partial r} - \mathrm{i}kv) = 0. \end{cases} \tag{6.5}$$

It is well known that there exists a unique solution $v = v(f)$ to (6.5). Let v^∞ be the far-field pattern for $v(f)$. Next we define an operator $G : H^{1/2}(\partial D) \to L^2(\mathbb{S}^{\mathrm{d}-1})$ by $G(f) = v^\infty$. If $z \in \mathbb{R}^{\mathrm{d}} \in D$, then in $\mathbb{R}^{\mathrm{d}} \setminus \overline{D}$, $G(x,z)$ satisfies the Helmholtz equation, and $G^\infty(\cdot,z)$ is in the range of G. But if $z \in \mathbb{R}^{\mathrm{d}} \setminus \overline{D}$, then because of the singularity of $G(x,z)$ at z, it doest satisfy the Helmholtz equation and $G^\infty(\cdot,z)$ lies outside the range of G. In summary, $G^\infty(\cdot,z) \in G(H^{1/2}(\partial D))$ if and only if $z \in D$. Hence one can determine the scatterer D by testing sampling points $z \in \mathbb{R}^{\mathrm{d}}$ using the Picard criterion, if the range of the operator G is known. Kirsch [199] established the following characterization of the range of G in terms of the far-field operator F:

$$\mathcal{R}(G) = \mathcal{R}((F^*F)^{1/4}).$$

Therefore, the range of G can be determined from the eigenvalues $\{\lambda_j\}$ and eigenfunctions $\{v_j\}$ of the operator F^*F. In particular, the range of $(F^*F)^{1/4}$ is given by

$$\mathcal{R}((F^*F)^{1/4}) = \left\{ f : \sum_j \lambda_j^{-1/2} |\langle v_j, f\rangle|^2 < \infty \right\}.$$

Following Picard criterion, the factorization method is to plot, at each sampling point z, the indicator function

$$\Phi(z) = \frac{1}{\sum_j |\lambda_j|^{-1/2} |\langle v_j, G^\infty(\cdot, z)\rangle|^2}.$$

It will be identically zero whenever z is outside the obstacle D, and nonzero whenever z is inside the obstacle D.

We only briefly describe the linear sampling method and factorization method for inverse scattering. There are several other closely related sampling methods, e.g., singular source methods. We refer to the surveys [252, 74] for comprehensive introduction as well as historical remarks.

6.1.3 *Direct sampling method*

Now we describe a direct sampling method for the inverse medium scattering problem of determining the shape of the scatterers/inhomogeneities from near-field scattering data, recently developed in [157]. The derivation of the method is carried out for a circular curve/spherical surface Γ, but the resulting index function is applicable to general geometries. It uses the fundamental solution $G(x, x_p)$ (see Section 2.2.3) associated with the Helmholtz equation in the homogeneous background:

$$\Delta G(x, x_p) + k^2 G(x, x_p) = -\delta_{x_p}(x), \tag{6.6}$$

where $\delta_{x_p}(x)$ refers to the Dirac delta function located at the point $x_p \in \Omega_\Gamma$ (the domain enclosed by Γ). By multiplying both sides of (6.6) by the conjugate $\overline{G}(x, x_q)$ of the fundamental solution $G(x, x_q)$ and then integrating over the domain Ω_Γ, we derive

$$\int_{\Omega_\Gamma} (\Delta G(x, x_p) + k^2 G(x, x_p)) \overline{G}(x, x_q) dx = -\overline{G}(x_p, x_q).$$

Next we consider equation (6.6) with $x_q \in \Omega_\Gamma$ in place of x_p, and take its conjugate. Then by multiplying both sides of the resulting equation by $G(x, x_p)$ and integrating over the domain Ω_Γ, we obtain

$$\int_{\Omega_\Gamma} (\Delta \overline{G}(x, x_q) + k^2 \overline{G}(x, x_q)) G(x, x_p) dx = -G(x_p, x_q).$$

Using integration by parts for the terms involving Laplacians in the preceding two equations, we readily deduce the Helmoltz-Kirchoff identity

$$\int_\Gamma \left[\overline{G}(x, x_q) \frac{\partial G(x, x_p)}{\partial n} - G(x, x_p) \frac{\partial \overline{G}(x, x_q)}{\partial n} \right] ds$$
$$= G(x_p, x_q) - \overline{G}(x_p, x_q) = 2\mathrm{i}\Im(G(x_p, x_q)).$$

Next we approximate the right hand side by means of the Sommerfeld radiation condition for the Helmholtz equation, i.e.,

$$\frac{\partial G(x, x_p)}{\partial n} = \mathrm{i}kG(x, x_p) + \text{h.o.t.}$$

Thus we use the following approximations:

$$\begin{cases} \dfrac{\partial G(x, x_p)}{\partial n} \approx \mathrm{i}kG(x, x_p), \\ \dfrac{\partial \overline{G}(x, x_q)}{\partial n} \approx -\mathrm{i}k\overline{G}(x, x_q), \end{cases}$$

which are valid if the points x_p and x_q are not close to the boundary Γ. Consequently, we arrive at the following approximate relation

$$\int_\Gamma G(x, x_p)\overline{G}(x, x_q) ds \approx k^{-1}\Im(G(x_p, x_q)). \tag{6.7}$$

The relation (6.7) represents one of the most crucial identities for deriving direct sampling method: in view of the decay behavior of the fundamental solution $G(x, y)$, it indicates that the fundamental solutions $G(x, x_p)$ and $G(x, x_q)$ are roughly orthogonal on the surface Γ.

The second essential ingredient of the derivation is the scattered field representation by means of the fundamental solution $G(x, y)$. This is rather straightforward for the Helmholtz equation in view of the Lipmann-Schwinger integral equation (2.5):

$$u^s = \int_\Omega I(y)G(x, y)dy, \tag{6.8}$$

where $I(x) = \eta(x)u(x)$ is the induced current.

With the two ingredients, we are now ready to derive an index function. To this end, we consider a sampling domain $\widetilde{\Omega}$ containing the support Ω. and divide the domain $\widetilde{\Omega}$ into a set of small elements $\{\tau_j\}$. Then by using the rectangular quadrature rule, we arrive at the following simple approximation from the integral representation (6.8):

$$u^s(x) = \int_{\widetilde{\Omega}} G(x, y)I(y)dy \approx \sum_j w_j\, G(x, y_j), \tag{6.9}$$

where the point $y_j \in \tau_j$, and the weight $w_j = |\tau_j| I(y_j)$ with $|\tau_j|$ being the measure of τ_j. Since the induced current I vanishes outside Ω, the summation in (6.9) is actually only over those elements intersecting with Ω. By elliptic regularity theory, the induced current $I = \eta u$ is smooth in each subregion, as long as the coefficient η is piecewise smooth. The relation (6.9) serves only the goal of motivating our method, and is not needed in the implementation. Physically, (6.9) also admits an interesting interpretation: the scattered field u^s at any fixed point $x \in \Gamma$ can be regarded as a weighted average of that due to the point scatterers located at $\{y_j\}$ within the true scatterer Ω.

By multiplying both sides of (6.9) by $\overline{G}(x, x_p)$ for any point x_p that lies in the sampling domain $\widetilde{\Omega}$, then integrating over the boundary Γ and using (6.7), we obtain the following approximate relation

$$\int_\Gamma u^s(x)\,\overline{G}(x, x_p)\,ds \approx k^{-1} \sum_j w_j \Im(G(y_j, x_p)). \tag{6.10}$$

The relation (6.10) is valid under the premises that the point scatterers $\{y_j\}$ and the sampling points $\{x_p\}$ are far apart from the surface Γ, and the elements $\{\tau_j\}$ are sufficiently small. The relation (6.10) underlies the essential idea of the direct sampling method. In particular, if a point x_p is close to some physical point scatterer $y_j \in \Omega$, then $G(y_j, x_p)$ is nearly singular and takes a very large value, hence it contributes significantly to the summation in (6.10). Conversely, if x_p is far away from all physical point scatterers, then the sum will be small due to the decay property of the fundamental solution $G(x, y)$.

These facts lead us to the following index function for any point $x_p \in \widetilde{\Omega}$:

$$\Phi(x_p) = \frac{|\langle u^s(x), G(x, x_p)\rangle_{L^2(\Gamma)}|}{\|u^s(x)\|_{L^2(\Gamma)} \|G(x, x_p)\|_{L^2(\Gamma)}}.$$

In practice, if a point x_p satisfies $\Phi(x_p) \approx 1$, then it is most likely within the scatterer support Ω according to the preceding discussions; whereas if $\Phi(x_p) \approx 0$, then the point x_p likely lies outside the support.

The direct method has the following distinct features:

(a) The index $\Phi(x_p)$ provides the likelihood of the point x_p lying within Ω and hence the coefficient distribution η.
(b) The index Φ involves only evaluating the free-space fundamental solution and its inner product with the measured data u^s, thus it is computationally cheap and embarrassingly parallel.

(c) The data noise enters the index Φ via the integral on Γ. The energy of the noise is expected to be equally distributed in all Fourier modes, whereas the fundamental solution $G(x, x_p)$ is very smooth on the curve/surface Γ and concentrates on only low-frequency Fourier modes. Hence the high-frequency modes (i.e., noises) in the noisy data is roughly orthogonal to $G(x, x_p)$, and contributes little to the index Φ. Hence the method is robust with respect to the noise.

Now we illustrate the performance of the method with an example.

Example 6.3. In this example, we consider the reconstruction of a ring-shape square located at the origin, with the outer and inner side lengths being 0.6 and 0.4, respectively. Two incident directions $d_1 = \frac{1}{\sqrt{2}}(1,1)^{\mathrm{t}}$ and $d_2 = \frac{1}{\sqrt{2}}(1,-1)^{\mathrm{t}}$ are considered. The results are shown in Fig. 6.3. One observes that one incident field is insufficient for recovering the ring structure, and only some parts of the ring can be resolved, depending on the incident direction. With two incident waves, the direct sampling method can provide a quite reasonable estimate of the ring shape. The estimate remains very stable for up to 20% relative noise in the data.

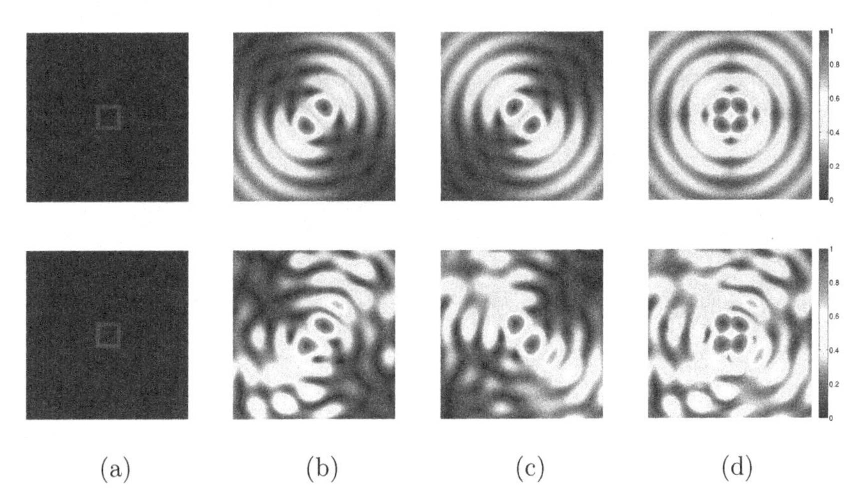

Fig. 6.3 Numerical results: (a) true scatterer; (b), (c) and (d), reconstructions respectively for the incident direction d_1, d_2 and both directions. The first and second rows are respectively for the exact data and noisy data with 20% noise in the data.

The estimate can be further refined by more expensive inversion techniques. In particular, it allows screening the computational domain for the indirect method, thereby bringing significant computational benefits. In Fig. 6.4, we present one refinement via a multi-parameter Tikhonov regularization model (cf. Section 3.6 for the general mathematical theory):

$$\tfrac{1}{2}\|K\eta - g^\delta\|^2_{L^2(\Gamma)} + \alpha\|\eta\|_{L^1(\Omega)} + \frac{\beta}{2}\|\nabla\eta\|^2_{L^2(\Omega)},$$

where the linear operator K is the Lippman-Schwinger integral operator, cf. (2.5), linearized around the first estimate $\Phi|_D$. The purpose of the first and second penalty terms in the multi-parameter model is to encourage a groupwise sparsity structure: localized yet smooth within the groups. The multi-parameter model can be efficiently solved using a semismooth Newton method, cf. Section 5.2.5. The index Φ identifies a sharp estimate of the scatterer support, and hence one may restrict the computation to a small subdomain D, which represents the region of interest, and seeks refinements only in the region D. The refined reconstructions are very appealing: the locations and magnitudes are accurately resolved, especially after taking into account that only two incident fields have been used.

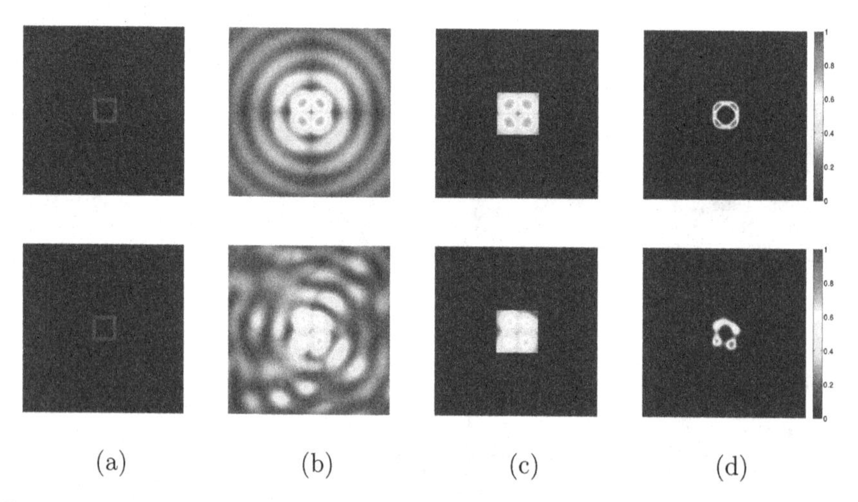

Fig. 6.4 Numerical results with further refinement: (a) true scatterer; (b) index Φ, (c) index $\Phi|_D$ and (d) sparse reconstruction. The first and second rows are respectively for the exact data and noisy data with 20% noise in the data.

The promising numerical results motivate the extension to more complex scattering scenarios, e.g., Maxwell system and elastic wave system. The key to such extension is to design probing functions that satisfy the following important properties: (1) the capability to represent scattered field representation, and (2) the approximate orthogonality relation. For the acoustic and electromagnetic scattering, the fundamental solution is a natural candidate, and numerically have been very successful [157, 158]. However, in general, a systematic strategy is still not available.

The extension of the direct sampling method for far-field scattered data u^∞ is straightforward in view of the asymptotic for the fundamental solution $G(x,y)$, i.e., (6.4). The index function $\Phi(x_p)$ is then given by

$$\Phi(x_p) = \frac{|\langle u^\infty, G^\infty(\cdot, x_p)\rangle_{L^2(\mathbb{S}^{d-1})}|}{\|u^\infty\|_{L^2(\mathbb{S}^{d-1})}\|G^\infty(\cdot, x_p)\|_{L^2(\mathbb{S}^{d-1})}}.$$

Up to a normalizing constant, it is equivalent to

$$\Phi(x_p) = |\int_{\mathbb{S}^{d-1}} u^\infty(\hat{x})e^{ik\hat{x}\cdot x_p}d\hat{x}|.$$

It is known as orthogonality sampling in the literature, first introduced by Potthast [253] and then analyzed in [120] for the multi-frequency case.

6.2 Point source identification

In this part, we describe a direct method for identifying point sources in the Laplace equation from the Cauchy data on the whole boundary Γ, developed by El Badia and Ha-Duong [86]; see Section 2.2.2 for an introduction to the inverse source problem. The mathematical model is as follows: given the Cauchy data ($f = u|_\Gamma, g = \frac{\partial u}{\partial n}|_\Gamma$) of a solution u to the Poisson's equation

$$-\Delta u = F \quad \text{in } \Omega,$$

we look for a source F is from the admissible set of monopoles

$$\mathcal{A} = \left\{ F = \sum_{k=1}^{m} \lambda_k \delta_{S_k} \right\},$$

i.e., identifying the unknowns m, $\{\lambda_k\}_{k=1}^m$ and $\{S_k\}_{k=1}^m$. Clearly, this is a nonlinear inverse problem.

The starting point of the direct inversion method is the reciprocity gap functional defined by

$$R(v) = \langle v, g\rangle_{H^{\frac{1}{2}}(\Gamma), H^{-\frac{1}{2}}(\Gamma)} - \left\langle f, \frac{\partial v}{\partial n}\right\rangle_{H^{\frac{1}{2}}(\Gamma), H^{-\frac{1}{2}}(\Gamma)},$$

where v is a harmonic function in Ω, i.e., $v \in H(\Omega) = \{v \in H^1(\Omega) : \Delta v = 0\}$. Using Green's formula, it is easy to see

$$R(v) = \sum_{k=1}^{m} \lambda_k v(S_k), \quad \forall v \in H(\Omega). \tag{6.11}$$

Therefore, the inverse problem is reduced to determining the unknowns (m, $\{\lambda_k\}$, $\{S_k\}$) when the sum in (6.11) is known for all $v \in H(\Omega)$.

The algorithm relies on a clever choice of harmonic test functions. In the two-dimensional case, the real and imaginary parts of any polynomial of the complex $z = x + iy$ are both harmonic. Clearly, the formula (6.11) remains valid for complex-valued harmonic functions v. Hence one may introduce the complex numbers

$$\alpha_j =: R(z^j) = \sum_{k=1}^{m} \lambda_k (S_k)^j,$$

where the point source S_k is identified with its complex affix.

Next we introduce matrices $A_j = [(A_j)_{kl}] \in \mathbb{C}^{M\times m}$ ($j \in \mathbb{N}$) with its klth entry given by $(A_j)_{kl} = S_l^{j+k-1}$, where M is an upper bound on the number of monopoles, and $D = \mathrm{diag}(S_1, S_2, \ldots, S_m)$, and the vectors

$$\mu_j = (\alpha_j, \ldots, \alpha_{M+j-1})^{\mathrm{t}},$$
$$\Lambda = (\lambda_1, \lambda_2, \ldots, \lambda_m)^{\mathrm{t}}.$$

The matrices A_j and D and vectors μ_j and Λ have the following property.

Lemma 6.1. *The vectors μ_j and Λ and matrices A_j and D satisfy*

(a) For all $j \in \mathbb{N}$, the vectors μ_j and matrices A_j verify the relations

$$\mu_j = A_j\Lambda \quad \textit{and} \quad A_{j+1} = A_j D.$$

(b) The rank of the family $(\mu_0, \mu_1, \ldots, \mu_{M+1})$ is $r = m$, and the vectors $(\mu_0, \mu_1, \ldots, \mu_{m-1})$ are linearly independent.

Proof. Part (a) of the lemma is immediate. The vectors μ_j are images of vectors Λ, $D\Lambda, \ldots, D^{M-1}\Lambda \in \mathbb{C}^m$ by the same matrix A_0, and hence $r \leq m$. Now let $\{c_j\}$ be complex constants such that

$$\sum_{j=0}^{m-1} c_j\mu_j = A_0\left(\sum_{j=0}^{m-1} c_j D^j \Lambda\right) = 0.$$

Since the first m rows and columns of A_0 form an invertible Vandermonde matrix (because S_k are distinct), there holds

$$\sum_{j=0}^{m-1} c_j D^j \Lambda = 0,$$

i.e., a polynomial of degree at most $m-1$ with m distinct roots $\{S_k\}_{k=1}^m$. The desired assertion in part (b) follows directly. □

Now we can present a uniqueness theorem, whose proof is constructive and can be used to develop a direct method.

Theorem 6.2. *Let M be a known upper bound for the number m of the point sources. Then the unknowns m, $\{\lambda_k\}$ and $\{S_k\}$ are uniquely determined by $2M$ numbers $\{\alpha_j\}_{j=0}^{2M-1}$.*

Proof. By replacing M by m in the matrices A_j and D and vectors μ_j, Λ, the square matrix A_0 is invertible. Further, Lemma 6.1 shows that there exists a unique matrix T satisfying the relations

$$\mu_{j+1} = T\mu_j \quad 0 \leq j \leq m-1.$$

Again from Lemma 6.1, one has for all j:

$$\mu_{j+1} = A_0 D^{j+1}\Lambda = A_0 D A_0^{-1} A_0 D^j \Lambda = A_0 D A_0^{-1}\mu_j,$$

and thus the matrix T is given by $T = A_0 D A_0^{-1}$. Therefore,

(1) the affixes $\{S_k\}$ are the eigenvalues of T,
(2) the intensity Λ is the solution of the system $\mu_0 = A_0\Lambda$.

This shows directly the desired assertion. □

The proof of Lemma 6.1 and Theorem 6.2 contains a constructive method for finding the locations and strengths of the poles. The complete procedure is listed in Algorithm 6.2. Numerically, one observes that the algorithm is relatively sensitive to the noise in the Cauchy data, since the matrix H in part (b) of Lemma 6.1 is a Hankel matrix, which is known be highly ill-conditioned even if the matrix size is fairly small. This is reminiscent of the ill-posed nature of the Cauchy problem for the Laplace equation. A rank-revealing procedure, such as singular value decomposition (cf. Appendix A), is often employed to determine the effective number m of poles. We note that by the proof of Theorem 6.2, the matrix T at step 5 is a companion matrix, and can be found directly by solving a linear system of size m. We remark that the severe ill-conditioning of the Hankel matrix indicates that the algorithm can work well only for a small number of point sources.

Remark 6.1. Algorithm 6.2 can be extended to identify dipole sources:

$$\mathcal{A} = \left\{ F = -\sum_{k=1}^{m} p_k \cdot \nabla\delta_{S_k} \right\}$$

Algorithm 6.2 El Badia-Ha Duong algorithm

1: Given the Cauchy data (f, g), and upper bound M;
2: Compute the harmonic moments α_j, $j = 0, \ldots, 2M-1$;
3: Form the matrix $H = [\mu_0 \ \mu_1 \ \cdots \ \mu_M]$;
4: Determine the rank m of H by singular value decomposition;
5: Find the eigenvalues of reduced matrix T;
6: Solve for $\Lambda = [\lambda_k]$ by $\mu_0 = A_0 \Lambda$.

where the integer m, the centers S_k, and the moments p_k of the dipoles are unknown. It straightforward to see that the reciprocity gap functional $R(v)$ for the monomials $v_j = z^j$ is given by

$$R(v_j) = \sum_{k=1}^{m} j(p_{k,1} + \mathrm{i}p_{k,2}) S_k^{j-1}.$$

Upon letting $\mu_j = R(v_j)/j$, then Algorithm 6.2 can be applied directly.

Algorithm 6.2 can be used efficiently for other more complicated linear or nonlinear inverse problems. For example, this idea has been pursued in [126] for the inverse source problem for the Laplace equation, where they first find the locations and size of the distributed source from the point source locations and strengths. The rough estimates are then refined by e.g., Newton method. A similar idea might be applied to the transmission problem of determining the interfaces or cracks from Cauchy data.

6.3 Numerical unique continuation

In this part, we describe one direct method due to the first author and M. Yamamoto for numerically continuing a harmonic function from one line (or line segment) to the domain. Let the domain Ω be $\Omega = \{(x, y) \in \mathbb{R}^2 : x > 0\}$, and its boundary $\Gamma = \{(x, y) \in \mathbb{R}^2 : x = 0\}$. There holds the Laplace equation, i.e.,

$$u_{xx} + u_{yy} = 0 \quad \text{in } \Omega \tag{6.12}$$

and the Cauchy data

$$u = g \quad \text{and} \quad u_x = h \ \text{ on } \Gamma.$$

The direct method builds on the fact that the function $u(x, y)$ is harmonic and real analytic. Therefore, it can be represented as

$$u(x,y) = u(0,y) + u_x(0,y)x + \frac{1}{2}u_{xx}(0,y)x^2 + \frac{1}{3!}u_{xxx}(0,y)x^3 + ...$$
$$= \sum_{i=0}^{\infty} \frac{x^i}{i!}\frac{\partial^i u(0,y)}{\partial x^i}.$$

Therefore, it suffices to determine the sequence of functions $\left\{\frac{\partial^i u(0,y)}{\partial x^i}\right\}_{i=0}^{\infty}$. According to the Cauchy data, $u(0,y) = g(y)$ and $u_x(0,y) = h(y)$. To determine the higher-order derivatives, we utilize the fundamental relation (6.12). To illustrate the point, we compute $u_{xx}(0,y)$ and $u_{xxx}(0,y)$. It follows from (6.12) that

$$u_{xx}(0,y) = -u_{yy}(0,y) = -\frac{\partial^2}{\partial y^2}(u(0,y))$$
$$= -\frac{\partial^2}{\partial y^2}g(y) = -g''(y),$$

and by differentiating (6.12) with respect to x, we arrive at

$$u_{xxx}(0,y) = -u_{yyx}(0,y) = -\frac{\partial^2}{\partial y^2}(u_x(0,y))$$
$$= -\frac{\partial^2}{\partial y^2}h(y) = -h''(y).$$

Analogously to the preceding derivations, we rewrite (6.12) as

$$u_{xx}(x,y) = -u_{yy}(x,y).$$

Applying the identity repeatedly yields that for any $i > 2$

$$\frac{\partial^i}{\partial x^i}u(0,y) = \frac{\partial^{i-2}}{\partial x^{i-2}}\left[\frac{\partial^2}{\partial x^2}u(0,y)\right] = \frac{\partial^{i-2}}{\partial x^{i-2}}\left[-u_{yy}(0,y)\right]$$
$$= \frac{\partial^{i-4}}{\partial x^{i-4}}\frac{\partial^2}{\partial x^2}\left[-u_{yy}(0,y)\right] = (-1)^2\frac{\partial^{i-4}}{\partial x^{i-4}}\frac{\partial^4}{\partial y^4}u(0,y)$$
$$= \cdots = \begin{cases} (-1)^{\frac{i-1}{2}}h^{(i-1)}(y) & \text{odd } i, \\ (-1)^{\frac{i}{2}}g^{(i)}(y) & \text{even } i. \end{cases}$$

It is easy to verify that the formula holds also for $i \le 2$. Consequently, the solution $u(x,y)$ can be represented by

$$\begin{aligned} u(x,y) &= \sum_{i=0}^{\infty}\frac{x^i}{i!}\frac{\partial^i u(0,y)}{\partial x^i} \\ &= \sum_{i=0,\text{even}}^{\infty}\frac{x^i}{i!}\frac{\partial^i}{\partial x^i}u(0,y) + \sum_{i=0,\text{odd}}^{\infty}\frac{x^i}{i!}\frac{\partial^i}{\partial x^i}u(0,y) \qquad (6.13) \\ &= \sum_{i=0,\text{even}}^{\infty}\frac{(-1)^{\frac{i}{2}}x^i g^{(i)}(y)}{i!} + \sum_{i=0,\text{odd}}^{\infty}\frac{(-1)^{\frac{i-1}{2}}x^i h^{(i-1)}(y)}{i!}. \end{aligned}$$

The representation (6.13) enables evaluating the value of the function $u(x, y)$ at any point in the half-plane Ω, if the Cauchy data are sufficiently differentiable. The method is strictly direct, and only involves computing the derivatives of the Cauchy data. In practice, the Cauchy data are noisy and do not allow very high-order differentiation, and thus a truncation of the expansion order would be needed to stabilize the inherent ill-posedness of unique continuation. Further, we note that the method is also applicable to rectangular domains with the Cauchy data specified on one side of the domain or cubic domains with data on one surface.

6.4 Gel'fand-Levitan-Marchenko transformation

Now we describe an elegant approach for inverse Sturm-Liouville problem based on the Gel'fand-Levitan-Marchenko transformation [226, 100], and numerically realized efficiently in [266].

6.4.1 *Gel'fand-Levitan-Marchenko transformation*

We first recall the transformation due to Gel'fand and Levitan [100], which in itself originates from an earlier work of Marchenko [226] and has found applications in many different areas. In essence, it transforms the solution of one ordinary differential equation into that of the other, via a Volterra integral operator of the second kind, with the kernel defined by a "wave"-type equation. The key point of deriving the transformation is to construct the adjoint operator of the leading part of the elliptic operator, with suitable boundary conditions.We illustrate the derivation of the transformation for three distinct cases, (i) the classical Sturm-Liouville problem (SLP) in the potential form, (ii) polar coordinate counterpart, and (iii) the discrete analogue.

case (i): SLP in potential form. Let $u(x)$ and $v(x)$ respectively be the solutions to

$$\begin{aligned} -u'' + q(x)u &= \lambda u, \\ -v'' + p(x)v &= \lambda v, \end{aligned}$$

and $u(0) = v(0) = 0$ and $u'(0) = v'(0) = 1$, where $\lambda \in \mathbb{R}$ is a parameter. The condition $u'(0) = v'(0) = 1$ represents a normalization condition. Given a solution v, we define a new function u via the integral transformation

$$u(x) = v(x) + \int_0^x K(x,t)v(t)dt,$$

where the kernel function $K(x,t)$ depends on the potentials q and p, but not on the parameter λ, such that $u(x)$ is the solution to the preceding problem. To find the kernel $K(x,t)$, we compute

$$u''(x) = v''(x) + (K(x,x)v(x))' + K_x(x,x)v(x) + \int_0^x K_{xx}(x,t)v(t)dt. \quad (6.14)$$

Now we define a hyperbolic equation

$$L_x K(x,t) - L_t^* K(x,t) - (p(t) - q(x))K(x,t) = 0,$$

where the differential operator L is given $Lv = -v''$ and L^* is its adjoint, and L_x and L_t refer to the differential operator in x and t variable, respectively.

Then by integration by parts and using the boundary conditions $v(0) = 0$ and $v'(0) = 1$, we can rewrite the term $\int_0^x K_{xx}(x,t)v(t)dt$ as

$$\begin{aligned}\int_0^x K_{xx}(x,t)v(t)dt &= \int_0^x (q(x) - p(t))K(x,t)v(t)dt + \int_0^t K_{tt}(x,t)v(t)dt \\ &= \int_0^x q(x)K(x,t)v(t)dt + K_t(x,x)v(x) - K(x,x)v'(x) \\ &\quad + K(x,0) - \int_0^x K(x,t)(L_t + p(t))v(t)dt,\end{aligned}$$

where the subscripts x and t refer to taking derivative with respect to x and t, respectively. Now substituting this back to (6.14) and noting the relations

$$(K(x,x)v(x))' = K_x(x,x)v(x) + K_t(x,x)v(x) + K(x,x)v'(x)$$

and

$$(L_t + p(t))v(t) = \lambda v(t),$$

the definition of u, and the total derivative $\frac{d}{dx}K(x,x) = K_x(x,x) + K_t(x,x)$ yields

$$\begin{aligned}u''(x) &= v''(x) + 2v(x)\frac{d}{dx}K(x,x) + \int_0^x q(x)K(x,t)v(t)dt \\ &\quad + K(x,0) - \lambda \int_0^x K(x,t)v(t)dt, \\ &= p(x)v(x) - \lambda u(x) + 2v(x)\frac{d}{dx}K(x,x) \\ &\quad + \int_0^x q(x)K(x,t)v(t)dt + K(x,0).\end{aligned}$$

Consequently, we have

$$-u''(x) + q(x)u(x) = \lambda u(x) + v(x)[q(x) - p(x) - 2\frac{d}{dx}K(x,x)] + K(x,0).$$

So it suffices to set

$$2\frac{d}{dx}K(x,x) = q(x) - p(x) \quad \text{and} \quad K(x,0) = 0$$

so that u does solve the initial value problem.

(ii) SLP in polar coordinates. Here the model is reduced from radial symmetry, and the differential operator L is given by $Lv = -\frac{1}{r}(rv')'$. Accordingly, the Sturm-Liouville problem is given by

$$\begin{aligned} Lu + q(r)u &= \lambda u, \\ Lv + p(r)v &= \lambda v. \end{aligned}$$

Like before, we define the transformation via

$$u(r) = v(r) + \int_0^r K(r,t)v(t)dt,$$

and K satisfies the relation

$$L_r K(r,t) - L_t^* K(r,t) + (q(r) - p(t))K(r,t) = 0,$$

where the adjoint L_t^* reads $L_t^* v = -(t(\frac{v}{t})')'$. Then there hold

$$\begin{aligned} L_r u(r) = L_r v(r) &- \frac{1}{r}(rK(r,r))'v(r) - K(r,r)v'(r) - K_r(r,r)v(r) \\ &+ \int_0^r L_r K(r,t)v(t)dt. \end{aligned}$$

By applying integration by parts and the relation $L_r v + p(r)v = \lambda v$, the integral term $\int_0^r L_r K(r,t)v(t)dt$ can be rewritten as

$$\begin{aligned} \int_0^r L_r K(r,t)v(t)dt &= \int_0^r (-q(r) + p(t))K(r,t)v(t)dt + \int_0^r L_t^* K(r,t)v(t)dt \\ &= -\int_0^r q(r)K(r,t)v(t)dt + \lambda \int_0^t K(r,t)v(t)dt \\ &\qquad - t(\frac{K(r,t)}{t})_t|_{t=r}v(r) + K(r,r)v'(r). \end{aligned}$$

Combining the above two identities with the relation $L_r v + p(r)v = \lambda v$ and the definition of u yields

$$L_r u(r) + q(r)u(r) = \lambda u(r) + v(r)[-2\frac{d}{dr}K(r,r) + q(r) - p(r)].$$

Upon setting $2\frac{d}{dr}K(r,r) = q(r) - p(r)$, we arrive at the desired assertion.

(iii) Discrete SLP. Now we present a discrete analogue of the Sturm-Liouville problem, i.e.

$$-\delta u + qu = \lambda u,$$
$$-\delta v + pv = \lambda v,$$

where $u(0) = v(0) = 0$, and

$$\delta u_i = u_{i+1} - 2u_i + u_{i-1}$$

is the discrete Laplacian. This arises, e.g., in the finite difference discretization of the classical SLP. Then we seek a discrete integral transformation such that the function u can be expressed in terms of v as

$$u_i = v_i + \sum_{j=1}^{i} k_{ij} v_j,$$

where the lower-triangle kernel matrix $K = [k_{ij}]$ depends on the (discrete) potentials q and p, but not on the parameter λ. To find the kernel k_{ij}, we compute

$$\delta u_i = \delta v_i + k_{i+1,i+1} v_{i+1} - k_{i-1,i} v_i + \sum_{j=1}^{i} \delta_1 K_{ij} v_j,$$

where δ_1 refers to the discrete Laplacian in the first index. Now we define a discrete hyperbolic equation

$$\delta_1 k_{ij} - \delta_2 k_{ij} + (p_j - q_i) k_{i,j} = 0,$$

where δ_2 refers to the discrete Laplacian in the second index. Then upon invoking the summation by parts formula, the discrete summation $\sum_{j=1}^{i} \delta_1 K_{ij} v_j$ can be rewritten as

$$\begin{aligned}
\sum_{j=1}^{i} \delta_1 k_{ij} v_j &= \sum_{j=1}^{i} (q_i - p_j) k_{i,j} v_j + \delta_2 k_{i,j} v_j \\
&= \sum_{j=1}^{i} (q_i - p_j) k_{ij} v_j + \sum_{j=1}^{i} k_{i,j} \delta v_j + k_{i,i+1} v_i - k_{ii} v_{i+1} + k_{i0} v_1 \\
&= q_i \sum_{j=1}^{i} k_{ij} v_j - \lambda \sum_{j=1}^{i} k_{ij} v_j + k_{i,i+1} v_i - k_{ii} v_{i+1} + k_{i0} v_1.
\end{aligned}$$

Now by substituting this back and using the definition of u, we derive

$$\begin{aligned}
-\delta u_i + q_i u_i = \lambda u_i &+ (q_i - p_i) v_i - k_{i+1,i+1} v_{i+1} + k_{i-1,i} v_i \\
&- k_{i,i+1} v_i + k_{i,i} v_{i+1} - k_{i0} v_1.
\end{aligned}$$

We note that asymptotically, both $k_{i+1,i+1}-k_{i,i}$ and $k_{i,i+1}-k_{i-1,i}$ represent a discrete analogue of the total derivative $\frac{d}{dx}K(x,x)$ in the sense that

$$(k_{i+1,i+1}-k_{i,i})v_{i+1}+(k_{i,i+1}-k_{i-1,i})v_i \to 2\frac{d}{dx}K(x,x)v(x).$$

Hence, the discrete transformation is consistent with the continuous counterpart. Meanwhile, since the values $k_{i,i+1}$ and $k_{i-1,i}$ are not needed in the definition of the transformation, and never used in the numerical implementation, one may choose them such that

$$(k_{i+1,i+1}-k_{i,i})v_{i+1}+(k_{i,i+1}-k_{i-1,i})v_i = 2(k_{i+1,i+1}-k_{i,i})v_i.$$

This simplifies the formula to:

$$-\delta u_i + q_i u_i = \lambda u_i + (q_i - p_i)v_i - 2(k_{i+1,i+1}-k_{i,i})v_i - k_{i0}v_1.$$

Consequently, we arrive at the following discrete transformation

$$\begin{aligned}
&\delta_1 k_{ij} - \delta_2 k_{ij} + (p_j - q_i)k_{i,j} = 0,\\
&k_{i0} = 0, \quad q_i - p_i = 2(k_{i+1,i+1}-k_{i,i}).
\end{aligned}$$

6.4.2 *Application to inverse Sturm-Liouville problem*

Now we illustrate the use of the Gel'fand-Levitan-Marchenko transformation for the inverse Sturm-Liouville problem. To this end, we consider an odd reflection of the differential equation into the domain $t < 0$ and take $p(x) = 0$. It is easy to see that the solution v to the initial value problem is given by $v = \frac{\sin\sqrt{\lambda}x}{\sqrt{\lambda}}$. Hence for any given potential $q \in L^2(0,1)$ there exists a function $K(x,t)$ defined on the triangle $\{0 \le |t| \le x \le 1\}$ such that

$$u(x,\lambda) = \frac{\sin\sqrt{\lambda}x}{\sqrt{\lambda}} + \int_0^x K(x,t)\frac{\sin\sqrt{\lambda}t}{\sqrt{\lambda}}dt, \tag{6.15}$$

and the kernel $K(x,t) = K(x,t;q)$ solves the characteristic boundary value problem

$$\begin{aligned}
&K_{tt} - K_{xx} + q(x)K = 0 \quad 0 \le |t| \le x \le 1,\\
&K(x,\pm x) = \pm\frac{1}{2}\int_0^x q(s)ds \quad 0 \le x \le 1.
\end{aligned} \tag{6.16}$$

On the basis of (6.15) and (6.16), an efficient direct method for the inverse Sturm-Liouville problem was developed in [266]. We briefly sketch the idea of the method for the case of two spectra, i.e., the Dirichlet spectrum $\{\lambda_j\}$ and Dirichlet-Neumann spectrum $\{\mu_j\}$. It consists of the following two steps.

(a) Recovery of Cauchy data for K The first step in the method is to use the given spectral data to determine the Cauchy data for K on $\{x = 1\}$, i.e., the pair of functions $\{K(1,t), K_x(1,t)\}$ for $0 \le t \le 1$, upon noting their oddness in t. This can be derived from (6.15). Evaluating (6.15) with $\lambda = \lambda_j$ and $x = 1$ gives

$$\int_0^1 K(1,t)\sin\sqrt{\lambda_j}t\,dt = -\sin\sqrt{\lambda_j}.$$

Next differentiating (6.15) with respect to x and evaluating at $\lambda = \mu_j$ and $x = 1$ yield

$$\int_0^1 K_x(1,t)\sin\sqrt{\mu_j}t\,dt = -\sqrt{\mu_j}\cos\sqrt{\mu_j} - \frac{1}{2}\sin\sqrt{\mu_j}\int_0^1 q(s)ds.$$

The mean $\int_0^1 q(s)ds$ can be extracted from either the sequence $\{\lambda_j\}_{j=1}^\infty$ or $\{\mu_j\}_{j=1}^\infty$, using the known asymptotics of these spectral data. Hence the unique recovery of the Cauchy data for K follows immediately from the fact that the sequences $\{\sin\sqrt{\lambda_j}t\}$ and $\{\sin\sqrt{\mu_j}t\}$ are complete in the space $L^2(0,1)$ (under minor conditions on the potential q). We note that this step is well-posed, and numerically stable.

(b) Recovery of q from Cauchy data With the Cauchy data for $K(x,t)$ on the line $\{x = 1\}$, the coefficient $q(x)$ is then determined from the system (6.16). It follows from d'Alambert's formula for the inhomogeneous wave equation with the Cauchy data $\{K(1,t), K_x(1,t)\}$ on the line $\{x = 1\}$ that the solution $K(x,t)$ is given by

$$\begin{aligned} K(x,t) = {} & \frac{1}{2}(K(1,1+t-x) + K(x+t-1)) - \frac{1}{2}\int_{t+x-1}^{1+t-x} K_x(1,s)ds \\ & - \frac{1}{2}\int_x^1\int_{t+y-x}^{t+x-y} q(y)K(y,s)dsdy. \end{aligned}$$

Hence on the line $x = t$, there holds

$$\begin{aligned} K(x,x) = {} & \frac{1}{2}\left(K(1,2x-1) + K(1,1)\right) - \frac{1}{2}\int_{2x-1}^1 K_x(1,s)ds \\ & - \frac{1}{2}\int_x^1\int_y^{2x-y} q(y)K(y,s)dsdy. \end{aligned}$$

Differentiating the equation with respect to x and using (6.16) gives

$$q(x) = 2\left(K_t(1,2x-1) + K_x(1,2x-1)\right) - 2\int_x^1 q(y)K(y,2x-y)dy, \quad (6.17)$$

which is a nonlinear equation in q, since $K = K(x, t; q)$. The term in the bracket can be computed from the Cauchy data $\{K(1,t), K_x(1,t)\}$ directly. The nonlinear equation (6.17) can be solved by the successive approximation method. It can be shown that the fixed point map defined by the right-hand side of (6.17) is contractive in $L^\infty(0,1)$, and hence there exists at most one solution [266]. Numerically, one can observe that a few iterations suffice accurate reconstructions.

Bibliographical notes

The design of direct inversion methods represent an important ongoing theme in numerical methods for inverse problems, and there are many such methods in the literature. We only briefly describe a few direct methods to give a flavor of such methods. It is worth mentioning that for electrical impedance tomography, there are several direct inversion methods, including layer stripping method [279], Ikehata's complex geometric method [149], Brühl-Hanke method [45, 46], and D-bar method [277, 239] etc.

Chapter 7

Bayesian Inference

Now we have developed efficient algorithms for finding a Tikhonov minimizer. A natural question is how plausible it is. Hence, it is highly desirable to have tools for assessing the quality of the inverse solution. Bayesian inference is one principled and systematic statistical framework for this purpose. In this chapter, we describe the basic idea and computational techniques of Bayesian inference. The starting point of Bayesian inference is Bayes' formula, i.e., for two random variables X and Y the conditional probability of X given Y is given by

$$p_{X|Y}(x|y) = \frac{p_{Y|X}(y|x)p_X(x)}{p_Y(y)}, \tag{7.1}$$

where the probability density functions $p_{Y|X}(y|x)$ and $p_X(x)$ are known as the likelihood function, i.e., the conditional density of Y given X, and the prior distribution of X, respectively. Consider the following inverse problem

$$F(X) = Y, \tag{7.2}$$

where X and Y denote the unknown coefficient and the noisy data, respectively, and the forward map $F : \mathbb{R}^m \mapsto \mathbb{R}^n$ is in general nonlinear and stands for the mathematical model. To apply Bayesian inference, we have regarded both the unknown X and the data Y as random variables, and encode the prior knowledge in a probability distribution. For example, if we assume that given $X = x$, Y follows a Gaussian distribution with mean $F(x)$ and variance $\sigma^2 I$, then

$$p_{Y|X}(y|x) = \frac{1}{(2\pi\sigma^2)^{n/2}} e^{-\frac{\|F(x)-y\|^2}{2\sigma^2}}.$$

In practice, the noise covariance may be more complex, and it can be incorporated into the likelihood function $p_{Y|X}(y|x)$ directly by changing the

Euclidean norm $\|\cdot\|$ into a weighted norm. The prior $p_X(x)$ encodes the a priori knowledge of the unknown. For example, if the solution X is sparse, i.e., it has many entries close to zero but also a few large entries, then a Laplace distribution on X is appropriate:

$$p_X(x) = (\tfrac{\lambda}{2})^m e^{-\lambda\|x\|_1}.$$

We use the unnormalized posteriori $p(x, y)$ defined by

$$p(x, y) = p_{Y|X}(y|x)p_X(x),$$

and shall often write

$$p_{X|Y}(x|y) \propto p(x, y)$$

to indicate the posterior probability density $p_{X|Y}(x|y)$ up to a multiplicative constant. Similar notation applies to the likelihood and the prior distribution. The posterior distribution $p_{X|Y}(x|y)$ holds the full information about the inverse problem. That is, one can evaluate the conditional distribution of the parameter X from the observation $Y = y$ using the posterior distribution $p_{X|Y}(x|y)$, which in principle can be used to calibrate the uncertainties associated with the inverse solutions. We shall describe computational methods for exploring the posterior state space, including Markov chain Monte Carlo (MCMC) and approximate inference methods. MCMC methods are exact but expensive, whereas the others are approximate but less expensive. Apart from reliability assessment, we also aim to improve Tikhonov estimates by means of Bayesian modeling, especially model selection and hyperparameter estimation.

For the model based inverse problem we have

$$h(X, \Theta) = Y \text{ subject to constraint } E(X, \Theta) = 0$$

where h is a measurement map and E is a PDE constraint for the state X and the parameter Θ. We may formulate the model errors for the constraint E and the measurement errors jointly as

$$p_{Y|X,\Theta}(y|x, \theta) \propto e^{-\frac{\|E(x,\theta)\|^2}{2\sigma_1^2} - \frac{\|h(x,\theta)-y\|^2}{2\sigma_2^2}}$$

where the constants σ_1 and σ_2 describe both model uncertainties and data uncertainties, respectively, if we assume Gaussian distributions for errors. We note that for the first term in the exponential, the (discrete) norm should be consistent with the natural choice at the continuous level, which often invokes the Riesz mapping (or Sobolev gradient). This can be regarded as preconditioning, and will be essential for the success of the model and its efficient exploration. In addition to the prior distribution $p_\Theta(\theta)$ on the parameter θ, we may also select a prior distribution $p_X(x)$ on the state X. Thus, we have in general the following posterior distribution

$$p(x,\theta|y) \propto e^{-\frac{\|E(x,\theta)\|^2}{2\sigma_1^2} - \frac{\|h(x,\theta)-y\|^2}{2\sigma_2^2}} p_\Theta(\theta)p_X(x). \tag{7.3}$$

The denominator $p_Y(y)$ in (7.1) is defined as

$$p_Y(y) = \int p(x,y)\,dx \text{ or } p_Y(y) = \int p(x,\theta,y)\,dx\,d\theta \tag{7.4}$$

and it is known as the marginal density of Y. We note that the marginalized parameters are parameters for a specific model, and the only remaining variable is the identity of the model itself. In this case, the marginalized likelihood $p_Y(y)$ is the probability of the data y given the model type, not assuming any particular model parameters. In this context, it is often known as the model evidence [224], and indicates the plausibility of a given model for explaining the data y. It forms the basis of model selection via Bayes factor [191]. That is, one can rank the prior models $p_\Theta^k(\theta)$ using the corresponding power $p_Y^k(y)$ (7.4) among a family of possible models expressed in terms of the posterior distributions $p_\Theta^k(\theta)$ of the parameter. For example, one can choose between two candidate models, e.g., the Gaussian model

$$p_\Theta^1(\theta) \propto e^{-\frac{\lambda_1}{2}\|\theta-\bar{\theta}\|^2}$$

and the Laplace model

$$p_\Theta^2(\theta) \propto e^{-\lambda_2\|\theta-\bar{\theta}\|_1}$$

for the parameter Θ using the model evidence. Also, we have used the augmented Tikhonov method in Section 3.5, which represents the joint maximizer of the posterior distribution over the Tikhonov solution and regularization parameters. In general we may maximize the model evidence over all prior models

$$p^k(x,\theta|y) \propto p_{Y|X,\Theta}^k(y|x,\theta)p_\Theta^k(\theta)p_X(x)$$

to select the "best" model.

7.1 Fundamentals of Bayesian inference

The likelihood function $p_{Y|X}(y|x)$ and prior distribution $p_X(x)$ are the two basic building blocks in Bayesian modeling. The former contains the information in the data y, or more precisely the statistics of the noise in the data y, whereas the latter encodes a prior knowledge available about the problem before collecting the data. Bayes' formula (7.1) provides an information fusion mechanism by integrating the prior knowledge together with the information in the data y.

Below we discuss the basic building blocks, and the application of Bayesian inference in the context of inverse problems. The more practical aspects of efficient implementations will be discussed in detail in Sections 7.3 and 7.4.

Noise model The presence of noise in the data y represents one of most salient features of any practical inverse problem. The likelihood function $p_{Y|X}(y|x)$ is completely determined by the noise statistics (and its dependence on the unknown x). In practice, there are diverse sources of errors in the model F and the data y:

(a) Modeling errors δF, i.e., $F = F^\dagger + \delta F$, where $F^\dagger$ is the true physics model. The modeling error δF can be attributed to replacing complex models with reduced ones, e.g., the continuum model in place of the complete electrode model in electrical impedance tomography, and diffusion approximation for the radiative transport model in optical tomography; see Sections 2.3.3 and 2.3.4.
(b) Discretization errors δF_h, i.e., $F = F^\dagger + \delta F_h$. In practical inversion algorithms, the forward model must be approximated by a finite-dimensional one. This can be achieved by the finite element method, finite difference method, finite volume method or boundary element method. The discretization error δF_h can often be rigorously estimated in terms of the mesh size h, under certain regularity conditions on the forward solution.
(c) Data errors δy, i.e., $y = y^\dagger + \delta y$, where $y^\dagger$ is the exact data. One source of data error is attributed to inherently limited precision of experimental instruments like thermal sensors. Further, the instruments occasionally may fail and if not carefully calibrated, can also induce significant bias. The data error can also come from transmission. For example, data transmitted in noisy channels are susceptible to corruptions, typically with pixels either completely corrupted or staying intact. Similar errors can occur in storage, due to hardware failure.

Often in practice, all sources of errors (e.g., these for the forward model F) are lumped into the data y. As was mentioned earlier, a careful modeling and account of all errors in the data y is essential for extracting useful information using the classical Tikhonov regularization. This remains crucial in the Bayesian setting. The most popular and convenient noise model is the additive Gaussian model, i.e.,

$$y = y^\dagger + \xi,$$

where $\xi \in \mathbb{R}^n$ is a realization of the random variable, with its components following an independent identically distributed (i.i.d.) Gaussian distribution with zero mean and variance σ^2. Further, if the noise ξ is independent of the true data $y^\dagger$ (and hence the sought-for solution x), then the likelihood $p_{Y|X}(y|x)$ is given by

$$p_{Y|X}(y|x) = (2\pi\sigma^2)^{-\frac{n}{2}} e^{-\frac{1}{2\sigma^2}\|F(x)-y\|^2}.$$

We shall frequently use the alternative representation $\tau = 1/\sigma^2$ below, and refer to Appendix B for further discussions on the Gaussian model and other noise models.

Prior modeling The prior $p_X(x)$ encodes the prior knowledge about the sought-for solution x in a probabilistic manner. The prior knowledge can be either qualitative or quantitative, including smoothness, monotonicity and features (sparsity). Generally, the prior knowledge can be based on expert opinion, historical investigations, statistical studies and anatomical knowledge etc. Since inverse problems are ill-posed due to lack of information, the careful incorporation of all available prior knowledge is of utmost importance in any inversion technique. So is the case for Bayesian modeling, and the prior plays the role of regularization in a stochastic setting [97]. Hence, prior modeling stays at the heart of Bayesian model construction, and crucially affects the interpretation of the data. One very versatile prior model is Markov random field, i.e.,

$$p_X(x) \propto e^{-\lambda\psi(x)},$$

where $\psi(x)$ is a potential function dictating the interaction energy between the components of the random field x, and can be any regularization functional studied earlier, e.g., sparsity, total variation or smoothness penalty. The scalar λ is a scale parameter, determining the strength of the local/global interactions. It plays the role of a regularization parameter in classical regularization theory, and hence its automated determination is very important. We shall discuss strategies for choosing the parameter λ (and other nuance parameters) in Section 7.2.

Hierarchical models Both likelihood $p_{Y|X}(y|x)$ and the prior $p_X(x)$ can contain additional unknown parameters, e.g.,

$$p_{Y|X}(y|x) = p_{Y|X,\Upsilon}(y|x,\tau) \quad \text{and} \quad p_X(x) = p_{X|\Lambda}(x|\lambda)$$

where τ and λ are the precision (inverse variance) and the scale parameter, respectively. These parameters are generically known as hyperparameters. In Bayesian estimation, the scaling parameter λ of the Markov random field prior, which acts as a regularization parameter, affects the posterior distribution $p_{X|Y}(x|y)$, and more explicitly it substantially affects the posterior point estimates and credible intervals. Hierarchical Bayesian modeling provides an elegant approach to choose these parameters automatically [303, 267], in addition to being a flexible computational tool. Specifically, we view the scale parameter λ and the inverse variance τ as random variables with their own priors (a.k.a., hyperpriors) and determine them from the data y. A convenient choice of these priors is their conjugate distribution, which leads to a posterior distribution in the same family as the prior distribution and often lends itself to computational convenience, while remaining sufficiently flexible. For both λ and τ, the conjugate distribution is given by a Gamma distribution:

$$\begin{aligned} p_\Lambda(\lambda) &= G(\lambda; a_0, b_0) = \frac{b_0^{a_0}}{\Gamma(a_0)}\lambda^{a_0-1}e^{-b_0\lambda}, \\ p_\Upsilon(\tau) &= G(\tau; a_1, b_1) = \frac{b_1^{a_1}}{\Gamma(a_1)}\tau^{a_1-1}e^{-b_1\tau}. \end{aligned} \tag{7.5}$$

Here the parameter pairs (a_0, b_0) and (a_1, b_1) determines the range of the prior knowledge on the parameters λ and τ. For the scale parameter λ, a noninformative prior is often adopted, which roughly amounts to setting a_0 to 1 and b_0 close to zero. As to the inverse variance τ, often a quite reasonable estimate is available by repeated experiments and thus one can apply a narrowly peaked prior distribution; otherwise one can also employ a noninformative prior. Once we have selected the prior parameter pairs (a_0, b_0) and (a_1, b_1), we can apply Bayes' formula (7.1) to form the posterior distribution $p_{X,\Lambda,\Upsilon|Y}(x, \lambda, \tau|y)$ which determines automatically these parameters along with the inverse solution x:

$$p_{X,\Lambda,\Upsilon|Y}(x,\lambda,\tau|y) \propto p_{Y|X,\Upsilon}(y|x,\tau)p_{X|\Lambda}(x|\lambda)p_\Lambda(\lambda)p_\Upsilon(\tau).$$

Connection with Tikhonov regularization Before proceeding to the use and interpretation of Bayesian formulations, let us illustrate its connection with the classical Tikhonov regularization, and see how the Bayesian formulation can complement the (classical) Tikhonov regularization with new ingredients. We illustrate this with the simplest Gaussian noise model and the sparsity constraint with a Laplace prior, i.e.,

$$p_{Y|X,\Upsilon}(y|x,\tau) \propto \tau^{-\frac{n}{2}} e^{-\frac{\tau}{2}\|F(x)-y\|^2},$$
$$p_{X|\Lambda}(x|\lambda) \propto \lambda^m e^{-\lambda\|x\|_1}.$$

In case of known λ and τ, a popular rule of thumb is to consider the maximum a posteriori (MAP) estimate x_{map}, i.e.,

$$\begin{aligned} x_{\mathrm{map}} &= \arg\max_x p_{X,\Lambda,\Upsilon|Y}(x,\lambda,\tau|y) \\ &= \arg\min_x \left\{ \tfrac{\tau}{2}\|F(x)-y\|^2 + \lambda\|x\|_1 \right\}. \end{aligned}$$

We note that the functional in the curly bracket can be rewritten as

$$\tfrac{1}{2}\|F(x)-y\|^2 + \lambda\tau^{-1}\|x\|_1,$$

which is nothing other than Tikhonov regularization with sparsity constraint and a regularization parameter $\alpha = \lambda\tau^{-1}$. This establishes the following important connection between Tikhonov regularization and Bayesian estimation

A Tikhonov minimizer is an MAP estimate of some Bayesian formulation.

If the parameters λ and τ are not known a prior, then it is natural to adopt a hierarchical model. Specifically, together with a Gamma distribution on both λ and τ, cf. (7.5), Bayes' formula gives rise to the following posterior distribution

$$\begin{aligned} p_{X,\Lambda,\Upsilon|Y}(x,\lambda,\tau|y) \propto\ & \tau^{\frac{n}{2}+a_1-1} e^{-\frac{\tau}{2}\|F(x)-y\|^2} \\ & \cdot \lambda^{m+a_0-1} e^{-\lambda\|x\|_1} \cdot e^{-b_1\tau} \cdot e^{b_0\lambda}. \end{aligned}$$

There are many different ways of handling the posterior distribution $p_{X,\Lambda,\Upsilon|Y}(x,\lambda,\tau|y)$. One simple idea is to consider the joint maximum a posteriori estimate $(x,\lambda,\tau)_{\mathrm{map}}$, which leads to the following minimization problem (with $\tilde{a}_0 = m + a_0 - 1$ and $\tilde{a}_1 = \frac{n}{2} + a_1 - 1$)

$$(x,\lambda,\tau)_{\mathrm{map}} = \arg\min_{x,\lambda,\tau} J(x,\lambda,\tau),$$

where the functional $J(x,\lambda,\tau)$ is given by

$$J(x,\lambda,\tau) = \tfrac{\tau}{2}\|F(x)-y\|^2 + \lambda\|x\|_1 - \tilde{a}_0 \ln\lambda + b_0\lambda - \tilde{a}_1 \ln\tau + b_1\tau.$$

The functional $J(x,\lambda,\tau)$ is the augmented Tikhonov regularization for sparsity constraint: the first two terms recover Tikhonov regularization, whereas the rest provides the mechanism for automatically determining the regularization parameter. The underlying mechanism essentially follows from the Gamma prior distributions on the parameters, and it is a balancing principle; see Section 3.5 for relevant analysis. We note that the augmented approach does select the hyperparameters λ and τ automatically, but it remains a point estimate and completely ignores the statistical fluctuations around the (joint) mode. Hence, the picture is incomplete from a Bayesian perspective. We shall discuss two alternative principled approaches, i.e., Bayesian information criterion and maximum likelihood method, for selecting hyperparameters in Section 7.2.

What is new in Bayesian inference? The posterior distribution $p_{X,\Lambda,\Upsilon|Y}(x,\lambda,\tau|y)$ represents the complete Bayesian solution to the inverse problem (7.2), and holds all the information about the problem. In comparison with the more conventional Tikhonov regularization, the Bayesian approach has the following distinct features.

First, the Bayesian solution $p_{X,\Lambda,\Upsilon|Y}(x,\lambda,\tau|y)$ is a probability distribution, and encompasses an ensemble of plausible solutions that are consistent with the given data y (to various extent). For example one can consider the mean μ and the covariance C

$$\mu = \int x p_{X|Y}(x|y)dx,$$

$$C = \int (x-\mu)(x-\mu)^{\mathrm{t}} p_{X|Y}(x|y)dx.$$

In particular, the mean μ can be regarded as a representative inverse solution, whereas the covariance C enables one to quantify the uncertainty of a specific solution, e.g., with credible intervals. For example, the 95% credible interval for the ith component μ_i is given by the interval I centered around the mean μ_i, i.e., $I = [\mu_i - \Delta\mu_i, \mu_i + \Delta\mu_i]$, such that

$$\int_{\mathbb{R}^{m-1}} \int_I p_{X|Y}(x|y)dx_i dx_{-i} = 95\%,$$

where x_{-i} denotes $(x_1, \ldots, x_{i-1}, x_{i+1}, \ldots, x_m)^{\mathrm{t}}$. In the presence of hyperparameters λ and τ, one can either fix them at some selected values, e.g., by the model selection rules described in Section 7.2 below, or marginalize them out. In contrast, deterministic techniques generally content with singling out one solution from the ensemble. For example, augmented Tikhonov regularization only looks for the point that maximizes the posterior distribution $p_{X,\Lambda,\Upsilon|Y}(x,\lambda,\tau|y)$. Since it completely ignores the statistical fluctuations, it may not be representative of the ensemble, and the conclusion drawn with only one solution at hand can be misleading.

Second, it shows clearly the crucial role of statistical modeling in designing appropriate regularization formulations for practical problems. Usually, the statistical assumptions of these models are not explicitly spelt out in classical regularization models, which consider only the worst-case scenario, and has not receive due attention, despite their obvious importance in extracting all available information. However, it is now well accepted that a proper statistical modeling is essential for fully extracting the information contained in the data. For example, for impulsive noise, an appropriate likelihood function is the Laplace distribution or t-distribution, since they are robust to the presence of a significant amount of outliers in the data. An inadvertent choice of the computationally convenient Gaussian distribution can significantly compromise the reconstruction accuracy [168].

Third, it provides a flexible regularization since hierarchical modeling can partially resolve the nontrivial issue of choosing an appropriate regularization parameter. The versatile framework, augmented Tikhonov regularization, discussed in detail in Section 3.5, was derived from hierarchical modeling, from which a variety of useful parameter choice rules, including the balancing principle, discrepancy principle and balanced discrepancy principle etc. Further, many other parameter choice rules can be deduced from Bayesian modeling; see Section 7.2 for further discussions.

Last, it allows seamlessly integrating structural/multiscale features of the problem through careful prior modeling. The prior knowledge is essential for solving many practical inverse problems, e.g., in petroleum engineering, biomedical imaging and geophysics. For example, in medical imaging, one can use anatomical information, e.g., segmented region from magnetic resonance imaging or x-ray computerized tomography can assist emission tomography image reconstruction. The idea of using one imaging modality to inform others is now well established.

Because of these distinct features, Bayesian inference has attracted much attention in a wide variety of applied disciplines in recent years, including inverse problems, imaging, machine learning and signal processing, to name a few. However, these nice features do come with a price: it brings huge computational challenges. For systematic methodological developments, we refer to the monographs [282, 186]. Our discussion shall focus on model selection and hyperparameter estimation, and the efficient numerical realization via stochastic and deterministic approaches.

7.2 Model selection

In practice, we generally have a multitude of models which can explain the observed data y to various extents. Here a model M generally can refer to either different priors, e.g., smoothness prior v.s. sparsity prior, as was indicated earlier. It can also refer to a family of priors that depend on a parameter, e.g., $p_{X|\Lambda}(x|\lambda)$, or their discrete analogues. Hence, an immediate practical question is which model we shall select to explain the data y. The marginal $p_Y(y)$, which crucially depends on the choice of the model M, contains valuable information about a given model M. In particular, it measures the plausibility/capability of the model M for explaining the data y. To indicate the dependence of the marginal $p_Y(y)$ on the model M, we shall write $p_{Y|\mathcal{M}}(y|M)$. The Bayesian approach to model selection is to maximize the posterior probability $p_{Y|\mathcal{M}}(y|M)$ of a model M, given the noisy data y. By Bayes' formula (7.1), the posterior probability $p_{\mathcal{M}|Y}(M|y)$ of the model M is given by

$$p_{\mathcal{M}|Y}(M|y) = \frac{p_{Y|\mathcal{M}}(y|M)p_{\mathcal{M}}(M)}{p_Y(y)},$$

where $p_Y(y)$ is the marginal likelihood (over all models) of the data y, and $p_{\mathcal{M}}(M)$ is the prior probability of the model M.

We begin our discussion with the simple case of two models, denoted by M_1 and M_2, and use Bayes factor to select between the two models M_1 and M_2 in question. The Bayes factor is defined as the ratio of the prior and posterior odds:

$$\begin{aligned} BF &= \frac{p_{\mathcal{M}}(M_1)/p_{\mathcal{M}}(M_2)}{p_{\mathcal{M}|Y}(M_1|y)/p_{\mathcal{M}|Y}(M_2|y)} \\ &= \frac{p_{\mathcal{M}}(M_1)p_{\mathcal{M}}(M_2)p_{Y|\mathcal{M}}(y|M_2)}{p_{\mathcal{M}}(M_2)p_{\mathcal{M}}(M_1)p_{Y|\mathcal{M}}(y|M_1)} \\ &= \frac{p_{Y|\mathcal{M}}(y|M_2)}{p_{Y|\mathcal{M}}(y|M_1)}. \end{aligned}$$

This number BF serves as an indicator of which of the models M_j is more supported by the data y. If the factor $BF > 1$, this indicates that M_2 is more strongly supported by the data y; Otherwise, the model M_1 is more strongly supported. Furthermore, the magnitude of the Bayes factor is a measure of how strong the evidence is for or against the model M_1. According to Kass and Raftery [191], when the Bayes factor exceeds 3, 20 and 150, one can say that, accordingly, a positive, strong and overwhelming evidence exists that the model M_2 is preferred.

Now we turn to the case of multiple models. If all candidate models are equally likely, then maximizing the posterior probability of a model M_i given the data y is equivalent to maximizing the marginal likelihood

$$p_{Y|\mathcal{M}}(y|M_i) = \int p_{Y|\Theta}(y|\theta_i)p_\Theta(\theta_i)d\theta_i,$$

where θ_i denotes the parameter vector of the model M_i, and $p_\Theta(\theta_i)$ is the (prior) probability distribution of θ_i, which might vary for different models. In order to arrive at a concise rule, we approximate the marginal likelihood $p_{Y|\mathcal{M}}(y|M_i)$ by Laplace's method. To this end, we first note that

$$\begin{aligned} p_{Y|\mathcal{M}}(y|M_i) &= \int p_{Y|\Theta}(y|\theta_i)p_\Theta(\theta_i)d\theta_i \\ &= \int e^{\ln p_{Y|\Theta}(y|\theta_i)p_\Theta(\theta_i)}d\theta_i. \end{aligned}$$

Next we expand $\ln p_{Y|\Theta}(y|\theta_i)p_\Theta(\theta_i)$ around the posterior mode $\tilde{\theta}_i$, i.e., the MAP estimate, where the posterior distribution $p_{Y|\Theta}(y|\theta_i)p_\Theta(\theta_i)$ achieves its strict maximum (assuming it has one) and so does the log posteriori $Q := \ln p_{Y|\Theta}(y|\theta_i)p_\Theta(\theta_i)$. Then we can approximate Q by its second-order Taylor expansion around $\tilde{\theta}_i$

$$Q \approx Q|_{\tilde{\theta}_i} + (\theta_i - \tilde{\theta}_i, \nabla_{\theta_i} Q|_{\tilde{\theta}_i}) + \tfrac{1}{2}(\theta_i - \tilde{\theta}_i)^{\mathrm{t}} H_{\theta_i}(\theta_i - \tilde{\theta}_i),$$

where $H_{\tilde{\theta}_i}$ is the Hessian matrix. By the optimality of $\tilde{\theta}_i$, $H_{\tilde{\theta}_i}$ is negative definite, and $\nabla_{\theta_i} Q|_{\tilde{\theta}_i} = 0$. We denote by $\tilde{H}_{\theta_i} = -H_{\theta_i}|_{\theta_i=\tilde{\theta}_i}$ and then approximate $p_{Y|\mathcal{M}}(y|M_i)$ by

$$\begin{aligned} p_{Y|\mathcal{M}}(y|M_i) &\approx \int e^{Q|_{\tilde{\theta}_i} - \frac{1}{2}(\theta_i-\tilde{\theta}_i)^{\mathrm{t}}\tilde{H}_{\theta_i}(\theta_i-\tilde{\theta}_i)}d\theta_i \\ &= e^{Q|_{\tilde{\theta}_i}} \int e^{-\frac{1}{2}(\theta_i-\tilde{\theta}_i)^{\mathrm{t}}\tilde{H}_{\theta_i}(\theta_i-\tilde{\theta}_i)}d\theta_i \\ &= p_{Y|\Theta}(y|\tilde{\theta}_i)p_\Theta(\tilde{\theta}_i)\frac{(2\pi)^{|\theta_i|/2}}{|\tilde{H}_{\theta_i}|^{1/2}}. \end{aligned}$$

This procedure is known as Laplace's method in the literature. The approximation is reasonable if there are a large number of data points, for which the posterior distribution is narrowly peaked around the mode $\tilde{\theta}_i$. Consequently, we arrive at

$$2\ln p_{Y|\mathcal{M}}(y|M_i) \approx 2\ln p_{Y|\Theta}(y|\tilde{\theta}_i) + 2\ln p_\Theta(\tilde{\theta}_i) + |\theta_i|\ln 2\pi + \ln|\tilde{H}_{\theta_i}^{-1}|.$$

To shed further insight into the approximation, we consider the case of $p_\Theta(\theta_i) = 1$, a noninformative prior. Then the posterior mode $\tilde{\theta}_i$ is identical with the maximum likelihood (ML) estimate $\hat{\theta}_i$. Further, each entry in the Hessian $\tilde{H}_{\theta_i}$ can be expressed as

$$\tilde{H}_{jk} = -\frac{\partial^2 \ln p_{Y|\Theta}(y|\theta_i)}{\partial\theta_{i,j}\partial\theta_{i,k}}|_{\theta_i=\hat{\theta}_i}.$$

That is, the Hessian $\tilde{H}_{\theta_i}$ is the observed Fisher information matrix. Moreover, if the observed data y is i.i.d., and the number n of data points is large, the weak law of large numbers implies that

$$\begin{aligned}\ln p_{Y|\Theta}(y|\theta_i) &= \frac{1}{n}\sum_{l=1}^{n} n \ln p_{Y|\Theta}(y_l|\theta_i) \\ &\to E[n \ln p_{Y|\Theta}(y_l|\theta_i)] \text{ in probability.}\end{aligned}$$

Consequently, the observed Fisher information matrix $\tilde{H}_{jk}$ simplifies to

$$\begin{aligned}\tilde{H}_{jk} &= -\frac{\partial^2 E[n \ln p_{Y|\Theta}(y_1|\theta_i)]}{\partial\theta_{i,j}\partial\theta_{i,k}}|_{\theta_i=\hat{\theta}_i} \\ &= -n\frac{\partial^2 E[\ln p_{Y|\Theta}(y_1|\theta_i)]}{\partial\theta_{i,j}\partial\theta_{i,k}}|_{\theta_i=\hat{\theta}_i} = nI_{jk},\end{aligned}$$

so $|\tilde{H}_{\theta_i}| = n^{|\theta_i|}|I_{\theta_i}|$, where $|\theta_i|$ is the number of parameters, and I_{θ_i} coincides with the Fisher information matrix for a single data point y_1. In summary, we get

$$\begin{aligned}\ln p_{Y|\mathcal{M}}(y|M_i) \approx& \ln p_{Y|\Theta}(y|\tilde{\theta}_i) + \ln p_{\Theta}(\tilde{\theta}_i) + \tfrac{|\theta_i|}{2}\ln 2\pi \\ &- \tfrac{|\theta_i|}{2}\ln n - \tfrac{1}{2}\ln|I_{\theta_i}|.\end{aligned}$$

For large n, keeping only the terms involving n and ignoring the rest yields

$$\ln p_{Y|\mathcal{M}}(y|M_i) = \ln p_{Y|\Theta}(y|\tilde{\theta}_i) - \tfrac{|\theta_i|}{2}\ln n,$$

which is known as the Bayesian information criterion for the model M_i. It is derived under a number of approximations, e.g., the Gaussian approximation around the mode $\tilde{\theta}_i$. The term $|\theta_i|\ln n$ is a complexity measure of the model M_i, and penalizes models of high complexity, thereby favoring models of low complexity.

The consistency of model selection procedures (as n tends to infinity or equivalently noise level tends to zero) is well studied in the statistical literature. We shall not dwell on this important issue, but refer interested readers to [305, 52] and references therein.

Now we give two examples illustrating the usage of the model selection idea: the first is on sparsity constraints, and the second is a probabilistic "MUSIC" algorithm.

Example 7.1. In this example, we consider a linear inverse problem $Fx = y$, $F \in \mathbb{R}^{n\times m}$, with sparsity constraints, i.e.,

$$p_{X|Y}(x|y) \propto e^{-\frac{\tau}{2}\|Fx-y\|^2} \cdot e^{-\lambda\|x\|_1}.$$

We regard the parametric family of priors $p_{X|\Lambda}(x|\lambda)$ as models, $p_{X|\Lambda}(x|\lambda) \propto \lambda^m e^{-\lambda\|x\|_1}$, with λ being the parameter. Surely, the normalizing constant $p_Y(y)$ depends on the parameter value λ, i.e., $p_{Y|\Lambda}(y|\lambda)$. By means of model selection, a natural idea is to look for the maximizer of the normalizing constant $p_{Y|\Lambda}(y|\lambda)$, and with the above derivation, we arrive at the following Bayesian information criterion for selecting the parameter λ by maximizing

$$\Psi(\lambda) = -\frac{\tau\|Fx_\lambda - y\|^2}{2n} - \frac{df(x_\lambda)}{2n}\ln n,$$

where x_λ is the posterior mode, and $df(x_\lambda)$ is the degree of freedom of the estimate x_λ. It is known that the support size of x_λ, i.e., $\|x_\lambda\|_0$, is a good estimate of the degree of freedom for sparsity regularization [314].

Example 7.2. In this example, we consider inverse scattering with one single point scatterer, i.e.,

$$-\Delta u - k^2 u = \delta_x.$$

Let y represent a measurement of u on a surface Γ, and $F(x)$ denote the forward map. We aim to determine the location of the point scatterer using the model evidence. Under the assumption that the data y is contaminated by i.i.d. Gaussian noise with mean zero and variance σ^2, the likelihood function $p_{Y|X}(y|x)$ is given by

$$p_{Y|X}(y|x) \propto e^{-\frac{1}{2\sigma^2}\|y-F(x)\|^2}.$$

Let $\{x_i\}$ be a set of sampling points in the domain and far from the measurement surface Γ. We consider the collection of prior models $p_X^i(x) = \delta_{x_i}$. For the prior model $p_X^i(x)$, the model evidence is given by

$$\int p_{Y|X}(y|x)p_X^i(x)dx = e^{-\frac{1}{2\sigma^2}\|y-F(x_i)\|^2}.$$

According to the model selection principle, we select the prior model maximizing the model evidence. Note that

$$e^{-\frac{1}{2\sigma^2}\|y-F(x_i)\|^2} = e^{-\frac{1}{2\sigma^2}[\|y\|^2 - 2\Re(y,F(x_i)) + \|F(x_i)\|^2]}.$$

Now under the assumption x_is are far from Γ, we may assume the term $\|F(x_i)\|^2$ is small for all x_i or their magnitudes are commeasurable for all sampling points $\{x_i\}$. Further, under Born's approximation, $F(x_i) \approx$

$G(x_i, z)$. Hence, maximizing the model evidence is equivalent to maximizing $\Re(y, G(x_i, z))$, which is almost identical with the MUSIC algorithm (and the direct sampling method) in Section 6.1. This provides a probabilistic derivation of these algorithms. However, the derivation suggests a higher-order correction, i.e., $\|F(x_i)\|^2$, if the preceding assumption is invalid. Further, the likelihood function can be of alternative form in order to achieve other nice properties, e.g., robustness. In particular, the Cauchy likelihood leads directly to the classical MUSIC indicator function. We observe that here the prior models can be more general, e.g., small circular/spherical regions centered at x_i instead of the point scatterer x_i alone or regions of any other geometrical shapes. The collection of regions can be viewed as a dictionary for plausible scatterers.

In addition to Bayesian information criterion, one can also employ the maximum likelihood method. We illustrate this on a Markov random field prior $p_{X|\Lambda}(x|\lambda) \propto e^{-\lambda\psi(x)}$. Given the observational data y, the maximum likelihood estimate of λ can be found by maximizing the marginal $p_{Y|\Lambda}(y|\lambda)$ defined by

$$\begin{aligned} p_{Y|\Lambda}(y|\lambda) &= \int p_{X|Y,\Lambda}(x|y,\lambda)dx = \int p_{Y|X}(y|x)p_{X|\Lambda}(x|\lambda)dx \\ &= \frac{\int e^{\ln p_{Y|X}(y|x)-\lambda\psi(x)}dx}{\int e^{-\lambda\psi(x)}dx} = \frac{Z(y,\lambda)}{Z(\lambda)}, \end{aligned}$$

where $Z(y, \lambda) = \int e^{\ln p_{Y|X}(y|x)-\lambda\psi(x)}dx$ is the partition function of the posterior density $p_{X|Y,\Lambda}(x|y, \lambda)$, and $Z(\lambda)$ is the partition function of the prior distribution. Therefore,

$$\ln p_{Y|\Lambda}(y|\lambda) = \ln Z(y, \lambda) - \ln Z(\lambda)$$

and the ML estimate of hyperparameter λ is a root of the equation

$$\frac{\partial \ln Z(y,\lambda)}{\partial \lambda} = \frac{\partial \ln Z(\lambda)}{\partial \lambda}.$$

It is straightforward to verify that

$$\frac{\partial \ln Z(\lambda)}{\partial \lambda} = -E_{p_{X|\Lambda}(x|\lambda)}[\psi] \quad \text{and} \quad \frac{\partial \ln Z(y,\lambda)}{\partial \lambda} = -E_{p_{X|Y,\Lambda}(x|y,\lambda)}[\psi].$$

In short, the ML estimate of λ is a root of the following nonlinear equation

$$E_{p_{X|\Lambda}(x|\lambda)}[\psi] = E_{p_{X|Y,\Lambda}(x|y,\lambda)}[\psi].$$

Statistically, the maximum likelihood method chooses the regularization parameter λ by balances the expectation of the penalty ψ under the prior distribution $p_{X|\Lambda}(x|\lambda)$ and that under the posterior distribution $p_{X|Y,\Lambda}(x|y, \lambda)$.

This equation can in principle be solved using an expectation-maximization (EM) algorithm [80], cf. Algorithm 7.1. Steps 3 and 4 are known as the E-step (Expectation) and M-step (maximization), respectively. An exact implementation of the EM algorithm is often impractical due to the complexity and high-dimensionality of the posterior distribution $p_{Y|X,\Lambda}(x|y,\lambda)$ (and possibly also the prior $p_{X|\Lambda}(x|\lambda)$). There are several different ways to arrive at an approximation, e.g., variational method in Section 7.4.

Algorithm 7.1 EM algorithm for hyperparameter estimation.

1: Initialize λ^0 and set $K \in \mathbb{N}$;
2: **for** $k = 0 : K$ **do**
3: estimate the complete-data sufficient statistics

$$\psi^k := E_{p_{X|Y,\Lambda}(x|y,\lambda^k)}[\psi];$$

4: determine λ^{k+1} as the solution to

$$E_{p_{X|\Lambda}(x|\lambda)}[\psi(x)] = \psi^k;$$

5: check the stopping criterion.
6: **end for**

We illustrate the maximum likelihood method with one simple example.

Example 7.3. In this example, we apply the maximum likelihood method to the Gaussian model for a linear inverse problem, i.e.,

$$p_{X|Y,\Lambda}(x|y,\lambda) \propto e^{-\frac{\tau}{2}\|Ax-y\|^2} \cdot \lambda^{\frac{m}{2}} e^{-\frac{\lambda}{2}\|Lx\|^2},$$

where the matrix $L \in \mathbb{R}^{m\times m}$ is of full rank. Here the potential function $\psi(x)$ is given by $\psi(x) = \frac{1}{2}\|Lx\|^2$. For fixed λ^k, the posterior $p_{X|Y,\Lambda}(x|y,\lambda^k)$ is Gaussian with covariance $C^k = (\tau A^{\mathrm{t}}A + \lambda^k L^{\mathrm{t}}L)^{-1}$ and mean $m^k = C^k \tau A^{\mathrm{t}} y$, and hence the E-step of the EM algorithm is given by

$$\psi^k = E_{p_{X|Y,\Lambda}(x|y,\lambda)}[\psi(x)] = \tfrac{1}{2}\|Lx^k\|^2 + \tfrac{1}{2}\mathrm{tr}(C^k L^{\mathrm{t}}L).$$

Meanwhile, the prior expectation is given by

$$E_{p_{X|\Lambda}(x|\lambda)}[\psi(x)] = \tfrac{1}{2}\mathrm{tr}((\lambda L^{\mathrm{t}}L)^{-1}L^{\mathrm{t}}L) = \tfrac{m}{2\lambda}.$$

Consequently, the M-step of the EM algorithm reads

$$\lambda^{k+1} = m(\|Lx^k\|^2 + \mathrm{tr}(C^k L^{\mathrm{t}}L))^{-1}.$$

Clearly, the example can be extended to estimate the precision τ simultaneously with the scale parameter λ. A straightforward computation shows that the EM update for τ is given by

$$\tau^{k+1} = n(\|Kx^k - y\|^2 + \mathrm{tr}(C^k A^{\mathrm{t}}A))^{-1}.$$

We observe that these updates are closely related to the fixed-point algorithm for balancing principle, cf. Algorithm 3.2. Especially, the maximum-likelihood approach can be regarded as a statistical counterpart of the (deterministic) balancing principle discussed in Section 3.5. Further, statistical priors on the hyperparameters can also be incorporated straightforwardly.

Remark 7.1. In general, the E-step and M-step are nontrivial to compute, and hence approximations must be sought [312]. In particular, one can apply the approximate inference methods in Section 7.4 below.

7.3 Markov chain Monte Carlo

The posterior distribution $p(x)$ (we have suppressed the conditional on y and the subscript X for notational simplicity) lives in a very high-dimensional space, and the distribution itself is not directly informative. Therefore, it is necessary to compute summarizing statistics, e.g., mean μ and covariance C defined by

$$\mu = \int xp(x)dx \quad \text{and} \quad C = \int (x-\mu)(x-\mu)^{t}p(x)dx.$$

Below we shall denote μ and C by $E_p[x]$ and $\mathrm{Var}_p[x]$, respectively. These are very high-dimensional integrals, and standard quadrature rules are inefficient due to the notorious curse of dimensionality. For example, for a probability density $p(x)$ living on one-hundred dimensional space, a tensor product type quadrature rule with only two quadrature points in each direction (thus a very crude approximation) would require evaluating $2^{100} \approx 1.27 \times 10^{30}$ points, which is an astronomically large number. A now standard approach to circumvent the computational issue is to use Monte Carlo methods, especially Markov chain Monte Carlo (MCMC) methods [109, 217]. In this part, we describe fundamentals of MCMC methods and acceleration techniques.

7.3.1 *Monte Carlo simulation*

The basic procedure of Monte Carlo simulation is to draw a large set of i.i.d. samples $\{x^{(i)}\}_{i=1}^{N}$ from the target distribution $p(x)$ defined on a high-dimensional space $\mathbb{R}^m$, which maybe only have implicit form, e.g., in PDE constrained parameter identifications. One can then approximate the expectation $E_p[f]$ of any function $f:\mathbb{R}^m \to \mathbb{R}$ by the sample mean $E_N[f]$ as follows

$$E_N[f] \equiv \frac{1}{N}\sum_{i=1}^{N} f(x^{(i)}) \to E_p[f] = \int f(x)p(x)dx \quad \text{as } N \to \infty.$$

More specifically, define the Monte Carlo integration error $e_N[f]$ by

$$e_N[f] = E_p[f] - E_N[f],$$

so that the bias is $E_p[e_N[f]]$ and the root mean square error is $E_p[e_N[f]^2]^{1/2}$. Then the central limit theorem asserts that for large N, the Monte Carlo integration error

$$e_N[f] \approx \mathrm{Var}_p[f]^{\frac{1}{2}} N^{-1/2}\nu,$$

where ν is a standard normal random variable, and $\mathrm{Var}_p[f]$ is the variance of f with respect to p. A more precise statement is

$$\begin{aligned}&\lim_{N\to\infty} \mathrm{Prob}\left(a < \frac{\sqrt{N}}{\sqrt{\mathrm{Var}_p[f]}} e_N[f] < b\right)\\ &= \mathrm{Prob}(a < \nu < b) = \int_a^b (2\pi)^{-1/2} e^{-\frac{x^2}{2}} dx.\end{aligned}$$

Hence the error $e_N[f]$ in Monte Carlo integration is of order $O(N^{-1/2})$ with a constant depending only on the variance of the integrand f. Further, the statistical distribution of the error $e_N[f]$ is roughly a normal random variable. We note that the estimate is independent of the dimensionality of the underlying space $\mathbb{R}^m$.

Generating a large set of i.i.d. samples from an implicit and high-dimensional joint distribution, which is typical for nonlinear inverse problems and nongaussian models of our interest, is highly nontrivial. Therefore, there have been intensive studies on advanced Monte Carlo methods. One useful idea is importance sampling. Suppose $q(x)$ is an easy-to-sample probability density function and it is close to the posterior distribution $p(x)$. Then we can approximate the expectation of the function f with respect to $p(x)$ by

$$\begin{aligned}\int f(x)p(x)dx &= \int f(x)\frac{p(x)}{q(x)}q(x)dx\\ &\approx \frac{1}{N}\sum_{i=1}^{N} f(x^{(i)})w_i,\end{aligned} \tag{7.6}$$

where the i.i.d. samples $\{x^{(i)}\}_{i=1}^N$ are drawn from the auxiliary distribution $q(x)$, and the importance weights w_i are given by $w_i = \frac{p(x^{(i)})}{q(x^{(i)})}$. In the approximation (7.6), the density $p(x)$ is assumed to be normalized, otherwise the normalizing constant should also be computed:

$$\int f(x)p(x)dx \approx \frac{\sum_{i=1}^N f(x^{(i)})w_i}{\sum_{i=1}^N w_i}.$$

The efficiency of importance sampling relies crucially on the quality of the approximation $q(x)$ to the true posterior distribution $p(x)$ in the main probability mass region. In inverse problems, a natural idea is to use the Jacobian approximation of the forward model around the mode, i.e., the Tikhonov minimizer, which can be efficiently computed (or has already been computed within the iterative solution process), using the adjoint technique for PDE parameter identifications. We illustrate this with one example.

Example 7.4. Consider a nonlinear forward model $F(x) = y$, with a Gaussian noise model and a smoothness prior, i.e.,

$$p(x) \propto e^{-\frac{\tau}{2}\|F(x)-y\|^2 - \frac{\lambda}{2}\|Lx\|^2},$$

where τ and λ are fixed parameters, and L is either the identity operator or the differential operator, i.e., the prior distribution imposes the smoothness on the solution. This model arises in many nonlinear inverse problems for differential equations. A natural candidate model $q(x)$ is a Gaussian approximation around the mode x^*, i.e., a Tikhonov minimizer. One approach is to linearize the forward model $F(x)$ around the mode x^*, i.e.,

$$F(x) = F(x^*) + F'(x^*)(x - x^*) + \text{h.o.t.},$$

which gives the following Gaussian approximation

$$q(x) \propto e^{-\frac{\tau}{2}\|F'(x^*)(x-x^*)-(y-F(x^*))\|^2 - \frac{\lambda}{2}\|Lx\|^2}.$$

We note that the Jacobian $F'(x^*)$ is required for iterative methods like Gauss-Newton iteration, and can be computed efficiently via the adjoint technique. A more refined (and more expensive) approach is to consider the full Hessian, i.e.,

$$\begin{aligned}\|F(x) - y\|^2 \approx {} & \|F(x^*) - y\|^2 + 2\langle F'(x^*)^*(F(x^*) - y), x - x^*\rangle \\ & + \langle F'(x^*)(x - x^*), F'(x^*)(x - x^*)\rangle \\ & + \langle F''(x^*)(F(x^*) - y)(x - x^*), x - x^*\rangle.\end{aligned}$$

In comparison with the approximation based on the linearized model, the full Hessian approach contains the term involving the second-derivative

$F''(x^*)$ and thus it represents a more faithful approximation, but at the expense of much increased computational efforts. Generally, the Hessian representation involving the last term is not necessarily positive definite, and may render the approximation a degenerate or even invalid Gaussian. One remedy is to threshold the small negative eigenvalues of the Hessian by some small positive values. Meanwhile, for nondifferentiable priors, e.g., Laplace prior $p_X(x) \propto e^{-\lambda\|x\|_1}$ or t distribution (see Appendix B), a straightforward definition of the Hessian can be problematic, since the Hessian for the prior term is then not well defined, and alternative approximations, e.g., via variational approximation or expectation propagation, cf. Section 7.4, must be sought after.

Remark 7.2. For many inverse problems for differential equations, the operator for the linearized model is compact, and under certain circumstances, its singular values even decay exponentially. Hence the Hessian can be well approximated with a few dominant singular vectors, and consequently the approach can be realized efficiently, e.g., via Lanczos' method. We refer interested reader to [228] for one such example.

7.3.2 *MCMC algorithms*

MCMC is the most popular general-purposed approach for exploring the posterior distribution $p(x)$. In this part, we provide an introduction to basic MCMC algorithms and will briefly discuss convergence analysis and acceleration techniques in Sections 7.3.3 and 7.3.4, respectively.

The idea of MCMC sampling was first introduced by Metropolis [233] as a method for efficiently simulating the energy levels of atoms in a crystalline structure, and was later adapted and generalized by Hastings [129] to focus on statistical problems. The basic idea is simple. Given a complex target distribution $p(x)$, we construct an aperiodic and irreducible Markov chain on the state space such that its stationary distribution is $p(x)$. By running the chain for sufficiently long, simulated values from the chain can be regarded as dependent samples from the target distribution $p(x)$, and used for computing summarizing statistics.

The Metropolis-Hastings algorithm is the most basic MCMC method, cf. Algorithm 7.2. Here u is a random number generated from the uniform distribution $U(0,1)$, $p(x)$ is the target distribution and $q(x,x')$ is an easy-to-sample proposal distribution. Having generated a new state x' from the distribution $q(x,x')$, we then accept this point as the new state of the chain with probability $\alpha(x, x')$ given by

$$\alpha(x, x') = \min\left\{1, \frac{p(x')q(x', x)}{p(x)q(x, x')}\right\}. \tag{7.7}$$

However, if we reject the proposal x', then the chain remains in the current state x. Note that the target distribution $p(x)$ enters the algorithm only through α via the ratio $p(x')/p(x)$, so a knowledge of the distribution only up to a multiplicative constant is sufficient for implementation. Further, in the case where the proposal distribution q is symmetric, i.e., $q(x, x') = q(x', x)$, the acceptance probability function $\alpha(x, x')$ reduces to

$$\alpha(x, x') = \min\left\{1, \frac{p(x')}{p(x)}\right\}. \tag{7.8}$$

The Metropolis-Hastings algorithm guarantees that the Markov chain converges to the target distribution $p(x)$ for any reasonable proposal distribution $q(x)$ (see Section 7.3.3 for details). There are many possible choices for the proposal distribution $q(x, x')$, the defining ingredient of the algorithm, and we list a few common choices below.

Algorithm 7.2 Metropolis-Hastings algorithm.

1: Initialize $x^{(0)}$ and set N;
2: **for** $i = 0 : N$ **do**
3: sample $u \sim U(0, 1)$;
4: sample $x^{(*)} \sim q(x^{(i)}, x^{(*)})$
5: **if** $u < \alpha(x^{(i)}, x^{(*)})$, c.f., (7.7) **then**
6: $x^{(i+1)} = x^{(*)}$;
7: **else**
8: $x^{(i+1)} = x^{(i)}$;
9: **end if**
10: **end for**

random walk updating If the proposal distribution $q(x, x') = f(x' - x)$ for some probability density f, then the candidate proposal is of the form $x^{(*)} = x^{(i)} + \xi$, where $\xi \sim f$. The Markov chain is driven by a random walk. There are many common choices for the density f, including uniform distribution, multivariate normal distribution or a t-distribution. These distributions are symmetric, so the acceptance probability $\alpha(x, x')$ is of the simple form (7.8). In particular, with the components being i.i.d. and following the Gaussian distribution $N(0, \sigma^2)$, $x_j^{(*)} = x_j^{(i)} + \xi$, with $\xi \sim$

$N(\xi; 0, \sigma^2)$. The variance σ^2 of the proposal distribution f controls the size of the random walks, and should be carefully tuned to improve the MCMC convergence and estimation efficiency.

The size of the move (like σ^2 in the random walk sampler) can significantly affect the convergence and mixing of the chain, i.e., exploration of the state space. It is necessary to tune the scale parameter σ^2 carefully so as to achieve good mixing. Heuristically, the optimal acceptance ratio should be around 0.25 for some model problems [102]. Theoretically, the efficiency of the MCMC samples is determined by the correlation between the samples, cf. Theorem 7.2.

independence sampler If $q(x, x') = q(x')$, then the candidate proposal x' is drawn independently of the current state x of the chain. In this case, the acceptance probability $\alpha(x, x')$ can be written as

$$\alpha(x, x') = \min\{1, w(x')/w(x)\},$$

where $w(x) = p(x)/q(x)$ is the importance weight function used in importance sampling given observations generated from f. The ideas of importance sampling and independence samplers are closely related. The essential difference between the two is that the former builds probability mass around points with large weights, by choosing these points relatively frequently. In contrast, the latter builds up probability mass on points with high weights by remaining at those points for a long period of time. The independence sampler is attractive if the proposal distribution $q(x)$ is a good approximation to the target distribution $p(x)$. There are many different ways to generate the independent proposal distribution $q(x)$, e.g., Gaussian approximations from the linearized forward model, coarse-scale/reduced-order representation (cf. Section 7.3.4), and approximate inference methods in Section 7.4.

Remark 7.3. Each iteration of Algorithm 7.2 requires evaluating the acceptance probability $\alpha(x^{(i)}, x^{(*)})$, in order to determine whether the proposal $x^{(*)}$ should be accepted. It boils down to evaluating the likelihood $p_{Y|X}(y|x^{(*)})$ (and the prior $p_X(x^{(*)})$, which is generally much cheaper), and thus invokes the forward simulation $F(x^{(*)})$. For PDE-constrained parameter identifications, it amounts to solving one differential equation (system) for the proposed parameter vector $x^{(*)}$. However, one forward simulation can already be expensive, and thus a straightforward application of the algorithm might be impractical. We shall describe several ways to alleviate the associated computational challenges in Section 7.3.4 below.

In Algorithm 7.2, the initial guess $x^{(0)}$ can be faraway from the stationary distribution, and the first samples are poor approximations as samples from the target distribution $p(x)$. Hence, one often discards these initial samples, which is called the burning-in period. To this end, one needs to assess the convergence of the MCMC chains, which is a highly nontrivial issue. There are several heuristic rules in the literature. For example, Brooks and Gelman [42] proposed the following diagnostic statistics. Suppose that we have L Markov chains, each of N samples, with the ith sample from the jth chain denoted by $x_j^{(i)}$. Then we compute

$$\widehat{V} = \frac{N-1}{N}W + \left(1 + \frac{1}{L}\right)\frac{B}{N},$$

where

$$W = \frac{1}{L(N-1)}\sum_{j=1}^{L}\sum_{i=1}^{N}(x_j^{(i)} - \bar{x}_j)(x_j^{(i)} - \bar{x}_j)^{\mathrm{t}},$$

$$\frac{B}{N} = \frac{1}{L-1}\sum_{j=1}^{L}(\bar{x}_j - \bar{x})(\bar{x}_j - \bar{x})^{\mathrm{t}},$$

which represent respectively the within and between-sequence covariance matrix estimates. One can then monitor the distance between $\widehat{V}$ and W to determine the chain convergence.

Gibbs sampler If the state space is high dimensional, it is rather difficult to update the entire vector x in one single step since the acceptance probability $\alpha(x, x')$ is often very small. A better approach is to update a part of the components of x each time and to implement an updating cycle inside each step. This is analogous to block Gauss-Seidel iteration in numerical linear algebra. The extreme case is the Gibbs sampler due to Geman and Geman [103], which updates a single component each time. Specifically, suppose we want to update the ith component x_i of x, then we choose the full conditional as the proposal distribution $q(x, x')$, i.e.,

$$q(x, x') = \begin{cases} p(x_i'|x_{-i}) & x_{-i}' = x_{-i}, \\ 0 & \text{otherwise}, \end{cases}$$

where x_{-i} denotes $(x_1, \ldots, x_{i-1}, x_{i+1}, \ldots, x_m)^{\mathrm{t}}$. With this proposal, the acceptance probability $\alpha(x, x')$ is given by

$$\begin{aligned} \alpha(x, x') &= \frac{p(x')q(x', x)}{p(x)q(x, x')} = \frac{p(x')/p(x_i'|x_{-i})}{p(x)/p(x_i|x_{-i}')} \\ &= \frac{p(x')/p(x_i'|x_{-i}')}{p(x)/p(x_i|x_{-i})} = \frac{p(x_{-i}')}{p(x_{-i})} = 1, \end{aligned}$$

where the last two steps follow from the fact $x_{-i} = x'_{-i}$. Thus, at each step, the only possible jumps are to states x' that match x on all components other than the ith component, and these proposals are automatically accepted. Now we explicitly give the Gibbs algorithm in Algorithm 7.3. We note that the full conditional $p(x'_i|x_{-i})$ is only available for a limited class of posterior distributions, mainly for linear inverse problems.

Algorithm 7.3 Gibbs algorithm.

1: Initialize $x^{(0)}$ and set N.
2: **for** $i = 0 : N$ **do**
3: sample $x_1^{(i+1)} \sim p(x_1|x_2^{(i)}, x_3^{(i)}, \ldots, x_m^{(i)})$,
4: sample $x_2^{(i+1)} \sim p(x_2|x_1^{(i+1)}, x_3^{(i)}, \ldots, x_m^{(i)})$,
5: $\vdots$
6: sample $x_m^{(i+1)} \sim p(x_m|x_1^{(i+1)}, x_2^{(i+1)}, \ldots, x_{m-1}^{(i+1)})$,
7: **end for**

Now we give two examples to illustrate the Gibbs sampler for linear inverse problems, one with the Gaussian model and the other with sparsity constraints.

Example 7.5. We first illustrate the Gibbs sampler on a linear inverse problem $Ax = y$ with a Gaussian noise model, smoothness prior and a Gamma hyperprior $p(\lambda) \propto \lambda^{a_0-1}e^{-b_0\lambda}$ on the scale parameter λ, i.e., the posterior distribution

$$p(x, \lambda) \propto e^{-\frac{\tau}{2}\|Ax-y\|^2} \cdot \lambda^{\frac{m}{2}} e^{-\frac{\lambda}{2}x^tWx}\lambda^{a_0-1}e^{-b_0\lambda},$$

where the matrix W encodes the local interaction structure. In order to apply the Gibbs sampler, we need to derive the full conditional $p(x_i|x_{-i}, \lambda)$, which is given by

$$p(x_i|x_{-i}, \lambda) \sim N(\mu_i, \sigma_i^2), \quad \mu_i = \frac{b_i}{2a_i}, \quad \sigma_i = \frac{1}{\sqrt{a_i}},$$

with a_i and b_i given by

$$a_i = \tau \sum_{j=1}^{n} A_{ji}^2 + \lambda W_{ii} \quad \text{and} \quad b_i = 2\tau \sum_{j=1}^{n} \mu_j A_{ji} - \lambda\mu_p,$$

and $\mu_j = y_j - \sum_{k\neq i} A_{jk}x_k$ and $\mu_p = \sum_{j\neq i} W_{ji}x_j + \sum_{k\neq i} W_{ik}x_k$. Lastly, we deduce the full conditional for λ:

$$p(\lambda|x) \sim G\left(\lambda; \tfrac{m}{2} + a_0, \tfrac{1}{2}x^tWx + \beta_0\right).$$

Example 7.6. The Gibbs sampler can also be applied to posterior distributions that admit Gaussian scale mixtures, which can represent a rich class of densities, e.g., Laplace distribution and t-distribution. We illustrate its use with the Laplace prior, which enforces sparsity constraint. This leads to the posteriori

$$p(x) \propto e^{-\frac{\tau}{2}\|Ax-b\|^2} \cdot e^{-\lambda\|x\|_1}.$$

The posterior is not directly amenable with Gibbs sampling due to non-gaussian nature of the Laplace prior. To circumvent the issue, we recall the scale mixture representation of the Laplace distribution [6, 248]

$$\tfrac{\lambda}{2}e^{-\lambda|z|} = \int_0^\infty \tfrac{1}{\sqrt{2\pi s}}e^{-\frac{z^2}{2s}} \cdot \tfrac{\lambda^2}{2}e^{-\frac{\lambda^2}{2}s}ds.$$

This suggests the following hierarchical representation of the full model

$$p(x,w) \propto e^{-\frac{\tau}{2}\|Ax-b\|^2}\prod_i w_i^{-\frac{1}{2}}e^{-\frac{x_i^2}{w_i}}\tfrac{\lambda^2}{2}e^{-\frac{\lambda^2}{2}w_i}.$$

The full conditional $p(x_i|x_{-i},w)$ follows a Gaussian distribution, and can be derived as in Example 7.5. The full conditional $p(w_i|x,w_{-i})$ is given by

$$p(w_i|x,w_{-i}) \propto w_i^{-\frac{1}{2}}e^{-\frac{x_i^2}{2w_i}-\frac{\lambda^2}{2}w_i},$$

which follows the generalized inverse Gaussian distribution

$$p(t;\alpha,\beta,\delta) = \frac{(\frac{\beta}{\delta})^{\frac{\alpha}{2}}}{2K_\alpha(\sqrt{\beta\delta})}t^{\alpha-1}e^{-\frac{1}{2}(\frac{\delta}{t}+\beta t)} \quad t > 0.$$

The mean of the distribution $p(t;\alpha,\beta,\delta)$ is $\frac{\sqrt{\delta}K_{\alpha+1}(\sqrt{\beta\delta})}{\sqrt{\beta}K_\alpha(\sqrt{\beta\delta})}$, where K_α refers to modified Bessel function of order α. There is public domain software to generate samples from the generalized inverse Gaussian distribution.

Remark 7.4. The sparsity constraint is of much recent interest in inverse problems. There are several alternative Bayesian implementations, e.g., the relevance vector machine [293] or direct Gibbs sampling [223].

7.3.3 *Convergence analysis*

We now briefly review the convergence theory of the Metropolis-Hastings algorithm, following Chapter 7 of [262]. First, we introduce two sets

$$\mathcal{E} = \{x : p(x) > 0\},$$
$$\mathcal{D} = \{x' : q(x,x') > 0 \text{ for some } x \in \mathcal{E}\}.$$

The set $\mathcal{E}$ contains all parameter vectors having a positive probability. It is the set that Algorithm 7.2 should draw samples from. The set $\mathcal{D}$ consists of all samples which can be generated by the proposal distribution $q(x, x')$, and hence contains the set that Algorithm 7.2 will actually draw samples from. A full exploration of the target distribution requires $\mathcal{E} \subset \mathcal{D}$.

Lemma 7.1. *If the relation $\mathcal{E} \subset \mathcal{D}$ holds, then $p(x)$ is a stationary distribution of the chain $\{x^{(n)}\}$.*

Proof. The transition kernel $k(x, x')$, which denotes the probability density that the next state of the chain lies at x', given that the chain is currently at state x, is given by

$$k(x, x') = k(x, x')q(x, x') + r(x)\delta_x(x'),$$

where δ_x is the Dirac δ-function at x, and

$$r(x) = 1 - \int q(x, x')\alpha(x, x')dx',$$

which denotes the size of the point mass associated with a rejection. It is easy to verify that the transition kernel satisfies

$$\begin{aligned} \alpha(x, x')q(x, x')p(x) &= \alpha(x', x)q(x', x)p(x'), \\ (1 - r(x))\delta_x(x')p(x) &= (1 - r(x'))\delta_{x'}(x)p(x'), \end{aligned}$$

which together establish the detailed balance relation for the Metropolis-Hastings chain. Now the distribution $p(x)$ is a stationary distribution follows directly from classical theory on Markov chain [262]. □

Theorem 7.1. *If $E_p[|f|] < \infty$ and*

$$q(x, x') > 0 \quad \text{for all } (x, x') \in \mathcal{E} \times \mathcal{E}. \tag{7.9}$$

Then there holds

$$\lim_{N \to \infty} E_N(f) = E_p[f].$$

The condition (7.9) guarantees that the every set of $\mathcal{E}$ with a positive Lebesgue measure can be reached in a single step. Hence it is sufficient for the chain to be irreducible. Clearly, the condition is satisfied for the random walk sampler. Lemma 7.1 and Theorem 7.1 together imply that asymptotically, the sample average $E_N[f]$ computed with samples generated by Algorithm 7.2 does converge to the desired expected value $E_p[f]$.

Now we turn to the error bound of the MCMC estimate $E_N[f]$ in the root mean square error

$$e(E_N(f)) = (E_{\widehat{p}}[E_p[f] - E_N[f]]^2)^{\frac{1}{2}},$$

where $E_{\widehat{p}}$ denotes the expected value with respect to the empirical distribution by the samples $\{x^{(i)}\}$ generated by Algorithm 7.2. As usual, it can be decomposed into the variance of the estimate and the squared bias

$$e(E_N(f))^2 = \mathrm{V}_{\widehat{p}}[E_N[f]] + (E_p[f] - E_{\widehat{p}}[E_N[f]])^2.$$

The two terms correspond to two sources of error in the MCMC estimate: the first term is the error due to using a finite sample average, and the second term is due to the samples in the estimate not all being perfect (i.i.d.) samples from the target distribution $p(x)$.

Bounding the variance and bias of an MCMC estimate in terms of the number N of samples is not an easy task. The difficulty stems from the fact that the samples used in the MCMC estimate are correlated. Hence a knowledge of the covariance structure is required in order to bound the error of the estimate. Nonetheless, asymptotically, the behavior of the MCMC related errors can still be described by a central limit theorem.

To this end, we assume that the chain constructed by Algorithm 7.2 is stationary, i.e., $\tilde{x}^{(i)} \sim p(x)$ for all $i \geq 0$. The covariance structure for the empirical distribution $\tilde{p}$ (generated by $\{\tilde{x}^{(i)}\}$) is still implicitly defined. However, for all $i \geq 0$

$$\mathrm{Var}_{\tilde{p}}[f(\tilde{x}^{(i)})] = \mathrm{Var}_p[f(\tilde{x}^{(i)})],$$
$$E_{\tilde{p}}[f(\tilde{x}^{(i)})] = E_p[f(\tilde{x}^{(i)})],$$

and

$$\mathrm{Cov}_{p,p}[f(\tilde{x}^{(0)}), f(\tilde{x}^{(i)})] = E_{p,p}[(f(\tilde{x}^{(0)}) - E_p[f])(f(\tilde{x}^{(i)}) - E_p[f])],$$

where $E_{p,p}[z] = \int\int z(x,x')dxdx'$ for a random variable z depending on x and x'. Now we can define the asymptotic variance of the MCMC estimate

$$\sigma_f^2 = \mathrm{Var}_p[f] + 2\sum_{i=1}^{\infty} \mathrm{Cov}_{p,p}[f(\tilde{x}^{(0)}), f(\tilde{x}^{(i)})].$$

The stationarity of the chain is assumed only in the definition of σ_f^2, and it is not necessary for the samples actually used in the computation. Then we have the following central limit theorem (cf. Theorem 6.65 of [262]) for the MCMC estimate $E_N[f]$.

Theorem 7.2. *Suppose $\sigma_f^2 < \infty$, condition (7.9) holds, and*

$$\mathrm{Prob}(\alpha = 1) < 1. \tag{7.10}$$

Then we have $\frac{1}{\sqrt{N}}(E_N[f] - E_p[f]) \to N(0, \sigma_f^2)$ in distribution.

The condition (7.10) is sufficient for the chain to be aperiodic. It is difficult to show theoretically. In practice, it seems always satisfied, since not all proposals in Algorithm 7.2 agree perfectly with the observed data and thus are accepted. The above theorem shows that asymptotically the sampling error of the MCMC estimator decays at a similar rate to that based on i.i.d. samples, but the constant σ_f^2 is generally larger than in the i.i.d. case. Hence, in order to get high estimation efficiency, the correlation between MCMC samples should be minimized.

7.3.4 *Accelerating MCMC algorithms*

The Metropolis-Hastings algorithm described in Section 7.3.2 is very universal. However, each sample requires evaluating the likelihood function, which in turn calls for the evaluation of the forward model $F(x)$. For many inverse problems, the forward model $F(x)$ is described implicitly by a differential equation. Hence each evaluation requires one forward solve of the differential equation, which makes the application of MCMC algorithms prohibitively expensive, if not impossible at all. Therefore, there has been of immense interest to reduce the associated computational efforts. In this part, we describe several promising ideas, including preconditioning [66, 83], multilevel decomposition [195], reduced-order modeling [166, 167], and approximation error approach [13, 187]. These ideas have been attested numerically.

Preconditioned MCMC This method is applicable when a cheap (probably local) approximation $p_x^*(x)$, which might be dependent on the current state x, to the target distribution $p(x)$ is available. For inverse problems with differential equations, the approximation $p_x^*(x)$ can be either a local linear approximation or an approximation on coarse grids. The main idea is as follows. Consider a proposal distribution $q(x, x')$. To avoid calculating $p(x')$ for the proposals that are rejected, we first correct the proposal with the approximation $p_x^*(x')$ to create a second proposal distribution $q^*(x, x')$, to be used in a standard Metropolis-Hastings algorithm. The second proposal hopefully has a high acceptance probability. Thereby, we sample from $p(x)$, but avoid calculating $p(x')$ when proposals are rejected by $p_x^*(x')$ and hence gain the desired computational efficiency. The complete procedure is listed in Algorithm 7.4. Here the acceptance ratio $\alpha(x, x')$ is given by $\alpha(x, x') = \min\{1, \frac{q(x',x)p_x^*(x')}{q(x,x')p_x^*(x)}\}$, and the actual proposal

$q^*(x, x')$ is given by

$$q^*(x, x') = \alpha(x, x')q(x, x') + (1 - r(x))\delta_x(x'),$$

where $r(x) = \int \alpha(x, x')q(x, x')dx'$. The actual acceptance ratio $\rho(x, x')$ is given by

$$\rho(x, x') = \min\left\{1, \frac{q^*(x', x)p(x')}{q^*(x, x')p(x)}\right\}.$$

We observe that the evaluation of $\rho(x, x')$ does not require r.

Algorithm 7.4 Preconditioned Metropolis-Hastings algorithm

1: Initialize $x^{(0)}$ and set N;
2: **for** $i = 0 : N$ **do**
3:    Generate a proposal $x^{(*)}$ from $q(x^{(i)}, x^{(*)})$;
4:    $u \sim U(0, 1)$,
5:    **if** $u < \alpha(x^{(i)}, x^{(*)})$ **then**
6:       $u \sim U(0, 1)$
7:       **if** $u < \rho(x^{(i)}, x^{(*)})$ **then**
8:          $x^{(i+1)} = x^{(*)}$
9:       **end if**
10:    **else**
11:       $x^{(i+1)} = x^{(i)}$;
12:    **end if**
13: **end for**

Under minor regularity condition on the proposal distribution q and the approximation p_x^*, one can show that Algorithm 7.4 will converge, and furthermore, the more accurate is the approximation p_x^* to the target distribution $p(x)$, the closer is the acceptance ratio ρ to one [66]. The latter property essentially leverages the computational efforts on the expensive model $p(x)$ to the cheaper model $p_x^*(x)$.

Remark 7.5. Although we have described only a two-level algorithm, there is no essential difficulty in extending it to a general multilevel algorithm. Further, the number of unknowns might also be dependent on the level of the approximations.

Multilevel MCMC Multilevel MCMC is an extension of the classical multigrid method for numerical differential equations [195]. The main idea is to sample from several levels of approximations, and then combines these samples in the framework of multilevel Monte Carlo method

[136, 108, 107]. Specifically, let $\{N_l\}_{l=1}^{L} \subset \mathbb{N}$ be an increasing sequence, and $p = p^L$ is the level of interest. For each level l, we denote the parameter vector by x_l and the quantity of interest by f_l, and the posterior distribution by p^l. Like multigrid methods for discretizing PDEs, the key is to avoid estimating the expected value of f_l directly on level l but instead to estimate the correction with respect to the next lower level f_{l-1}. Since the target distribution p^l depends on the level l, a multilevel MCMC estimate has to be defined carefully. The basis of the multilevel estimate is the following telescopic sum

$$E_{p^L}(f_L) = E_{p^0}[f_0] + \sum_{l=1}^{L} E_{p^l,p^{l-1}}[f_l - f_{l-1}], \tag{7.11}$$

where the $E_{p^l,p^{l-1}}[f_l - f_{l-1}]$ is the expected value with respect to the joint distribution $(p^l(x_l), p^{l-1}(x_{l-1}))$, i.e.,

$$E_{p^l,p^{l-1}}[f_l - f_{l-1}] = \int\int (f_l(x_l) - f_{l-1}(x_{l-1}))p^l(x_l)p^{l-1}(x_{l-1})dx_l dx_{l-1}.$$

The identity (7.11) follows from the linearity of expectation

$$\begin{aligned} E_{p^l,p^{l-1}}[f_l - f_{l-1}] &= E_{p^l}[f_l]\int p^{l-1}dx_{l-1} - E_{p^{l-1}}[f_{l-1}]\int p^l dx_l \\ &= E_{p^l}[f_l] - E_{p^{l-1}}[f_{l-1}]. \end{aligned}$$

The idea of a multilevel estimate is to estimate each difference term on the right hand side of (7.11) independently so as to minimize the variance of the estimate for fixed computational effort. In particular, we can estimate each term by the Metropolis-Hastings algorithm. The first term can be estimated using the standard MCMC estimate described in Algorithm 7.2. The terms involving difference require an effective two-level version of the standard Metropolis-Hastings algorithm. For every $l \geq 1$, we denote $\tilde{f}_l = f_l - f_{l-1}$ and define the estimate on level l as

$$E_{N_l}[\tilde{f}_l] = \frac{1}{N_l}\sum_{i=0}^{N_l} f_l(x_l^{(i)}) - f_{l-1}(x_{l-1}^{(i)}).$$

The main ingredient is a judicious choice of samples $\{(x_l^{(i)}, x_{l-1}^{(i)})\}$ from the joint distribution (p^l, p^{l-1}). This can be constructed by Algorithm 7.5.

In Algorithm 7.5, we assume that the coarse level (level $l-1$) random variable x_{l-1} is nested into the one x_l on the fine level (level l), i.e., $x_l = [x_{l,c}\ x_{l,f}]$ with $x_{l,c} = x_{l-1}$. Hence the fine level proposal $q^l(x_l, x_l')$ should impose the constraint $x_{l,c}' = x_{l-1}'$ and generates only the remaining part

Algorithm 7.5 Two-level Metropolis-Hastings algorithm

1: Choose initial state $x_{l-1}^{(0)}$ and $x_l^{(0)} = [x_{l-1}^{(0)}, x_{l,f}^{(0)}]$
2: **for** $i = 0 : N$ **do**
3: $\quad x_{l-1}^{(*)} \sim q^{l,c}(x_{l-1}^{i}, x_{l-1}^{(*)})$
4: $\quad u \sim U(0,1)$
5: $\quad$ **if** $u < \alpha^{l,c}(x_{l-1}^{(i)}, x_{l-1}^{(*)})$ **then**
6: $\quad\quad x_{l-1}^{(i+1)} = x_{l-1}^{(*)}$
7: $\quad$ **else**
8: $\quad\quad x_{l-1}^{(i+1)} = x_{l-1}^{(i)}$
9: $\quad$ **end if**
10: $\quad x_l^{(*)} \sim q^l(x_l^{(i)}, x_l^{(*)})$ with $x_{l,c}^{(*)} = x_{l-1}^{(i+1)}$, $x_{l,f}^{(*)} \sim q^{l,f}(x_{l,f}^{(i)}, x_{l,f}^{(*)})$
11: $\quad u \sim U(0,1)$
12: $\quad$ **if** $u < \alpha^{l,f}(x_l^{(i)}, x_l^{(*)})$ **then**
13: $\quad\quad x_l^{(i+1)} = x_l^{(*)}$
14: $\quad$ **else**
15: $\quad\quad x_l^{(i+1)} = x_l^{(i)}$
16: $\quad$ **end if**
17: **end for**

$x'_{l,f}$ with a proposal distribution $q^{l,f}(x_{l,f}, x'_{l,f})$. Analogous to (7.7), the acceptance probabilities $\alpha^{l,c}(x_{l-1}, x'_{l-1})$ and $\alpha^{l,f}(x_l, x'_l)$ on level $l-1$ and level l are respectively given by

$$\alpha^{l,c}(x_{l-1}, x'_{l-1}) = \min\left\{1, \frac{p^{l-1}(x'_{l-1})q^{l,c}(x'_{l-1}, x_{l-1})}{p^{l-1}(x_{l-1})q^{l,c}(x_{l-1}, x'_{l-1})}\right\},$$

$$\alpha^{l,f}(x_l, x'_l) = \min\left\{1, \frac{p^l(x'_l)q^l(x'_l, x_l)}{p^l(x_l)q^l(x_l, x'_l)}\right\}.$$

The cost reduction for the multilevel estimate stems from two observations. First, samples from p^l for $l < L$ are cheaper to generate than that from p^L, reducing the cost of the estimates on the coarser levels for any fixed number of samples. Second, if the variance $\mathrm{Var}_{p^l,p^{l-1}}[f_l - f_{l-1}]$ is small, a small number of samples would suffice an accurate estimate of $E_{p^l,p^{l-1}}[f_l - f_{l-1}]$ on the fine grids, and so the computational effort on fine grids is also reduced. In view of the latter fact, multilevel MCMC can also be regarded as a multilevel control variate technique.

Reduced-order modeling The forward model $F(x)$ for inverse problems are often implicitly defined by differential equations, with the unknown

$x \in I$ (I is the parametric domain), i.e., find $u \in V_h$ such that

$$a(u(x), v; x) = (f, v) \quad \forall v \in V_h,$$

where a is a bilinear form on a suitable finite element space V_h, and f is the source term. The key observation is that albeit the solution $u(x)$ is high-dimensional, the parametrically induced manifold $\mathcal{M} = \{u(x), x \in I\}$ is often low dimensional. Hence, general-purposed methods are often unduly expensive in the many-query context, e.g., Bayesian inference. Instead, one should represent $u(x)$ in a problem-dependent basis spanning $\mathcal{M}$ [166]. This can be efficiently achieved via reduced order modeling. Typically, it consists of the following three stages.

The first step is to collect snapshots, i.e., solutions $\{s_j = u(x_j)\}$ at a set of parameter values $\{x_j\}_{j=1}^n \subset I$, which are obtained by solving the full-order system. The snapshots should be rich enough to capture essential characteristics of the manifold $\mathcal{M}$, and thus many snapshots are needed. In practice, ad hoc procedures are often invoked.

Then one extracts the essential information about the manifold $\mathcal{M}$ in the snapshots by the SVD, cf. Appendix A. Let $S = [s_1\, s_2 \ldots s_n]$. Then the SVD of the matrix S is given by

$$S = U\Sigma V^{\mathrm{t}}$$

where $U = [u_1\ u_2\ \ldots\ u_n] \in \mathbb{R}^{N\times n}$, with N being the dimension of the space V_h, and $V = [v_1\ v_2\ \ldots\ v_n]$ are column orthonormal matrices, and $\Sigma = \mathrm{diag}(\sigma_1, \sigma_2, \ldots, \sigma_n)$ is a diagonal matrix with nonnegative diagonal entries, i.e. singular values, in descending order. Then an optimal basis of cardinality p is given by $\{u_j\}_{j=1}^p$. The cardinality p required to accurately portray $\{s_j\}$ is the smallest integer such that

$$\frac{\sum_{j=1}^p \sigma_j^2}{\sum_{j=1}^n \sigma_j^2} \geq 1 - \epsilon,$$

where ϵ is a tolerance. Hence, the singular value σ_j dictates the significance of the basis u_j, and a fast decay indicates that a handful of basis functions can accurately represent the snapshots.

Finally, for a given parameter $x \in I$, we approximate the solution $u(x)$ in the basis $\{u_j\}_{j=1}^p$ e.g., by Galerkin procedure, which leads to a reduced-order model. The formulation for the reduced-order model is formally identical with that for the full-order model, but of a much smaller size, which significantly effects online cost reduction. We note that such a construction is efficient only for governing equations that are affine linear in the parameter x. For problems involving a general nonlinearity term, refinements, e.g., discrete empirical interpolation method [57], should be applied as well to reduced the computational efforts.

Remark 7.6. The accuracy of the reduced-order model depends very much on the parametric dimension $|I|$. It is especially suited to low-dimensional parameter spaces (preferably less than 10). Numerically, it can be used either as a preconditioner to the MCMC, or as a surrogate to the full model for cost reduction. If the solution $u(x)$ is smooth with respect to the parameter x and the dimensionality of the parameter x is low, one can also use stochastic Galerkin method or stochastic collocation method to generate a surrogate [105, 229].

Approximation error approach The approximation error approach [13, 187] adopts a statistical framework to model the discretization error so as to reducing its effect on the reconstruction resolution. In the preceding discussions, we assume that the finite-dimensional model $F(x)$ is exact within the measurement accuracy to the physical model, i.e.,

$$y = F(x) + e.$$

In the approach, the observation model is rewritten in the form

$$\begin{aligned} y &= F_H(x) + (F(x) - F_H(x)) + e \\ &= F_H(x) + \epsilon(x) + e, \end{aligned}$$

where $F_H(x)$ is the reduced model (e.g., obtained via coarse-level discretization), and $\epsilon(x)$ is the modeling error. The modeling error $\epsilon(x)$ describes the discrepancy between the high-resolution forward model and the reduced model.

Then, a Gaussian approximation is constructed for ϵ, i.e.,

$$\epsilon \sim N(\epsilon; \epsilon_*, C_\epsilon).$$

Together with a Gaussian model on the data error e, $e = N(e; e_*, C_e)$, the total error $\zeta = \epsilon + e$ is approximately a Gaussian distribution

$$\zeta \sim N(\zeta; \zeta_*, C_\zeta),$$

with $\zeta_* = \epsilon_* + e_*$ and $C_\zeta = C_\epsilon + C_e$. Further, if we assume that x and ϵ are mutually independent and the prior is given by $p_X(x)$, e.g., a Gaussian model $N(x; x^*, C_x)$ or Markov random field $p_X(x) \propto e^{-\lambda\psi(x)}$, we arrive at the enhanced error model

$$p_{X|Y}(x|y) \propto e^{-\frac{1}{2}\|y - F_H(x) - \epsilon_* - e_*\|^2_{C_\zeta^{-1}}} p_X(x).$$

It remains to construct the mean ϵ_* and the covariance C_ϵ for the approximation error ϵ, which can be simulated as follows. We draw a set of samples $\{x^{(l)}\}_{l=1}^L$ from the prior distribution $p_X(x)$, and compute the respective approximation error

$$\epsilon_H^{(l)} = F(x^{(l)}) - F_H(x^{(l)}).$$

The mean and covariance of the approximation error ϵ are estimated from the samples by

$$\epsilon_* = \frac{1}{L}\sum_{l=1}^{L} \epsilon_H^{(l)} \quad \text{and} \quad C_\epsilon = \frac{1}{L-1}\sum_{l=1}^{L} \epsilon_H^{(l)} \epsilon_H^{(l)^{\mathrm{t}}} - \epsilon_* \epsilon_*^{\mathrm{t}}.$$

We note that the mean ϵ_* and the covariance C_ϵ of the approximation error ϵ are computed before taking measurements and they are valid for the employed geometry and other parameters as long as the prior model is feasible. Therefore, the approximation error ϵ can be precomputed, and the online computational burden is essentially identical with the conventional likelihood model.

Remark 7.7. There are several different ways to exploit the approximation error model. Often it is used to derive an enhanced Tikhonov model, by considering the MAP estimate, in the hope of compensating the discretization error. Surely, it can also serve as either a preconditioner or a proposal distribution in the Metropolis-Hastings' algorithm. One interesting question is to extend the approach to the case of nongaussian data noise model, which seems open so far.

7.4 Approximate inference

For many real-world inverse problems involving differential equations, MCMC methods are often too expensive to apply. Therefore, there has been of immense interest in developing approximate inference methods, which can deliver reliable statistics of the posterior distribution $p(x)$ yet are far less expensive. In this part, we describe three deterministic approximate inference methods based on minimizing Kullback-Leibler divergence, i.e., variational Bayes, Gaussian lower bound and expectation propagation. They are especially attractive when the posterior distribution $p(x)$ exhibits certain structures.

7.4.1 *Kullback-Leibler divergence*

The main idea of approximate inference methods is to transform the inference problem of exploring the posterior distribution $p_{X|Y}(x|y)$ into an equivalent optimization problem, and solves the optimization problem approximately to arrive at an approximation. The distributional nature of the posteriori $p_{X|Y}(x|y)$ requires probabilistic metrics. There are many possible choices, including Hellinger metric, Kullback-Leibler divergence and L^1 metric etc. We shall mainly employ the Kullback-Leibler divergence [205]. Given two probability density functions $q(x)$ and $\tilde{q}(x)$, the Kullback-Leibler divergence $D_{KL}(q, \tilde{q})$ is defined by

$$D_{KL}(q, \tilde{q}) = \int q(x) \ln \frac{q(x)}{\tilde{q}(x)} dx.$$

The divergence $D_{KL}(q, \tilde{q})$ is asymmetric in q and $\tilde{q}$, and does not satisfy the triangle inequality. It is a special case of the α-divergence $D_\alpha(q, \tilde{q})$, $\alpha \in \mathbb{R}$ defined by

$$D_\alpha(q, \tilde{q}) = \frac{\int \alpha q(x) + (1-\alpha)\tilde{q}(x) - q(x)^\alpha \tilde{q}(x)^{1-\alpha} dx}{\alpha(1-\alpha)}$$

in the sense that $\lim_{\alpha \to 1} D_\alpha(q, \tilde{q}) = D_{KL}(q, \tilde{q})$. In view of Jensen's inequality for the convex function $\varphi(x) = -\ln x$, the divergence $D_{KL}(q, \tilde{q})$ is always nonnegative:

$$\begin{aligned} D_{KL}(q, \tilde{q}) &= -\int q(x) \ln \frac{\tilde{q}(x)}{q(x)} dx \geq -\ln \int q(x) \cdot \frac{\tilde{q}(x)}{q(x)} dx \\ &= -\ln \int \tilde{q}(x) dx = -\ln 1 = 0, \end{aligned} \tag{7.12}$$

and it vanishes if and only if $q = \tilde{q}$ almost everywhere. This fact will be used later, and hence we record it in a lemma.

Lemma 7.2. *Let $q(x)$ be a probability density function. Then for any probability density function $\tilde{q}(x)$, the divergence $D_{KL}(q, \tilde{q}) \geq 0$ and vanishes if and only if $q = \tilde{q}$.*

Example 7.7. The divergence between two normal distributions $q(x) = N(x; \mu, C)$ and $\tilde{q}(x) = N(x; \tilde{\mu}, \tilde{C})$, with $\mu, \tilde{\mu} \in \mathrm{R}^m$ and $C, \tilde{C} \in \mathbb{R}^{m \times m}$ being positive definite, is given by

$$
\begin{aligned}
&D_{KL}(q, \tilde{q}) \\
&= \frac{1}{2}\int q(x)\left[-\ln\frac{|C|}{|\tilde{C}|} - (x-\mu)^{\mathrm{t}}C^{-1}(x-\mu) + (x-\tilde{\mu})^{\mathrm{t}}\tilde{C}^{-1}(x-\tilde{\mu})\right]dx \\
&= \frac{1}{2}\left[-\ln\frac{|C|}{|\tilde{C}|} + \int N(z;0,I_m)\left[-\|z\|^2 + (\mu + C^{\frac{1}{2}}z - \tilde{\mu})^{\mathrm{t}}\tilde{C}^{-1}(\mu + C^{\frac{1}{2}}z - \tilde{\mu})\right]dz\right] \\
&= \frac{1}{2}\left[(\mu-\tilde{\mu})^{\mathrm{t}}\tilde{C}^{-1}(\mu-\tilde{\mu}) + \mathrm{tr}(C\tilde{C}^{-1}) - m - \ln\frac{|C|}{|\tilde{C}|}\right],
\end{aligned}
$$

where the second identity follows from the transformation $x = \mu + C^{\frac{1}{2}}z$. We note that in the special case $C = \tilde{C}$, the Kullback-Leibler divergence $D_{KL}(q, \tilde{q})$ simplifies to the squared C^{-1}-weighted Euclidean norm

$$D_{KL}(q, \tilde{q}) = \tfrac{1}{2}(\mu - \tilde{\mu})^{\mathrm{t}}C^{-1}(\mu - \tilde{\mu}),$$

which is symmetric and positive definite.

A direct consequence of Lemma 7.2 is the following lower bound on the normalizing constant $p_Y(y) = \int p(x,y)dx$.

Lemma 7.3. *For any probability density function $q(x)$, the normalizing constant $p_Y(y)$ satisfies*

$$\ln p_Y(y) \geq \int q(x)\ln\frac{p(x,y)}{q(x)}dx.$$

Proof. By (7.12), there holds

$$
\begin{aligned}
\ln p_Y(y) &= \int q(x)\ln p_Y(y)dx = \int q(x)\ln\frac{p(x,y)/q(x)}{p_{X|Y}(x|y)/q(x)}dx \\
&= \int q(x)\ln\frac{p(x,y)}{q(x)}dx + \int q(x)\ln\frac{q(x)}{p_{X|Y}(x|y)}dx \\
&\geq \int q(x)\ln\frac{p(x,y)}{q(x)}dx.
\end{aligned}
$$

□

The lower bound

$$\mathrm{LB}(y,q) := \int q(x)\ln\frac{p(x,y)}{q(x)}dx \tag{7.13}$$

in Lemma 7.3 serves the basis of model comparison, hyperparameter estimation and optimal experimental design. By the proof of Lemma 7.3, $D_{KL}(q(x), p_{X|Y}(x|y))$ is the gap between the margin $\ln p_Y(y)$ and the lower bound, and hence maximizing LB is equivalent to minimizing the divergence $D_{KL}(q(x), p_{X|Y}(x|y))$ with respect to q.

The approximate inference methods to be described below rely on minimizing the Kullback-Leibler divergence with respect to different arguments.

Theorem 7.3. *Let q be a probability distribution with a finite second moment, $\mu^* = E_q[x]$ and $C^* = \mathrm{Var}_q[x]$. Then there exists a minimizer to*

$$\min_{\tilde{q} \text{ Gaussian}} D_{KL}(q, \tilde{q})$$

over any compact set of Gaussians with a positive definite covariance. Further, if $N(x; \mu^*, C^*)$ belongs to the interior of the compact set, then it is a local minimizer.

Proof. The existence follows directly from the compactness assumption, and the continuity of Kullback-Leibler divergence in the second argument. Let $\tilde{q}(x) = N(x; \mu, C)$. The divergence $D_{KL}(q, \tilde{q})$ can be expanded into

$$D_{KL}(q, \tilde{q}) = \int q(x)\big(\ln(x) + \tfrac{d}{2}\ln 2\pi - \tfrac{1}{2}\ln|C^{-1}| + \tfrac{1}{2}(\mu - x)^t C^{-1}(\mu - x)\big)\,dx.$$

We first compute the gradient $\nabla_\mu D_{KL}(q, \tilde{q})$ of the divergence $D_{KL}(q, \tilde{q})$ with respect to the mean μ of the Gaussian factor $\tilde{q}(x)$, i.e.,

$$\nabla_\mu D_{KL}(q, \tilde{q}) = \int q(x) C^{-1}(\mu - x) dx$$

and set it to zero to obtain the first condition $\mu = E_q[x]$.

Likewise, we compute the gradient $\nabla_P D_{KL}(q, \tilde{q})$ of $D_{KL}(q, \tilde{q})$ with respect to the precision $P = C^{-1}$

$$\nabla_P D_{KL}(q, \tilde{q}) = \int q(x)\left(-\tfrac{1}{2}P^{-1} + \tfrac{1}{2}(\mu - x)(\mu - x)^t\right) dx,$$

where we have used the following matrix identities [82]

$$\frac{\partial}{\partial A}\log|A| = A^{-1}, \quad x^t A x = \mathrm{tr}(Axx^t) \quad \text{and} \quad \frac{\partial}{\partial A}\mathrm{tr}(AB) = B^t.$$

The gradient $\nabla_P D_{KL}(q, \tilde{q})$ vanishes if and only if the condition $C = E_q[(x - \mu)(x - \mu)^t]$ holds. Together with the condition on the mean μ, this is equivalent to $C = \mathrm{Var}_q[x]$. Now the second assertion follows from the convexity of the divergence in the second argument and the assumption that $N(x; \mu^*, C^*)$ is an interior point of the compact set. □

Remark 7.8. Using the matrix identity [82] $\frac{\partial A^{-1}}{\partial A} = -A^{-t} \otimes A^{-1}$, one can deduce that the Hessian $\nabla^2_{\mu^*,P} D_{KL}(q, \tilde{q})$ is given by

$$\nabla^2_{\mu,P} D_{KL}(q, \tilde{q}) = \begin{bmatrix} C^{-1} & \\ & \frac{1}{2} C \otimes C \end{bmatrix},$$

where $\otimes$ denotes matrix Kronecker product. Note that $C \otimes C$ is positive definite if and only if C is positive definite. This indicates that the minimizer $N(\mu^*, C^*)$ is locally unique.

7.4.2 *Approximate inference algorithms*

Based on the Kullback-Leibler divergence D_{KL}, we now describe three algorithms for delivering an approximation $q(x)$ to the posterior distribution $p_{X|Y}(x|y)$

(i) separable approximation: $q(x)$ factorizes into $\prod_i q_i(x_i)$ for some disjoint partition $\{x_i\}$ of x;
(ii) parametric approximation: $q(x)$ is a member of a parametric family of density functions;
(iii) expectation propagation: $p_{X|Y}(x|y) = \prod_i t_i(x)$ and $q(x) = \prod_i \tilde{t}_i(x)$, with $\tilde{t}_i(x)$ being a Gaussian approximation to $t_i(x)$.

In all three algorithms, the approximation $q(x)$ of the posterior distribution $p_{X|Y}(x|y)$ is achieved by first restricting it to a manageable subclass of densities with simpler structure (e.g., separability and Gaussian family), and then minimizing a certain Kullback-Leibler divergence. Now we describe the three approximations in sequel.

case (i). The separability assumption decouples the interdependence between different factors, and thus makes the approximation computable. It is often known as mean field approximation and variational Bayes in the literature.

The separability assumption on product densities yields explicit solutions for each product component in terms of the others. This lends itself to an alternating direction iterative scheme for obtaining the simultaneous solutions, cf. Algorithm 7.6. Now we derive the update in Algorithm 7.6 from (7.12). Under the assumption $q(x) = \prod q_i(x_i)$ with x_i being disjoint, we have

$$\begin{aligned}
\mathrm{LB} &= \int \prod_i q_i(x_i) \left\{ \ln p(x, y) - \sum_i \ln q_i(x_i) \right\} dx \\
&= \int q_j(x_j) \left\{ \int \ln p(x, y) \prod_{i \neq j} q_i(x_i) dx_{-j} \right\} dx_j \\
&\quad - \int q_j(x_j) \ln q_j(x_j) dx_j + T(x_{-j}),
\end{aligned}$$

where the term $T(x_{-j})$ depend only on x_{-j}. Next we define a joint density function $\tilde{p}(x_j, y)$ by

$$\tilde{p}(x_j, y) \propto e^{E_{q_{-j}(x_{-j})}[\ln p(x,y)]},$$

with $q_{-j}(x_{-j}) = \prod_{i \neq j} q_i(x_i)$. Then

$$\mathrm{LB} = \int q_j(x_j) \ln \frac{\tilde{p}(x_j, y)}{q_j(x_j)} dx_j + T(x_{-j}).$$

By inequality (7.12), the optimal $q_j(x_j)$ is then given by

$$q_j(x_j) = \tilde{p}(x_j|y) \propto e^{E_{q_{-j}(x_{-j})}[\ln p(x,y)]},$$

which represent optimality conditions for the optimal densities. Algorithm 7.6 takes a fixed-point iteration of alternating direction type to solve the optimality system, which is in the same spirit of the Gibbs sampler described in Section 7.3.

Algorithm 7.6 Variational Bayes under the product density restriction.

1: Initialize $q_i(x_i), \imath = 2, \ldots, M$.
2: **for** $k = 1 : K$ **do**
3: Update sequentially $q_j(x_j)$ by

$$q_j^{k+1}(x_j) = \frac{e^{E_{q_{-j}^k(x_{-j})}[\ln p(x,y)]}}{\int e^{E_{q_{-j}^k(x_{-j})}[\ln p(x,y)]}}$$

4: Check the stopping criterion.
5: **end for**

The convexity property of Kullback-Leibler divergence implies that the convergence of the algorithm to a local optimum is guaranteed [183, 301]. The algorithm is especially suited to conjugate families. Then the optimal densities belong to recognizable density families and the updates reduce to updating parameters in the conjugate family; see the examples below. In Algorithm 7.6, the stopping criterion is based on monitoring the convergence of the lower bound. The main computational effort lies in evaluating the expectations, which in general can be expensive and hence further approximations are often necessary. In particular, one can apply MCMC or control variate techniques to compute the expectations [139].

Now we illustrate the variational Bayesian method with two examples from Section 7.3.

Example 7.8. We revisit a linear inverse problem $Ax = y$ with a smoothness prior, i.e., $p_{X|\Lambda}(x|\lambda) \propto \lambda^{-\frac{m}{2}} e^{-\frac{\lambda}{2}x^tWx}$, and a Gamma hyperprior on the scale parameter λ, i.e., $q_\Lambda(\lambda) \propto \lambda^{a_0-1}e^{-b_0\lambda}$. This leads to the following posterior distribution

$$p(x,\lambda|y) \propto e^{-\frac{\tau}{2}\|Ax-y\|^2} \cdot \lambda^{\frac{m}{2}} e^{-\frac{\lambda}{x}{}^{\mathrm{t}}Wx} \cdot \lambda^{a_0-1}e^{-b_0\lambda}.$$

The posteriori $p(x,\lambda|y)$ does not admits closed-form solution. Hence an approximation must be sought. To apply the separable approximation, we decompose the unknown naturally into two blocks x and λ. Then we initialize $q(\lambda)$ to the Gamma distribution $q^0(\lambda) \propto \lambda^{a_0-1}e^{-b_0\lambda}$. Then the iterates in Algorithm 7.6 read

$$\begin{aligned}
\ln q^k(x) &\propto E_{q^{k-1}(\lambda)}[\ln p(x,\lambda,y)],\\
\ln q^k(\lambda) &\propto E_{q^k(x)}[\ln p(x,\lambda,y)].
\end{aligned}$$

Now we compute the two expectations

$$\begin{aligned}
E_{q^{k-1}(\lambda)}[\ln p(x,\lambda,y)] &= E_{q^{k-1}(\lambda)}[-\tfrac{\tau}{2}\|Ax-y\|^2 - \tfrac{\lambda}{2}x^{\mathrm{t}}Wx] + T(\lambda)\\
&= -\tfrac{\tau}{2}\|Ax-y\|^2 - \tfrac{E_{q^{k-1}(\lambda)}[\lambda]}{2}x^{\mathrm{t}}Wx + T(\lambda),
\end{aligned}$$

where $T(\lambda)$ is a constant depending only on λ, and similarly

$$\begin{aligned}
E_{q^k(x)}[\ln p(x,\lambda,y)] &= E_{q^k(x)}[(\tfrac{m}{2}+a_0-1)\ln\lambda - (b_0\lambda + \tfrac{\lambda}{2}x^{\mathrm{t}}Wx)] + T(x)\\
&= (\tfrac{m}{2}+a_0-1)\ln\lambda - [b_0 + \tfrac{1}{2}E_{q^k(x)}[x^{\mathrm{t}}Wx]]\lambda + T(x),
\end{aligned}$$

where $T(x)$ is a constant depending only on x. It follows from these two updates that the iterates $q^k(x)$ and $q^k(\lambda)$ would remain Gaussian and Gamma, respectively. We observe that this does not follow from the a prior assumption, but from the conjugate form of the posteriori.

Example 7.9. In this example, we revisit Example 7.6 with sparsity constraints, and adopt the approach of scale mixture representation for the

Laplace prior, which gives a hierarchical representation of the full Bayesian model

$$p_{X,W|Y}(x,w|y) \propto e^{-\frac{\tau}{2}\|Ax-y\|^2} \prod_i w_i^{-\frac{1}{2}} e^{-\frac{x_i^2}{2w_i}} \frac{\lambda^2}{2} e^{-\frac{\lambda^2}{2} w_i}.$$

The inference is obstructed by the strong coupling between x and w. Therefore, we decompose the unknowns into two groups, i.e., x and w. Following the variational method, it leads to the following iterates

$$\begin{aligned}
\ln q^k(x) &\propto E_{q^{k-1}(w)}[\ln p(x,w,y)], \\
\ln q^k(w) &\propto E_{q^k(x)}[\ln p(x,w,y)].
\end{aligned}$$

Now we evaluate two expectations, i.e.,

$$\begin{aligned}
E_{q^{k-1}(w)}[\ln p(x,\lambda,y)] &= E_{q^{k-1}(w)}[-\tfrac{\tau}{2}\|Ax-y\|^2 - \textstyle\sum \frac{1}{2w_i} x_i^2] + T(w) \\
&= -\tfrac{\tau}{2}\|Ax-y\|^2 - \tfrac{1}{2} x^t W x + T(w),
\end{aligned}$$

where $T(w)$ is a constant depending only on w and W is a diagonal matrix with the ith diagonal given by

$$[W]_{ii} = E_{q^{k-1}(w_i)}[\tfrac{1}{w_i}],$$

which can be evaluated in closed for a generalized inverse Gaussian distribution. Similarly,

$$\begin{aligned}
E_{q^k(x)}[\ln p(x,w,y)] &= \sum_i E_{q^k(x)}[-\tfrac{1}{2}\ln w_i - \tfrac{x_i^2}{2w_i} - \tfrac{\lambda^2}{2} w_i] + T(x) \\
&= \sum_i [-\tfrac{1}{2}\ln w_i - \tfrac{1}{2w_i} E_{q^k(x)}[x_i^2] - \tfrac{\lambda^2}{2} w_i] + T(x),
\end{aligned}$$

where $T(x)$ is a constant depending only on x. Hence, $q^k(w)$ follows a generalized inverse Gaussian distribution of product form, i.e.,

$$q^k(w) = \prod_i GIG\left(w_i; \tfrac{1}{2}, \lambda^2, E_{q^k(x)}[x_i^2]\right).$$

It follows from these two updates that the iterates $q^k(x)$ and $q^k(w)$ would remain Gaussian and (generalized) inverse Gaussian, respectively. We reiterate that this does not follow from the a prior assumption, but from the conjugate form of the posteriori $p_{X,W|Y}(x,w|y)$.

case (ii) Here we assume that the approximating density $q(x)$ belongs to a particular parametric family, e.g., Gaussians, and then optimize the Kullback-Leilber divergence $D_{KL}(q(x), p_{X|Y}(x|y))$ with respect to the parameters in $q(x)$. In some cases, this leads to a scalable approximation to the posterior distribution $p_{X|Y}(x|y)$. In particular, the objective function is convex for log concave posterior distribution [54].

Theorem 7.4. *Let the posterior distribution $p_{X|Y}(x|y)$ be log concave, and $q(x) = N(x;\mu, LL^t)$, $L \in \mathbb{R}^m$ being a positive lower triangular matrix. Then the functional $D_{KL}(q(x), p_{X|Y}(x|y))$ is jointly convex in μ and L.*

Proof. First, with the shorthand notation $p(x) = p_{X|Y}(x|y)$, straightforward computation shows

$$D_{KL}(q(x), p(x)) = -\frac{m}{2}\ln 2\pi - \frac{m}{2} - \ln|L| - \int N(x;\mu, LL^t)\ln p(x)dx.$$

It is known that the log determinant term $-\ln|L|$ is convex. Hence it suffices to show that the functional $J(\mu, L) = -\int N(x;\mu, LL^t)\ln p(x)dx$ is convex. By means of the transformation $x = Lz + \mu$, the functional $J(\mu, L)$ can be rewritten as

$$J(\mu, L) = -\int N(z; 0, I_m)p(L^{-1}z + \mu)dz.$$

Consequently, for all $t \in [0, 1]$, μ_1, $\mu_2 \in \mathbb{R}^m$ and positive L_1, $L_2 \in \mathbb{R}^{m\times m}$ there holds

$$\begin{aligned}
&J(t\mu_1 + (1-t)\mu_2, tL_1 + (1-t)L_2)\\
&= -\int N(z; 0, I_m)\ln p(t(\mu_1 + L_1 z) + (1-t)(\mu_2 + L_2 z))dz\\
&\le -t\int N(z; 0, I_m)\ln p(\mu_1 + L_1 z)\\
&\quad - (1-t)\int N(z; 0, I_m)\ln p(\mu_2 + L_2 z)dz\\
&= tJ(\mu_1, L_1) + (1-t)J(\mu_2, L_2),
\end{aligned}$$

where the second line follows from the log concavity of the density p. This shows the desired assertion. □

Now we illustrate the approach by one example on Poisson inverse problem; see [123] for statistical discussions.

Example 7.10. Consider the Poisson model

$$y_i \sim \mathrm{Pois}(e^{(a_i,x)}), \quad i = 1, \ldots, n,$$

where $a_i \in \mathbb{R}^m$ and $x \in \mathbb{R}^m$ represents the data formation mechanism and the unknown image, respectively, and $(\cdot,\cdot)$ refers to the Euclidean inner product on $\mathbb{R}^m$. The exponential Poisson rate $e^{(a_i,x)}$ can be regarded as an approximation to $1 + (a_i, x)$, where 1 and (a_i, x) are often referred to as exposure and offset, respectively. The prior distribution $p_X(x)$ on the unknown coefficient x takes the form $p_X(x) = N(x; \mu_0, C_0)$. Let $A = [a_i^t] \in \mathbb{R}^{n\times m}$. Then the likelihood $p_{Y|X}(y|x)$ is given by

$$p_{Y|X}(y|x) = e^{(Ax,y)-(e^{Ax},\mathbf{1}_n)-(\ln(y!),\mathbf{1}_n)},$$

where the vector $\mathbf{1}_n = [1 \ \ldots \ 1]^t \in \mathbb{R}^n$, and the exponential and logarithm of a vector are understood to be componentwise operation. The unnormalized posteriori $p_{X|Y}(x|y)$ is given by

$$p_{X|Y}(x|y) \propto (2\pi)^{-\frac{m}{2}}|C_0|^{-\frac{1}{2}}e^{(Ax,y)-(e^{Ax},\mathbf{1}_n)-(\log(y!),\mathbf{1}_n)-\frac{1}{2}(x-x_0)^tC_0^{-1}(x-x_0)}.$$

The marginal likelihood $p_Y(y)$ is given by

$$p_Y(y) = (2\pi)^{-\frac{m}{2}}|C_0|^{-\frac{1}{2}}\int e^{(Ax,y)-(e^{Ax},\mathbf{1}_n)-(\log(y!),\mathbf{1}_n)-\frac{1}{2}(x-x_0)^tC_0(x-x_0)}dx,$$

which involves an intractable integral over $\mathbb{R}^m$. To arrive at a lower bound with the variational method, we take $q(x)$ to be $N(x; \bar{x}, C)$. Then the lower bound LB defined in (7.13) admits an explicit expression

$$\begin{aligned}\text{LB} = y^tA\bar{x} - \mathbf{1}_n^t e^{A\bar{x}+\frac{1}{2}\text{diag}(ACA^t)} - \tfrac{1}{2}(\bar{x}-\mu_0)^tC_0^{-1}(\bar{x}-\mu_0)\\ - \tfrac{1}{2}\text{tr}(C_0^{-1}C) + \tfrac{1}{2}\ln|C| - \tfrac{1}{2}\ln|C_0| + \tfrac{m}{2} - \mathbf{1}^t\ln(y!).\end{aligned}$$

Then one chooses the variational parameters ($\bar{x}$ and C) to maximize the lower bound LB so as to make the approximation accurate. To this end, we compute the gradient of the lower bound LB with respect to $\bar{x}$ and C. It follows from the relation $\frac{\partial(\mathbf{1}_n,e^{A\bar{x}})}{\partial\bar{x}} = A^te^{A\bar{x}}$, that

$$\frac{\partial \text{LB}}{\partial \bar{x}} = A^ty - A^te^{A\bar{x}+\frac{1}{2}\text{diag}(ACA^t)} - \text{C}_0^{-1}(\bar{x}-\mu_0).$$

Next with the notation $D = \text{diag}(e^{A\bar{x}+\frac{1}{2}\text{diag}(ACA^t)})$, we have the following first-order Taylor expansion

$$\begin{aligned}&\mathbf{1}_n^te^{A\bar{x}+\frac{1}{2}\text{diag}(A(C+H)A^t)} - \mathbf{1}_n^te^{A\bar{x}+\frac{1}{2}\text{diag}(ACA^t)}\\ &\approx (e^{A\bar{x}+\frac{1}{2}\text{diag}(ACA^t)}, \tfrac{1}{2}\text{diag}(AHA^t))\\ &= \tfrac{1}{2}\text{tr}(DAHA^t) = \tfrac{1}{2}(A^tDA, H).\end{aligned}$$

Hence, we can deduce

$$\frac{\partial \text{LB}}{\partial C} = \tfrac{1}{2}\left[-A^t\text{diag}(e^{A\bar{x}+\frac{1}{2}\text{diag}(ACA^t)})A - C_0^{-1} + C^{-1}\right].$$

With the optimality condition at hand, a gradient descent or fixed point iteration algorithm may be applied to find the maximizer. In particular, one possible fixed point algorithm is given by

$$A^{t}e^{\frac{1}{2}\mathrm{diag}(ACA^{t})}e^{A\bar{x}^{k+1}}+C_0^{-1}\bar{x}^{k+1}=A^{t}y+C_0^{-1}\mu_0,$$
$$C^{-1}=C_0^{-1}+A^{t}\mathrm{diag}(y-A^{-t}C_0^{-1}(\bar{x}^{k+1}-\mu_0))A.$$

The first equation remains nonlinear and requires inner iteration (e.g., via Newton method or fixed-point iteration).

Remark 7.9. In parametric approximations such as Gaussian lower bound, the covariance C or its inverse can be structured, e.g., blockwise diagonal and tridiagonal. This can be enforced by incorporating a projection step, i.e., $(C^{k+1})^{-1}=P_{\mathcal{C}}(C^{k+1})^{-1}$, where $\mathcal{C}$ denotes the subspace of admissible covariance. Such structural conditions might be useful in capturing the essential features of the posterior distribution $p_{X|Y}(x|y)$, while maintaining reasonable computational efforts.

case (iii). In the expectation propagation (EP) algorithm [236, 237], one assumes the posteriori $p_{X|Y}(x|y)$ can be factorized into $p_{X|Y}(x|y)=\prod t_i(x)$, and then constructs a Gaussian approximation $\tilde{q}(x)$ by approximating each factor $t_i(x)$ by one Gaussian $\tilde{t}_i(x)$ via

$$\tilde{t}_i(x)=\arg\min_{\tilde{t}\text{ Gaussian}} D_{KL}\left(t_i(x)\prod_{j\neq i}\tilde{t}_j(x),\tilde{t}(x)\prod_{j\neq i}\tilde{t}_j(x)\right). \tag{7.14}$$

The goal of the update $\tilde{t}_i(x)$ is to "learn" the information contained in the factor $t_i(x)$, and by looping over all indices, hopefully the essential characteristics of the posterior distribution $p(x)$ are captured.

Remark 7.10. Cases (ii) and (iii) both assume a parametric density family, predominantly Gaussians. However, the minimization problems are different. First, case (ii) minimizes the Kullback-Leibler divergence with respect to the first argument, whereas (iii) in the second argument. Second, there is no explicit objective function for case (iii), although it can be regarded a hybridization of (i) and (ii) with a successive iteration method. Last, the arguments in $\{\tilde{t}_i(x)\}$ for the EP algorithm are not necessarily disjoint.

According to Theorem 7.3, minimizing the divergence involves computing integrals, i.e., mean and covariance. This can be numerically realized only if the integrals can be reduced to a low-dimensional case. One such case is that each factor involves a probability density function of the form $t_i(U_i x)$, i.e.,

$$p_{X|Y}(x|y) = \prod_i t_i(U_i x), \tag{7.15}$$

where the matrices U_i have full row rank and their ranges are of low-dimensionality, and represent linear combinations of the variables. This arises quite often in practice, e.g., robust Bayesian formulation with a Laplace likelihood. The result below provides the formula for transforming high-dimensional integrals into low-dimensional ones. The proof follows from a change of variable [99].

Theorem 7.5. *Let $\mu \in \mathbb{R}^n$, $C \in \mathbb{R}^{n\times n}$ be symmetric positive definite and $U \in \mathbb{R}^{l\times n}$ ($l \leq n$) be of full rank, $Z = \int t(Ux)N(x;\mu,C)dx$ be the normalizing constant. Then the mean $\mu^* = E_{Z^{-1}t(Ux)N(x;\mu,C)}[x]$ and covariance $C^* = \mathrm{Var}_{Z^{-1}t(Ux)N(x;\mu,C)}[x]$ are given by*

$$\begin{aligned}
\mu^* &= \mu + CU^t(UCU^t)^{-1}(\overline{s} - U\mu),\\
C^* &= C + CU^t(UCU^t)^{-1}(\overline{C} - UCU^t)(UCU^t)^{-1}UC,
\end{aligned}$$

where $\overline{s} \in \mathbb{R}^l$ and $\overline{C} \in \mathbb{R}^{l\times l}$ are respectively given by

$$\overline{s} = E_{Z^{-1}t(s)N(s;U\mu,UCU^t)}[s] \quad \text{and} \quad \overline{C} = \mathrm{Var}_{Z^{-1}t(s)N(s;U\mu,UCU^t)}[s].$$

The complete procedure is listed in Algorithm 7.7, where we have adopted the following representation of the approximation $\tilde{t}_i(U_i x)$:

$$\tilde{t}_i(U_i x) \propto e^{(U_i x)^t h_i - \frac{1}{2}(U_i x)^t K_i (U_i x)}, \tag{7.16}$$

where h_i and K_i are parameters to be updated iteratively within the inner loop. Note that with the representation (7.16), there holds

$$\begin{aligned}
\tilde{t}_i(x) = N(x,\mu_i,C_i) &\propto e^{-\frac{1}{2}(x-\mu_i)^t C_i^{-1}(x-\mu_i)}\\
&\propto e^{-\frac{1}{2}x^t C_i^{-1} x + x^t C_i^{-1}\mu_i} =: G(x,h_i,K_i),
\end{aligned}$$

with $h_i = C_i^{-1}\mu_i$ and $K_i = C_i^{-1}$. One distinct feature of the representation is that the tuple (K_i, h_i) is very low-dimensional.

In brevity, Steps 6–7 form the distribution $\tilde{p}^{\setminus i} = \prod_{j\neq i}\tilde{t}_j(U_j x)$, i.e., mean $\mu_{\setminus i}$ and covariance $C_{\setminus i}$, and the reduced representation $\widehat{C}_i = U_i C_{\setminus i} U_i^t$ and $\widehat{\mu}_i = U_i \mu_{\setminus i}$. By Theorem 7.5, the integrals of the nongaussian $t_i(U_i x)$ against the cavity distribution $\tilde{p}^{\setminus i}$ can be reduced to low-dimensional ones. Steps 8–10 update the parameter pair (h_i, K_i) for the ith gaussian factor $\tilde{t}_i(U_i x)$. Then steps 11–12 update the global approximation (h, K) by the

Algorithm 7.7 Serial EP algorithm with projection.

1: Initialize with $K_0 = C_0^{-1}$, $h_0 = C_0^{-1}\mu_0$, $K_i = I$ and $h_i = 0$ for $i \in I$
2: $K = K_0 + \sum_i U_i^t K_i U_i$
3: $h = h_0 + \sum_i U_i^t h_i$
4: **while** not converged **do**
5: **for** $i \in I$ **do**
6: $\widehat{C}_i^{-1} = (U_i K^{-1} U_i^t)^{-1} - K_i$
7: $\widehat{\mu}_i = (I - U_i K^{-1} U_i^t K_i)^{-1}(U_i K^{-1} h - U_i K^{-1} U_i^t h_i)$
8: $Z = \int t_i(s_i) N(s_i; \widehat{\mu}_i, \widehat{C}_i) ds_i$
9: $K_i = \mathrm{Var}_{Z^{-1}t_i(s_i)N(s_i;\widehat{\mu}_i,\widehat{C}_i)}[s_i]^{-1} - \widehat{C}_i^{-1}$
10: $h_i = \mathrm{Var}_{Z^{-1}t_i(s_i)N(s_i;\widehat{\mu}_i,\widehat{C}_i)}[s_i]^{-1} E_{Z^{-1}t_i(s_i)N(s_i;\widehat{\mu}_i,\widehat{C}_i)}[s_i] - \widehat{C}_i^{-1}\widehat{\mu}_i$
11: $K = K_0 + \sum_i U_i^t K_i U_i$
12: $h = h_0 + \sum_i U_i^t h_i$
13: **end for**
14: Check the stopping criterion.
15: **end while**
16: Output the covariance $C = K^{-1}$ and the mean $\mu = K^{-1}h$.

new values of the ith factor. A parallel variant is obtained by moving steps 11–12 behind the inner loop. It can be much faster on modern multi-core architectures, but it is less robust than the serial variant.

In general, the convergence of the EP algorithm remains fairly unclear, and nonconvergence can occur for non log-concave factors, e.g., t likelihood/prior. The next result, essentially due to [273] and slightly extended in [99], shows the well-definedness of the EP algorithm for the case of log-concave factors. The proof makes use of several auxiliary results on exponential families and log-concave functions in Appendix C.

Theorem 7.6. *Let the factors $\{t_i\}$ in the posteriori $\prod t_i(x)$ be log-concave, uniformly bounded and have support of positive measure. If at iteration k, the factor covariances $\{C_i\}$ of the approximations $\{\tilde{t}_i\}$ are positive definite, then the EP updates at $k+1$ step are positive semidefinite.*

Proof. By the positive definiteness of the factor covariances $\{C_i\}$, the covariance $C_{\setminus i}$ of the cavity distribution $\tilde{t}_{\setminus i} = \prod_{j \neq i} \tilde{t}_j$ is positive definite. Next we show that the partition function $Z = \int N(x; \mu_{\setminus i}, C_{\setminus i}) t_i(x) dx$ is log-concave in $\mu_{\setminus i}$. A straightforward computation yields that $N(x; \mu_{\setminus i}, C_{\setminus i})$ is log-concave in $(x, \mu_{\setminus i})$. By Lemma C.3 and log-concavity of $t_i(x)$, $N(x; \mu_{\setminus i}, C_{\setminus i}) t_i(x)$ is log-concave in $(x, \mu_{\setminus i})$, and thus Lemma C.4 implies that Z is log-concave in $\mu_{\setminus i}$, and $\nabla^2_{\mu_{\setminus i}} \log Z$ is negative semi-definite.

Now by Lemma C.1, the covariance C of the distribution $Z^{-1}N(x;\mu_{\setminus i}, C_{\setminus i})t_i(x)$ is given by

$$C = C_{\setminus i}\left(\nabla^2_{\mu_{\setminus i}} \log Z\right) C_{\setminus i} + C_{\setminus i}.$$

By the boundedness of factors t_i, the covariance

$$Z^{-1}\int N(x;\mu_{\setminus i}, C_{\setminus i})t_i(x)(x-\mu^*)(x-\mu^*)^t dx$$

(with $\mu^* = Z^{-1}\int N(x;\mu_{\setminus i}, C_{\setminus i})t_i(x)xdx$) exists, and further by Lemma C.2 and the assumption on t_i, it is positive definite. Hence, C is positive definite.

Since the Hessian $\nabla^2_{\mu_{\setminus i}} \log Z$ is negative semi-definite, $C - C_{\setminus i} = C_{\setminus i}\left(\nabla^2_{\mu_{\setminus i}} \log Z\right) C_{\setminus i}$ is also negative semi-definite, i.e., for any vector w, $w^tCw \leq w^tC_{\setminus i}w$. Now we let $\tilde{w} = C^{\frac{1}{2}}w$. Then $\|\tilde{w}\|^2 \leq \tilde{w}^tC^{-\frac{1}{2}}C_{\setminus i}C^{-\frac{1}{2}}\tilde{w}$ for any vector $\tilde{w}$. By the minmax characterization of eigenvalues of a Hermitian matrix, its minimum eigenvalue $\lambda_{\min}(C^{-\frac{1}{2}}C_{\setminus i}C^{-\frac{1}{2}}) \geq 1$. Consequently, $\lambda_{\max}(C^{\frac{1}{2}}C_{\setminus i}^{-1}C^{\frac{1}{2}}) \leq 1$, and equivalently $\|\tilde{w}\|^2 \geq \tilde{w}^tC^{\frac{1}{2}}C_{\setminus i}^{-1}C^{\frac{1}{2}}\tilde{w}$ for any vector $\tilde{w}$. With the substitution $w = C^{\frac{1}{2}}\tilde{w}$, we get $w^tC^{-1}w \geq w^tC_{\setminus i}^{-1}w$ for any vector w, i.e., $C^{-1} - C_{\setminus i}^{-1}$ is positive semidefinite. The conclusion follows from this and the fact that the inverse covariance of the factor approximation $\tilde{t}_i$ is given by $C^{-1} - C_{\setminus i}^{-1}$. □

Remark 7.11. Theorem 7.6 only ensures the well-definedness of the EP algorithm for one iteration. One might expect that in case of strictly log-concave factors, the well-definedness holds for all iterations. However, this would require a strengthened version of the Prékopa-Leindler inequality, i.g., preservation of strict log-concavity under marginalization.

Nonlinear problems. Now we extend algorithms to the case of a nonlinear operator F by the idea of recursive linearization

$$F(x) \approx F(\mu^k) + F'(\mu^k)(x - \mu^k).$$

See Algorithm 7.8 for EP. Like optimization algorithms based on linearization, step size control is often necessary. It improves robustness of the algorithm and avoids jumping back and forth between two states, as was observed for some practical problems. There are many possible rules, and we only briefly mention the Barzilai-Borwein rule here. It was first proposed for gradient type methods in [20]. It approximates by a scalar multiple of the identity matrix to the Hessian of the problem based on the last two iterates and gradient descent directions. In this manner, it incorporates the curvature of the potential to accelerate/stabilize the descent into the minimum. In our context, we adopt the following simple update rule

$$\mu^{k+1} = \mu^k + s^k(\mu_*^k - \mu^k),$$

and thus the difference $\mu_*^k - \mu^k$ serves as a "descent" direction and μ^k acts as the iterate in the Barzilai-Borwein rule. The step size s^k is computed by

$$s^k = \frac{\langle \mu^k - \mu^{k-1}, (\mu_*^k - \mu^k) - (\mu_*^{k-1} - \mu^{k-1})\rangle}{\langle \mu^k - \mu^{k-1}, \mu^k - \mu^{k-1}\rangle},$$

which approximately solves (in a least-squares sense)

$$s^k(\mu^k - \mu^{k-1}) \approx (\mu_*^k - \mu^k) - (\mu_*^{k-1} - \mu^{k-1}).$$

In practice, the step size s^k is necessarily constrained to $[0, 1]$.

Algorithm 7.8 EP for non-linear problem $F(x) = y$.

1: Initialize with $\mu = \mu^0$
2: **for** $k = 0 : K$ **do**
3: Linearize around μ^k;
4: Use EP Algorithm 7.7 to get $N(x; \mu_*^k, C_*^k)$;
5: Compute Barzilai-Borwein step size s^k
6: Update $\mu^{k+1} = \mu^k + s^k(\mu_*^k - \mu^k)$
7: Check the stopping criterion.
8: **end for**
9: Output the covariance C^* and the mean μ^*.

We illustrate the performance of EP with electrical impedance tomography taken from [99].

Example 7.11. In this example, we revisit the sparsity approach for EIT imaging with the complete electrode model, cf. Section 2.3.3. First we assume that the voltage measurements U^δ are contaminated by additive noises, i.e., $U^\delta = U(\sigma^\dagger) + \zeta$, and the noise components ζ_i follow an i.i.d. Gaussian distribution with zero mean and inverse variance τ. Then the likelihood function $p(U^\delta|\sigma)$ is given by

$$p(U^\delta|\sigma) \propto e^{-\frac{\tau}{2}\|F(\sigma)-U^\delta\|_2^2},$$

where the forward map $F(\sigma)$ is determined by the complete electrode model. The conductivity σ is pointwise non-negative due to physical constraints, and hence we impose $\chi_{[c_0,\infty]}(\sigma_i)$ on each component σ_i with a small $c_0 > 0$. Further, we assume that the conductivity consists of a known (but possibly inhomogeneous) background $\sigma^{\text{bg}} \in \mathbb{R}^m$ plus some small inclusions, enforced by using a Laplace prior on the difference $\sigma - \sigma^{\text{bg}}$:

$$(\tfrac{\lambda}{2})^m e^{-\lambda\|\sigma-\sigma^{\text{bg}}\|_{\ell^1}},$$

where $\lambda > 0$ is a scale parameter. We note that one may incorporate an additional smooth prior to enhance the cluster structure of typical inclusions. This leads to the following posterior distribution $p(\sigma|U^\delta)$

$$p(\sigma|U^\delta) \sim e^{-\frac{\alpha}{2}\|F(\sigma)-U^\delta\|_2^2} e^{-\lambda\|\sigma-\sigma^{\text{bg}}\|_{\ell^1}} \prod_{i=1}^m \chi_{[c_0,\infty]}(\sigma_i).$$

To apply the EP algorithm to the posterior $p(\sigma|U^\delta)$, we factorize it into

$$t_0(\sigma) = e^{-\frac{\alpha}{2}\|F(\sigma)-U^\delta\|_2^2},$$

$$t_i(\sigma) = e^{-\lambda|\sigma_i-\sigma_i^{\text{bg}}|}\chi_{[c_0,\infty]}(\sigma_i), \quad i = 1, \ldots, m.$$

Then the factor t_0 is Gaussian (after linearization, see Algorithm 7.8), and the remaining factors $t_1, \ldots, t_m$ each depend only on one component of σ. This enables us to rewrite the necessary integrals, upon appealing to Theorem 7.5, into one-dimensional ones. One reconstruction from experimental data is shown in Fig. 7.1 (see [99] for the experiment setup). The EP reconstructions using the serial algorithm took about 2 minutes with up to 7 outer iterations and a total of 22 inner iterations on a 2.6 GHz CPU. Numerically, we observe that EP and MCMC show a very good match on the mean. For the variance, EP usually captures the same structure as MCMC, but the magnitudes are slightly different. The variance generally gets higher towards the center of the watertank (due to less information encapsulated in the measurements), and at the edges of detected inclusions (due to uncertainty in the edge's exact position). The Bayesian reconstructions with the Laplace prior contain many close-to-zero components, however it is not truly sparse but only approximately sparse. This is very different from the Tikhonov formulation, which yields a genuinely sparse reconstruction, as long as the regularization parameter is large enough.

Remark 7.12. Generally, deterministic inference methods are overshadowed by Monte Carlo methods, especially MCMC, for performing approximate inference. They are much faster alternatives to MCMC, especially for large-scale problems. They are, however, limited in their approximation accuracy – as opposed to MCMC, which can be made arbitrarily accurate

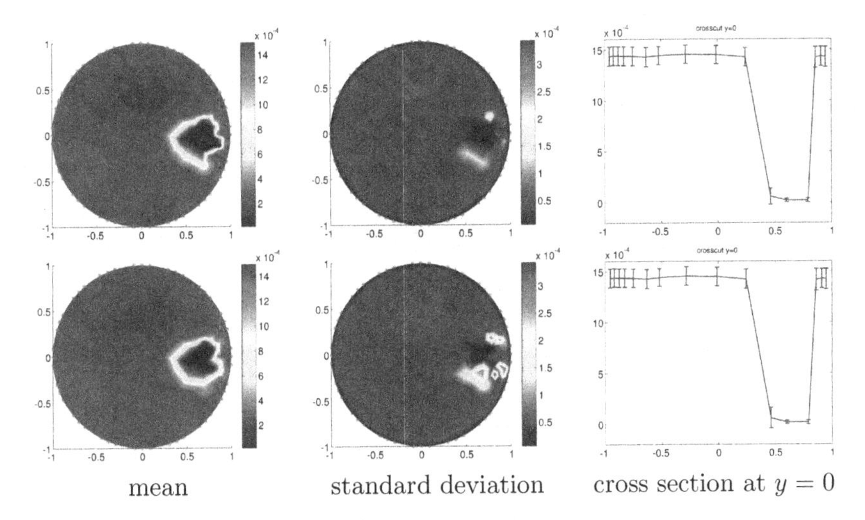

Fig. 7.1 Numerical results for one plastic inclusion: EP (top) v.s. MCMC (bottom).

by increasing the Monte Carlo sample sizes. The issue of accuracy is very delicate, and still poorly understood; see [302] for discussions on mean field approximation.

Bibliographical notes

Since the appearance of the seminal works [103, 28], Bayesian inference has gained great popularity in the inverse problem community. This is made possible partly by increasingly powerful computing facilities, and partly by many important algorithmic innovations. Hence, relevant literature on the topic is quite extensive. In this chapter we have described its basic ideas, fundamental algorithms and a few examples. Also we briefly mention several more advanced topics.

On the algorithmic side, Markov chain Monte Carlo methods remain the first choice. In particular, there are many advanced Markov chain Monte Carlo methods. We refer interested readers to comprehensive monographs [109, 262, 217]. In recent years, approximate inference methods also attracted considerable interest. The idea of variational Bayes in Section 7.4 was already contained in [15, 22], and we refer to [301] for an excellent survey. The book [186] contains many applications from the point of view of inverse problems; see also [75, 76, 280] for recent theoretical developments, where well-posedness of the posteriori and finite-dimensional approximation errors were discussed.

We have focused on discussions on the modeling and computational aspects of the Bayesian approach. There are a number of interesting theoretical results. We just mention a few below. In [140], the convergence of the posterior mean to the true solution in Ky Fan metric was studied, and a certain convergence rate was established, in the context of finite-dimensional linear inverse problems. The order-optimality was shown in [247]. However, there seems no analogous result for infinite-dimensional problems or nonlinear inverse problems.

Appendix A

Singular Value Decomposition

In this part, we recall the singular value system, and its discrete analogue, singular value decomposition. The latter is one of the most important tools for analyzing discrete ill-posed problems.

Singular value system One convenient tool for analyzing regularization methods in the Hilbert space setting is the singular value system. Let U and V be infinite-dimensional Hilbert spaces, and $K : U \to V$ be an injective compact linear operator. Then there exist an orthonormal basis $\{u_j\}$ in U, an orthonormal basis $\{v_j\}$ in V and a nonincreasing sequence $\{\sigma_j\}$ of nonnegative numbers with a limit 0 as $j \to \infty$ such that for all j

$$Ku_j = \sigma_j v_j.$$

We call the system $\{(\sigma_j, u_j, v_j)\}$ a singular system of the operator K. In other words, the operator K admits the decomposition

$$K = \sum \sigma_j u_j \otimes v_j.$$

Example A.1. Consider the integral operator $K : L^2(0,1) \to L^2(0,1)$ defined by $(Kf)(s) = \int_0^s f(t)dt$. Then one can verify directly that

$$\begin{aligned} u_j(s) &= \sqrt{2}\cos((j+\tfrac{1}{2})\pi s), \\ v_j(t) &= \sqrt{2}\sin((j+\tfrac{1}{2})\pi t), \\ \sigma_j &= \tfrac{1}{\pi(j+\frac{1}{2})}, \end{aligned}$$

is a singular value system.

With the help of the singular value system, one can state the Picard condition: in order that there exists a solution $f \in U$ to

$$Kf = g,$$

the right hand side g must satisfy

$$\sum_j \left(\frac{(u_j, g)}{\sigma_j}\right)^2 < \infty.$$

The Picard condition implies that for large j, the absolute value of the coefficients (u_j, g) must decay faster than the singular values $\sigma_j j^{-\frac{1}{2}}$ in order that a solution exists. This requirement is identical to $g \in \text{Range}(K)$. Range characterization and inclusion underlies many inversion algorithms, e.g, MUSIC algorithm and factorization method.

Singular value decomposition Let $A \in \mathbb{R}^{m\times n}$, $m \geq n$, be a rectangular or square matrix. Then the singular value decomposition (SVD) of A is a decomposition of the form

$$A = U\Sigma V^{\text{t}} = \sum_{j=1}^{n} u_j \sigma_j v_j^{\text{t}},$$

where $U = [u_1 \ \ldots \ u_n] \in \mathbb{R}^{m\times n}$ and $V = [v_1 \ \ldots \ v_n] \in \mathbb{R}^{n\times n}$ are matrices with orthonormal columns, $U^{\text{t}}U = V^{\text{t}}V = I$, and where the diagonal matrix $\Sigma = \text{diag}(\sigma_1, \ldots, \sigma_n)$ has nonnegative diagonal elements in nonincreasing order such that

$$\sigma_1 \geq \sigma_2 \geq \ldots \geq \sigma_n \geq 0.$$

The numbers σ_j are called the singular values of A, while the vectors u_j and v_j are the left and right singular vectors of A, respectively. Geometrically speaking, the SVD of A provides two sets of orthonormal basis vectors, namely, the columns of U and V, such that the matrix becomes diagonal when transformed into these two bases.

In connection with discrete ill-posed problems, two characteristic features of the SVD are very often found.

(a) The singular values σ_j decay gradually to zero with no particular gap in the spectrum. An increase of the dimension of A will increase the number of small singular values.
(b) The left and right singular vectors u_j and v_j tend to have more sign changes in their elements as the index j increases, i.e., as σ_i decreases.

Both features are consequences of the fact that the SVD of A is closely related to the singular value expansion of the underlying kernel K. In fact, the singular values σ_j of A are in many cases approximations to the singular values σ_j of K, and singular vectors u_j and v_j yield information about the singular functions of K.

The SVD also gives important insight into another aspect of discrete ill-posed problems, i.e., the smoothing effect typically associated with a square integrable kernel K. As σ_j decreases, the singular vectors u_j and v_j become more and more oscillatory. Now consider the mapping Ax of an arbitrary vector x. Using the SVD, we get

$$Ax = \sum_{j=1}^{n} \sigma_j (v_j^{\mathrm{t}} x) u_j.$$

The relation clearly show that due to the multiplication with σ_j, the high-frequency components of x are more damped in Ax than the low-frequency components. The inverse problem, i.e., that of computing x from $Ax = b$ or $\min \|Ax - b\|_2$, must have the opposite effect: it amplifies the high-frequency oscillations in the right-hand side b. In particular, the least-squares solution x_{ls} to $\|Ax - b\|_2$ is given by

$$x_{\mathrm{ls}} = \sum_{j=1}^{\mathrm{rank}(A)} \frac{u_j^{\mathrm{t}} b}{\sigma_i} v_j.$$

Clearly, it is the division by small singular values in the expression for the solution x_{ls} that amplifies the high-frequency components in b.

Appendix B

Noise Models

In this appendix, we describe several popular noise models, which underly the statistical ground of the fidelity functional in Tikhonov regularization and also the likelihood function in Bayesian modeling.

additive noise The simplest model for measurement errors is the additive noise model. Let $y^\dagger = F(x^\dagger) \in \mathbb{R}^n$ be the exact/noise-free data, and $\xi \in \mathbb{R}^n$ b a realization of certain random variable. If the measured data y is corrupted from the exact one $y^\dagger$ by

$$y = y^\dagger + \xi,$$

and the noise vector ξ is independent of $y^\dagger$, then the noise formation is called additive. One common noise model for the random variable ξ is Gaussian, and others include Laplace noise, Cauchy noise and uniform noise, which we shall explain briefly. If the true data $y^\dagger$ and the noise ξ are independent, then the likelihood function $p_{Y|X}(y|x)$ is given by

$$p_{Y|X}(y|x) = p(y - F(x)).$$

(i) **Gaussian noise** The Gaussian noise is definitely the most common model in practice. In the model, the noise ξ is taken to be a realization of a Gaussian random variable, typically with mean zero and covariance Σ, i.e.,

$$\xi \sim N(0, \Sigma),$$

where notation $N(\bar{\xi}, \Sigma)$ refers to a multivariate Gaussian distribution, with a mean $\bar{\xi}$ and covariance matrix Σ. The density function $p(\xi)$ is given by

$$p(\xi) = (2\pi)^{-\frac{n}{2}} |\Sigma|^{-\frac{1}{2}} e^{-\frac{1}{2}(\xi - \bar{\xi})^t \Sigma^{-1} (\xi - \bar{\xi})},$$

where $|\Sigma|$ denotes the determinant of the covariance Σ. Consequently, the likelihood function $p_{Y|X}(y|x)$ reads

$$p_{Y|X}(y|x) = \frac{1}{(2\pi)^{\frac{n}{2}} \sqrt{|\Sigma|}} e^{-\frac{1}{2}(y - F(x))^t \Sigma^{-1} (y - F(x))}.$$

A particularly simple case is that the covariance matrix Σ is a multiple of the identity matrix I_n, i.e., $\Sigma = \sigma^2 I_n$ for some covariance σ^2. In other words, the noise components follow an independent and identically distributed Gaussian distribution $N(0, \sigma^2)$. Then the likelihood $p_{Y|X}(y|x)$ is given by

$$p_{Y|X}(y|x) = \frac{1}{(2\pi\sigma^2)^{\frac{n}{2}}} e^{-\frac{1}{2\sigma^2}\|y-F(x)\|^2}.$$

We often employ the inverse covariance (precision) $\tau = \frac{1}{\sigma^2}$ in Bayesian modeling. This formula is used extensively in Chapter 7. The predominance of the Gaussian model in practice is primarily due to the analytical and computational tractability (at least for linear models): the mean and variance can be readily computed, and these statistics fully characterizes the solution. Theoretically, this is usually justified by appealing to the central limit theorem, which asserts that for data formed from the average of many independent random variables, the Gaussian model is a good approximation. This popularity of the model is also partially supported by its success in practical applications.

A well acknowledged limitation of the Gaussian model is its lack of robustness against the outliers, i.e., data points away from the bulk of the data, in the observations: one single aberrant data point can significantly influence all the parameters in the model, even for these with little substantive connection to the outlying observation [101].

(ii) **Laplace and Student-*t* noise** The Laplace noise arises in certain signal processing problems [3]. In comparison with the Gaussian model, one distinct feature of the these models is that the realizations are more likely to contain larger values, i.e., the distribution is more heavy tailed. Therefore, they are appropriate for data in the presence of significant outliers, and known to be robust with respect to outliers [147, 207].

Under the i.i.d. assumption on the components, the density function $p(\xi)$ for the Laplace distribution is given by

$$p(\xi) = \left(\tfrac{\sigma}{2}\right)^n e^{-\frac{\sigma}{2}\|\xi\|_1},$$

where σ is a scale parameter. Accordingly, the likelihood $p_{Y|X}(y|x)$ is given by

$$p_{Y|X}(y|x) = \left(\tfrac{\sigma}{2}\right)^n e^{-\frac{\sigma}{2}\|F(x)-y\|_1},$$

where $\|\cdot\|_1$ denotes the one-norm of a vector. The t model asserts that the noise components ξ_i are i.i.d. according to a centered t distribution, i.e.,

$$p(\xi;\nu,\sigma) = \frac{\Gamma(\frac{\nu+1}{2})}{\Gamma(\frac{\nu}{2})\sqrt{\pi\nu}\sigma}\left\{1 + \frac{1}{\nu}\left(\frac{\xi}{\sigma}\right)^2\right\}^{-\frac{\nu+1}{2}},$$

where ν is a degree of freedom parameter, σ is a scale parameter, and $\Gamma(\cdot)$ is Euler's Gamma function defined by $\Gamma(\alpha) = \int_0^\infty t^{\alpha-1}e^{-t}dt$. The t model is very flexible, and encompasses a number of other interesting models, e.g., the Cauchy distribution and the normal distribution as special cases, which correspond to the case of $\nu = 1$, and $\nu = \infty$, respectively. Accordingly, the likelihood function $p_{Y|X}(y|x)$ is given by

$$p_{Y|X}(y|x) = \left(\frac{\Gamma(\frac{\nu+1}{2})}{\Gamma(\frac{\nu}{2})\sqrt{\pi\nu}\sigma}\right)^n \prod_{i=1}^n \left\{1 + \frac{1}{\nu}\left(\frac{|(F(x)-y)_i|}{\sigma}\right)^2\right\}^{-\frac{\nu+1}{2}}.$$

The Laplace model and t-model are particularly attractive for handling data outliers. In practice, outliers can arise due to, e.g., erroneous recording and transmission in noisy channels.

multiplicative noise Multiplicative noise occurs as soon as one deals with active imaging systems, e.g., laser images, microscopic images, synthetic aperture radar images. The precise model is given by

$$y = f(y^\dagger, \xi),$$

where f denotes the noise formation mechanism, which often denotes pointwise multiplication and ξ refers to a realization of (non-negative, but not necessarily i.i.d.) random vector. There are several prominent noise models with a functional dependence of the noise on the signal, which is neither multiplicative nor additive. One special case of functional dependence is the multiplicative noise, for which we have the following representation.

Lemma B.1. *Assume that U and V are independent random variables, with continuous density functions $p_U(u)$ and $p_V(v)$, and $F = UV$. Then we have for $u > 0$*

$$p_V(\tfrac{f}{u})\tfrac{1}{u} = p_{F|U}(f|u).$$

Proof. Let $B \subset \mathbb{R}$ be an open subset. We have

$$\begin{aligned}\int p_{F|U}(f|u)1_{\{f\in B\}} = P(F \in B|U) &= \frac{P(F \in B, U)}{P(U)} \\ &= \frac{P((V = \frac{F}{U}) \in \frac{B}{U}, U)}{P(U)}.\end{aligned}$$

Using the fact that U and V are independent, we have

$$\begin{aligned}\frac{P((V = \frac{F}{U}) \in \frac{B}{U}, U)}{P(U)} &= P((V = \frac{F}{U}) \in \frac{B}{U}) \\ &= \int p_V(v)1_{\{v\in\frac{B}{u}\}}dv \\ &= \int p_v(\frac{f}{u})1_{f\in B}\frac{df}{u}.\end{aligned}$$

□

(v) **speckle noise** Synthetic aperture radar images are strongly corrupted by a noise called speckle. A radar sends a coherent wave which is reflected on the ground and then registered by the radar sensor. If the coherent wave is reflected on a coarse surface (compared to the radar wavelength), then the image processed by the radar is degraded by a noise with large amplitude: this gives a speckled aspect to the image, and this is the reason such a noise is called speckle noise. The classical approach [295] to model for such images is

$$y = y^\dagger \xi,$$

where y is the intensity of the observed image, $y^\dagger$ the reflectance of the scene (which is to be recovered), and ξ the speckle noise. ξ is assumed to follow a Gamma law with mean equal to 1, i.e.,

$$p(\xi) = \frac{L^L}{\Gamma(L)} \xi^{L-1} e^{-L\xi}.$$

Consequently, according Lemma B.1, the likelihood function $p_{Y|X}(y|x)$ is given

$$p_{Y|X}(y|x) = \prod_i \frac{L^L}{(Ax)_i^L \Gamma(L)} y_i^{L-1} e^{-\frac{-Ly_i}{(Ax)_i}}.$$

A variational approach taking into account such a noise model has been studied in [17].

The Gamma distribution is a versatile family of probability distributions for modeling noise, prior and parameter specification. The density function $p(t)$ takes the form

$$p(t) = G(t; a, b) = \frac{b^a}{\Gamma(a)} t^{a-1} e^{-bt},$$

where a and b are known as the shape parameter and the rate parameter, respectively. It covers the exponential distribution $a = 1$ and the χ^2 distribution ($a = 2/\nu$, $b = 2$) as special cases. It is the conjugate prior to many likelihood distributions: the Poisson, exponential, normal (with known mean), Pareto, gamma with known shape parameter, inverse gamma with known shape parameter etc. Therefore, it is very widely employed.

(vii) **salt-and-pepper noise** This model is especially common in image processing, and it reflects a wide variety of processes that result in the same image degradation: the corrupted data points (where $\xi_r \neq 0$) only take a fixed maximum ("salt") or minimum ("pepper") value. A simple model is as follows:

$$y_i = \begin{cases} y_i^\dagger, & \text{with probability } 1 - r - s, \\ y_{\max}, & \text{with probability } r, \\ y_{\min}, & \text{with probability } s, \end{cases}$$

where $y_{\max}$ and $y_{\min}$ are the maximum and minimum of the signal, respectively, and the parameter $r, s \in (0,1)$ (with $r + s < 1$) represents the percentage of corrupted data points. A closely related by more challenging noise is the so-called random-valued impulse noise, which allows arbitrary random values at the corrupted data points. Specifically, the noise model is given by

$$y_i = \begin{cases} y_i^\dagger, & \text{with probability } 1 - r, \\ y_i^\dagger + \xi, & \text{with probability } r, \end{cases}$$

where ξ is a random variable, e.g., normally distributed with mean zero and typically large variance. Clearly, this model is generated by the random variable ξ and reproduces the latter if $r = 1$. However, its characteristic is fundamentally different from that of ξ for $r < 1$: there exist data points which are not corrupted by noise, which carry a significant amount of information in the data. Salt-and-pepper noise and random-valued impulse noise are two prominent models for impulsive noise. The likelihood function $p_{Y|X}(y|x)$ is typically taken to be

$$p_{Y|X}(y|x) \propto e^{-\tau \|Ax - y\|_1},$$

for a suitable $\tau > 0$.

(vi) **Poisson noise** Poisson noise are typically employed for modeling photon counting errors produced by CCD sensors, e.g., florescence microscopy and optical/infrared astronomy. During the exposure, each sensor counts the number of incoming photons. The number of photons detected by the ith sensor can be modeled as a realization of a Poisson distributed random variable with mean $y_i^\dagger$, i.e., $y_i \sim \mathrm{Pois}(y_i^\dagger)$, where $\mathrm{Pois}(\lambda)$ is the Poisson distribution with a parameter λ. Specifically, the probability for measuring the value $k \in \mathbb{N} \cup \{0\}$ at the ith sensor is given by

$$y_i = k \quad \text{with } p(k) = \frac{(y_i^\dagger)^k}{k!} e^{-y_i^\dagger}, \quad k = 0, 1, 2, \ldots.$$

Under the i.i.d. assumption on the noise model, this gives rise to the following likelihood function

$$p_{Y|X}(y|x) = \prod_{i=1}^{n} \frac{e^{-(F(x))_i}(F(x))_i^{y_i}}{y_i!}.$$

Usually one approximates the unknown x with a continuous random variable to facilitate the subsequent model exploration [209].

Appendix C

Exponential Families

Now we recall exponential family [43]. We only need their specialization to normal distributions, but we follow the general framework for two reasons: it is the usual form found in the literature, and it allows direct generalization from normal distributions to other exponential families.

Definition C.1. An exponential family is a set $\mathcal{F}$ of distributions with density of the form

$$p(x|\theta) = e^{\theta^t \phi(x) - \Phi(\theta)},$$
$$\Phi(\theta) = \ln \int e^{\theta^t \phi(x)} d\nu(x),$$

for natural parameter θ from the natural parameter space Θ. The exponential family is fully characterized by the sufficient statistics ϕ and the base measure $d\nu$. The natural parameter space Θ is a convex set of all natural parameters such that $p(x|\theta)$ is a valid distribution.

The log partition function Φ allows computing the mean and variance of $\phi(x)$:

$$\nabla_\theta \Phi = E_{p(x|\theta)}[\phi(x)] \quad \text{and} \quad \nabla_\theta^2 \Phi = \mathrm{Var}_{p(x|\theta)}[\phi(x)]. \tag{C.1}$$

Let $f(x)$ be a positive function, and $\ln E_{p(x|\theta)}[f(x)] + \Phi(\theta)$ exist for every $\theta \in \Theta$. Then it defines a new exponential family with a base measure $f(x)d\nu(x)$, with the log partition function $\Phi_f(\theta)$ given by $\ln E_{p(x|\theta)}[f(x)] + \Phi(\theta)$.

We now consider the exponential family of normal distributions. We denote a normal distribution with mean μ and covariance C by $N(x; \mu, C) = (2\pi)^{-\frac{n}{2}} (\det C)^{-\frac{1}{2}} e^{-\frac{1}{2}(x-\mu)^t C^{-1}(x-\mu)}$. Hence, the sufficient statistics $\phi(x)$ and the natural parameter θ are given by $\phi(x) = (x, xx^t)$, and $\theta =: (h, K) = (C^{-1}\mu, -\frac{1}{2}C^{-1})$, respectively. This together with (C.1) and chain rule yields

$$E_{N(x;\mu,C)}[x] = C\nabla_\mu \Phi(\theta(\mu, C)) \quad \text{and} \quad \mathrm{Var}_{N(x;\mu,C)}[x] = C\nabla^2_\mu \Phi(\theta(\mu, C))C.$$

Now we can state a result on a tilted normal distribution $f(x)N(x, \mu, C)$.

Lemma C.1. *Let $N(x; \mu, C)$ be a normal distribution and f a positive function. Then the mean μ_f and covariance C_f of the tilted distribution $\tilde{p}(x) = Z^{-1} f(x)N(x; \mu, C)$ $(Z = \int f(x)N(x; \mu, C)$ are given by*

$$\begin{aligned}\mu_f &= C\left(\nabla_\mu \ln E_{N(x;\mu,C)}[f(x)]\right) + \mu, \\ C_f &= C\left(\nabla^2_\mu \ln E_{N(x;\mu,C)}[f(x)]\right) C + C.\end{aligned}$$

Proof. Let Φ and Φ_f be the log partition function of the normal distribution $p(x|\theta) = N(x; \mu, C)$ and the tilted distribution $\tilde{p}(x|\theta) = f(x)N(x; \mu, C)e^{-\Phi_f}$, respectively, with $\theta(\mu, C) = (h, K) = (C^{-1}\mu, -\frac{1}{2}C^{-1})$. Then $\Phi_f(\theta) = \ln E_{p(x|\theta)}[f(x)] + \Phi(\theta)$. Further, the first component of (C.1) reads

$$\nabla_h \Phi_f(\theta) = E_{\tilde{p}(x|\theta)}[x] \quad \text{and} \quad \nabla^2_h \Phi_f(\theta) = \mathrm{Var}_{\tilde{p}(x|\theta)}[x].$$

By chain rule there hold

$$\begin{aligned}\nabla_\mu \ln E_{N(x;\mu,C)}[f(x)] &= C^{-1}\nabla_h \ln E_{p(x|\theta)}[f(x)], \\ \nabla^2_\mu \ln E_{N(x;\mu,C)}[f(x)] &= C^{-1}\nabla^2_h \ln E_{p(x|\theta)}[f(x)]C^{-1}.\end{aligned}$$

Consequently, we deduce

$$\begin{aligned}E_{\tilde{p}(x|\theta)}[x] &= \nabla_h \Phi_f(\theta) = C\nabla_\mu \ln E_{N(x;\mu,C)}[f(x)] + \nabla_h \Phi(\theta), \\ \mathrm{Var}_{\tilde{p}(x|\theta)}[x] &= \nabla^2_h \Phi_f(\theta) = C\nabla^2_\mu \ln E_{N(x;\mu,C)}[f(x)]C + \nabla^2_h \Phi(\theta).\end{aligned}$$

Now the desired assertion follows directly from the relation $\nabla_h \Phi(\theta) = \mu$ and $\nabla^2_h \Phi(\theta) = C$ using (C.1). □

Lemma C.2. *Let $p : \mathbb{R}^n \to \mathbb{R}$ be a probability density function with a support of positive measure. If the covariance $\mathrm{Var}_p[x] = \int p(x)(E[x] - x)(E[x] - x)^t dx$ exists, then it is positive definite.*

Proof. It suffices to show $v^t \mathrm{Var}_p[x] v > 0$ for any nonzero vector $v \in \mathbb{R}^n$. Let $U = \{x \in \mathrm{supp}(p) : v^t(E[x] - x) \neq 0\}$. However, the complement set $\mathrm{supp}(p) \setminus U = \{x \in \mathrm{supp}(p) : v^t(E[x] - x) = 0\}$ lies in a co-dimensional one hyperplane in $\mathbb{R}^n$, thus it has only zero measure. This together with the positive measure assumption of the support $\mathrm{supp}(p)$, the set U has positive measure. Therefore, $v^t \mathrm{Var}_p[x] v \geq \int_U p(x)(v^t(E_p[x] - x))^2 dx > 0$. □

Finally we recall the concept of log-concavity. It plays a role in the theory of expectation propagation as convexity in classical optimization theory. A nonnegative function $f : V \to \mathbb{R}_0^+$ is log-concave if

$$f(\lambda x_1 + (1 - \lambda)x_2) \geq f(x_1)^\lambda f(x_2)^{1-\lambda}$$

holds for all elements x_1, x_2 from a real convex vector space V and for all $\lambda \in [0, 1]$.

Lemma C.3. *Let $f, g : V \to \mathbb{R}_0^+$ be log-concave. Then the product fg is log-concave.*

Log-concavity is preserved by marginalization by the Prékopa-Leindler inequality [30, 36].

Lemma C.4. *Let $f : \mathbb{R}^n \times \mathbb{R}^m \to \mathbb{R}_0^+$ be log-concave and bounded. Then the marginalized function $g(x) = \int_{\mathbb{R}^m} f(x, y)dy$ is log-concave.*

Bibliography

[1] Adams, R. A. and Fournier, J. J. F. (2003). *Sobolev Spaces*, 2nd edn. (Elsevier, Amsterdam).

[2] Alessandrini, G., Rondi, L., Rosset, E. and Vessella, S. (2009). The stability for the cauchy problem for elliptic equations, *Inverse Problems* **25**, 12, pp. 123004, 47 pp.

[3] Alliney, S. and Ruzinsky, S. A. (1994). An algorithm for the minimization of mixed l_1 and l_2 norms with application to Bayesian estimation, *IEEE Trans. Signal Proc.* **42**, 3, pp. 618–627.

[4] Ammari, H., Iakovleva, E. and Lesselier, D. (2005). A MUSIC algorithm for locating small inclusions buried in a half-space from the scattering amplitude at a fixed frequency, *Multiscale Model. Simul.* **3**, 3, pp. 597–628.

[5] Ammari, H., Iakovleva, E., Lesselier, D. and Perrusson, G. (2007). MUSIC-type electromagnetic imaging of a collection of small three-dimensional inclusions, *SIAM J. Sci. Comput.* **29**, 2, pp. 674–709.

[6] Andrews, D. F. and Mallows, C. L. (1974). Scale mixtures of normal distributions, *J. Roy. Statist. Soc. Ser. B* **36**, pp. 99–102.

[7] Anzengruber, S. W., Hofmann, B. and Mathe, P. (2014). Regularization properties of the sequential discrepancy principle for Tikhonov regularization in Banach spaces, *Appl. Anal.* **93**, 7, pp. 1382–1400.

[8] Anzengruber, S. W., Hofmann, B. and Ramlau, R. (2013). On the interplay of basis smoothness and specific range conditions occuring in sparsity regularization, *Inverse Problems* **29**, 12, pp. 125002, 21 pp.

[9] Anzengruber, S. W. and Ramlau, R. (2011). Convergence rates for Morozov's discrepancy principle using variational inequalities, *Inverse Problems* **27**, 10, pp. 105007, 18 pp.

[10] Arcangeli, R. (1966). Pseudo-solution de l'équation $Ax = y$, *C. R. Acad. Sci. Paris Sér. A-B* **263**, pp. A282–A285.

[11] Arens, T. (2004). Why linear sampling works, *Inverse Problems* **20**, 1, pp. 163–173.

[12] Arens, T. and Lechleiter, A. (2009). The linear sampling method revisited, *J. Integral Equations Appl.* **21**, 2, pp. 179–202.

[13] Arridge, S. R., Kaipio, J. P., Kolehmainen, V., Schweiger, M., Somersalo,

E., Tarvainen, T. and Vauhkonen, M. (2006). Approximation errors and model reduction with an application in optical diffusion tomography, *Inverse Problems* **22**, 1, pp. 175–195.

[14] Arridge, S. R. and Schotland, J. C. (2009). Optical tomography: forward and inverse problems, *Inverse Problems* **25**, 12, pp. 123010, 59 pp.

[15] Attias, H. (2000). A variational Bayesian framework for graphical models, in S. Solla, T. Leen and K.-R. Muller (eds.), *NIPS 12* (MIT Press, Cambridge, MA), pp. 209–215.

[16] Attouch, H., Buttazzo, G. and Michaille, G. (2006). *Variational Analysis in Sobolev and BV spaces* (SIAM, Philadelphia, PA).

[17] Aubert, G. and Aujol, J.-F. (2008). A variational approach to removing multiplicative noise, *SIAM J. Appl. Math.* **68**, 4, pp. 925–946.

[18] Bakushinskiĭ, A. B. (1984). Remarks on the choice of regularization parameter from quasioptimality and ratio tests, *Zh. Vychisl. Mat. i Mat. Fiz.* **24**, 8, pp. 1258–1259.

[19] Banks, H. T. and Kunisch, K. (1989). *Parameter Estimation Techniques for Distributed Systems* (Birkhäuser, Boston).

[20] Barzilai, J. and Borwein, J. M. (1988). Two-point step size gradient methods, *IMA J. Numer. Anal.* **8**, 1, pp. 141–148.

[21] Bauer, F. and Kindermann, S. (2008). The quasi-optimality criterion for classical inverse problems, *Inverse Problems* **24**, 3, pp. 035002, 20 pp.

[22] Beal, M. J. (2003). *Variational Algorithms for Approximate Bayesian Inference*, Ph.D. thesis, Gatsby Computational Neuroscience Unit, University College London.

[23] Beck, J. V., St. Clair, C. R. and Blackwell, B. (1985). *Inverse Heat Conduction: Ill-Posed Problems* (John Wiley, New York).

[24] Becker, S. M. A. (2011). Regularization of statistical inverse problems and the Bakushinskiĭ veto, *Inverse Problems* **27**, 11, pp. 115010, 22 pp.

[25] Belgacem, F. B. (2007). Why is the Cauchy problem severely ill-posed? *Inverse Problems* **23**, 2, pp. 823–836.

[26] Belge, M., Kilmer, M. E. and Miller, E. L. (2002). Efficient determination of multiple regularization parameters in a generalized L-curve framework, *Inverse Problems* **18**, 4, pp. 1161–1183.

[27] Benning, M. and Burger, M. (2011). Error estimates for general fidelities, *Electron. Trans. Numer. Anal.* **38**, pp. 44–68.

[28] Besag, J. (1986). On the statistical analysis of dirty pictures, *J. Roy. Statist. Soc. Ser. B* **48**, 3, pp. 259–302.

[29] Bissantz, N., Hohage, T., Munk, A. and Ruymgaart, F. (2007). Convergence rates of general regularization methods for statistical inverse problems and applications, *SIAM J. Numer. Anal.* **45**, 6, pp. 2610–2636.

[30] Bogachev, V. I. (1998). *Gaussian Measures* (AMS, Providence, RI).

[31] Bonesky, T. (2009). Morozov's discrepancy principle and Tikhonov-type functionals, *Inverse Problems* **25**, 1, pp. 015015, 11 pp.

[32] Bonesky, T., Kazimierski, K. S., Maass, P., Schöpfer, F. and Schuster, T. (2008). Minimization of Tikhonov functionals in Banach spaces, *Abstr. Appl. Anal.* pp. Art. ID 192679, 19 pp.

[33] Borcea, L. (2002). Electrical impedance tomography, *Inverse Problems* **18**, 6, pp. R99–R136.

[34] Borg, G. (1946). Eine Umkehrung der Sturm-Liouvilleschen Eigenwertaufgabe, *Acta Math.* **76**, 1, pp. 1–96.

[35] Boţ, R. I. and Hofmann, B. (2010). An extension of the variational inequality approach for obtaining convergence rates in regularization of nonlinear ill-posed problems, *J. Integral Equations Appl.* **22**, 3, pp. 369–392.

[36] Brascamp, H. J. and Lieb, E. H. (1976). On extensions of the Brunn-Minkowski and Prékopa-Leindler theorems, including inequalities for log concave functions, and with an application to the diffusion equation, *J. Fun. Anal.* **22**, 4, pp. 366–389.

[37] Bredies, K., Kunisch, K. and Pock, T. (2010). Total generalized variation, *SIAM J. Imag. Sci.* **3**, 3, pp. 492–526.

[38] Bredies, K. and Lorenz, D. A. (2009). Regularization with non-convex separable constraints, *Inverse Problems* **25**, 8, pp. 085011, 14 pp.

[39] Bredies, K. and Pikkarainen, H. K. (2013). Inverse problems in spaces of measures, *ESAIM Control Optim. Calc. Var.* **19**, 1, pp. 190–218.

[40] Bregman, L. M. (1967). The relaxation method of finding the common point of convex sets and its application to the solution of problems in convex programming, *USSR Comput. Math. Math. Phys.* **7**, 3, pp. 200–217.

[41] Brezinski, C., Redivo-Zaglia, M., Rodriguez, G. and Seatzu, S. (2003). Multi-parameter regularization techniques for ill-conditioned linear systems, *Numer. Math.* **94**, 2, pp. 203–228.

[42] Brooks, S. P. and Gelman, A. (1998). General methods for monitoring convergence of iterative simulations, *J. Comput. Graph. Statist.* **7**, 4, pp. 434–455.

[43] Brown, L. D. (1986). *Fundamentals of Statistical Exponential Families with Applications in Statistical Decision Theory* (Institute of Mathematical Statistics, Hayward, CA).

[44] Broyden, C. G. (1965). A class of methods for solving nonlinear simultaneous equations, *Math. Comp.* **19**, pp. 577–593.

[45] Brühl, M. (2001). Explicit characterization of inclusions in electrical impedance tomography, *SIAM J. Math. Anal.* **32**, 6, pp. 1327–1341.

[46] Brühl, M. and Hanke, M. (2000). Numerical implementation of two noniterative methods for locating inclusions by impedance tomography, *Inverse Problems* **16**, 4, pp. 1029–1042.

[47] Brune, C., Sawatzky, A. and Burger, M. (2011). Primal and dual Bregman methods with application to optical nanoscopy, *Int. J. Comput. Vis.* **92**, 2, pp. 211–229.

[48] Bukhgeĭm, A. L. and Klibanov, M. V. (1981). Uniqueness in the large of a class of multidimensional inverse problems, *Dokl. Akad. Nauk SSSR* **260**, 2, pp. 269–272.

[49] Burger, M., Flemming, J. and Hofmann, B. (2013). Convergence rates in ℓ^1-regularization if the sparsity assumption fails, *Inverse Problems* **29**, 2, pp. 025013, 16 pp.

[50] Burger, M. and Osher, S. (2004). Convergence rates of convex variational

regularization, *Inverse Problems* **20**, 5, pp. 1411–1421.

[51] Cakoni, F., Colton, D. and Monk, P. (2011). *The Linear Sampling Method in Inverse Electromagnetic Scattering* (SIAM, Philadelphia, PA).

[52] Casella, G., Girón, F. J., Martínez, M. L. and Moreno, E. (2009). Consistency of Bayesian procedures for variable selection, *Ann. Statist.* **37**, 3, pp. 1207–1228.

[53] Chaabane, S., Feki, I. and Mars, N. (2012). Numerical reconstruction of a piecewise constant Robin parameter in the two- or three-dimensional case, *Inverse Problems* **28**, 6, pp. 065016, 19 pp.

[54] Challis, E. and Barber, D. (2011). Concave Gaussian variational approximations for inference in large-scale Bayesian linear models, in *Proc. 14th Int. Conf. Artif. Int. Stat. (AISTATS)* (Fort Lauderdale, FL).

[55] Chambolle, A. and Lions, P.-L. (1997). Image recovery via total variation minimization and related problems, *Numer. Math.* **76**, 2, pp. 167–188.

[56] Chan, T. F. and Esedoḡlu, S. (2005). Aspects of total variation regularized L^1 function approximation, *SIAM J. Appl. Math.* **65**, 5, pp. 1817–1837.

[57] Chaturantabut, S. and Sorensen, D. C. (2010). Nonlinear model reduction via discrete empirical interpolation, *SIAM J. Sci. Comput.* **32**, 5, pp. 2737–2764.

[58] Chavent, G. and Kunisch, K. (1994). Convergence of Tikhonov regularization for constrained ill-posed inverse problems, *Inverse Problems* **10**, 1, pp. 63–76.

[59] Chavent, G. and Kunisch, K. (1996). On weakly nonlinear inverse problems, *SIAM J. Appl. Math.* **56**, 2, pp. 542–572.

[60] Chen, J., Chen, Z. and Huang, G. (2013). Reverse time migration for extended obstacles: acoustic waves, *Inverse Problems* **29**, 8, pp. 085005, 17 pp.

[61] Chen, X., Nashed, Z. and Qi, L. (2000). Smoothing methods and semismooth methods for nondifferentiable operator equations, *SIAM J. Numer. Anal.* **38**, 4, pp. 1200–1216.

[62] Chen, Z., Lu, Y., Xu, Y. and Yang, H. (2008). Multi-parameter Tikhonov regularization for linear ill-posed operator equations, *J. Comput. Math.* **26**, 1, pp. 37–55.

[63] Cheney, M. (2001). The linear sampling method and the MUSIC algorithm, *Inverse Problems* **17**, 4, pp. 591–595.

[64] Cheney, M., Isaacson, D. and Newell, J. C. (1999). Electrical impedance tomography, *SIAM Rev.* **41**, 1, pp. 85–101.

[65] Cheng, J. and Yamamoto, M. (2000). On new strategy for a priori choice of regularization parameters in Tikhonov's regularization, *Inverse Problems* **16**, 4, pp. L31–L38.

[66] Christen, J. A. and Fox, C. (2005). Markov chain Monte Carlo using an approximation, *J. Comput. Graph. Statist.* **14**, 4, pp. 795–810.

[67] Chu, M. T. and Golub, G. H. (2005). *Inverse Eigenvalue Problems* (Oxford University Press, Oxford, New York).

[68] Chung, E. T., Chan, T. F. and Tai, X.-C. (2005). Electrical impedance tomography using level set representation and total variational regulariza-

tion, *J. Comput. Phys.* **205**, 1, pp. 357–372.
[69] Cioranescu, I. (1990). *Geometry of Banach Spaces, Duality Mappings and Nonlinear Problems* (Kluwer, Dordrecht).
[70] Clason, C., Jin, B. and Kunisch, K. (2010). A semismooth Newton method for L^1 data fitting with automatic choice of regularization parameters and noise calibration, *SIAM J. Imaging Sci.* **3**, 2, pp. 199–231.
[71] Clason, C. and Kunisch, K. (2011). A duality-based approach to elliptic control problems in non-reflexive Banach spaces, *ESAIM Control Optim. Calc. Var.* **17**, 1, pp. 243–266.
[72] Colton, D. and Kirsch, A. (1996). A simple method for solving inverse scattering problems in the resonance region, *Inverse Problems* **12**, 4, pp. 383–393.
[73] Colton, D. and Kress, R. (1998). *Inverse Acoustic and Electromagnetic Scattering Theory* (Springer-Verlag, Berlin).
[74] Colton, D. and Kress, R. (2006). Using fundamental solutions in inverse scattering, *Inverse Problems* **22**, 3, pp. R49–R66.
[75] Cotter, S. L., Dashti, M., Robinson, J. C. and Stuart, A. M. (2009). Bayesian inverse problems for functions and applications to fluid mechanics, *Inverse Problems* **25**, 11, pp. 115008, 43 pp.
[76] Cotter, S. L., Dashti, M. and Stuart, A. M. (2010). Approximation of Bayesian inverse problems for PDEs, *SIAM J. Numer. Anal.* **48**, 1, pp. 322–345.
[77] Daubechies, I., Defrise, M. and De Mol, C. (2004). An iterative thresholding algorithm for linear inverse problems with a sparsity constraint, *Comm. Pure Appl. Math.* **57**, 11, pp. 1413–1457.
[78] Davison, M. E. (1981). A singular value decomposition for the Radon transform in n-dimensional Euclidean space, *Numer. Funct. Anal. Optim.* **3**, 3, pp. 321–340.
[79] de Hoop, M. V., Qiu, L. and Scherzer, O. (2012). Local analysis of inverse problems: Hölder stability and iterative reconstruction, *Inverse Problems* **28**, 4, pp. 045001, 16 pp.
[80] Dempster, A. P., Laird, N. M. and Rubin, D. B. (1977). Maximum likelihood from incomplete data via the EM algorithm, *J. Roy. Stat. Soc. Ser. B* **39**, 1, pp. 1–38.
[81] Devaney, A. J. (1999). Super-resolution processing of multi-static data using time-reversal and MUSIC, available at `http://www.ece.neu.edu/faculty/devaney/preprints/paper02n_00.pdf`.
[82] Dwyer, P. S. (1967). Some applications of matrix derivatives in multivariate analysis, *J. Amer. Stat. Assoc.* **62**, 318, pp. 607–625.
[83] Efendiev, Y., Hou, T. and Luo, W. (2006). Preconditioning Markov chain Monte Carlo simulations using coarse-scale models, *SIAM J. Sci. Comput.* **28**, 2, pp. 776–803.
[84] Efron, B., Hastie, T., Johnstone, I. and Tibshirani, R. (2004). Least angle regression, *Ann. Statist.* **32**, 2, pp. 407–499, with discussion, and a rejoinder by the authors.
[85] Egger, H. and Schlottbom, M. (2010). Analysis and regularization of prob-

lems in diffuse optical tomography, *SIAM J. Math. Anal.* **42**, 5, pp. 1934–1948.

[86] El Badia, A. and Ha-Duong, T. (2000). An inverse source problem in potential analysis, *Inverse Problems* **16**, pp. 651–663.

[87] Engl, H. W. (1987). On the choice of the regularization parameter for iterated Tikhonov regularization of ill-posed problems, *J. Approx. Theory* **49**, 1, pp. 55–63.

[88] Engl, H. W., Hanke, M. and Neubauer, A. (1996). *Regularization of Inverse Problems* (Kluwer Academic, Dordrecht).

[89] Engl, H. W., Kunisch, K. and Neubauer, A. (1989). Convergence rates for Tikhonov regularisation of nonlinear ill-posed problems, *Inverse Problems* **5**, 4, pp. 523–540.

[90] Engl, H. W. and Zou, J. (2000). A new approach to convergence rate analysis of Tikhonov regularization for parameter identification in heat equation, *Inverse Problems* **16**, 6, pp. 1907–1923.

[91] Evans, L. C. and Gariepy, R. F. (1992). *Measure Theory and Fine Properties of Functions* (CRC Press, Boca Raton, FL).

[92] Flemming, J. (2010). Theory and examples of variational regularization with non-metric fitting functionals, *J. Inverse Ill-Posed Probl.* **18**, 6, pp. 677–699.

[93] Flemming, J. (2012). *Generalized Tikhonov Regularization and Modern Convergence Rate Theory in Banach Spaces* (Shaker Verlag, Aachen).

[94] Flemming, J. and Hofmann, B. (2010). A new approach to source conditions in regularization with general residual term, *Numer. Funct. Anal. Optim.* **31**, 1-3, pp. 254–284.

[95] Flemming, J. and Hofmann, B. (2011). Convergence rates in constrained Tikhonov regularization: equivalence of projected source conditions and variational inequalities, *Inverse Problems* **27**, 8, pp. 085001, 11 pp.

[96] Fornasier, M., Naumova, V. and Pereverzyev, S. V. (2013). Parameter choice strategies for multi-penalty regularization, preprint.

[97] Franklin, J. N. (1970). Well-posed stochastic extensions of ill-posed linear problems, *J. Math. Anal. Appl.* **31**, pp. 682–716.

[98] Frick, K., Marnitz, P. and Munk, A. (2012). Shape-constrained regularization by statistical multiresolution for inverse problems: asymptotic analysis, *Inverse Problems* **28**, 6, pp. 065006, 31 pp.

[99] Gehre, M. and Jin, B. (2014). Expectation propagation for nonlinear inverse problems – with an application to electrical impedance tomography, *J. Comput. Phys.* **259**, pp. 513–535.

[100] Gel'fand, I. M. and Levitan, B. M. (1955). On the determination of a differential equation from its spectral function, *Amer. Math. Soc. Transl. (2)* **1**, pp. 253–304.

[101] Gelman, A., Carlin, J. B., Stern, H. S. and Rubin, D. B. (2004). *Bayesian Data Analysis* (Chapman and Hall/CRC, Boca Raton, Florida).

[102] Gelman, A., Roberts, G. O. and Gilks, W. R. (1996). Efficient Metropolis jumping rules, in J. M. Bernardo, J. O. Berger, A. P. Dawid and A. F. M. Smith (eds.), *Bayesian Statistics 5* (Oxford University Press), pp. 599–607.

[103] Geman, S. and Geman, D. (1984). Stochastic relaxation, Gibbs distributions, and the Bayesian restoration of images, *IEEE Trans. Patt. Anal. Mach. Int.* **6**, 6, pp. 721–741.

[104] Gfrerer, H. (1987). An a posteriori parameter choice for ordinary and iterated Tikhonov regularization of ill-posed problems leading to optimal convergence rates, *Math. Comp.* **49**, 180, pp. 507–522, S5–S12.

[105] Ghanem, R. G. and Spanos, P. D. (1991). *Stochastic Finite Elements: a Spectral Approach* (Springer-Verlag, New York).

[106] Gilbarg, D. and Trudinger, N. S. (2001). *Elliptic Partial Differential Equations of Second Order* (Springer-Verlag, Berlin).

[107] Giles, M. (2008a). Improved multilevel Monte Carlo convergence using the Milstein scheme, in *Monte Carlo and quasi-Monte Carlo methods 2006* (Springer, Berlin), pp. 343–358.

[108] Giles, M. B. (2008b). Multilevel Monte Carlo path simulation, *Oper. Res.* **56**, 3, pp. 607–617.

[109] Gilks, W. R., Richardson, S. and Spielgelhalter, D. J. (1996). *Markov Chain Monte Carlo in Practice* (Chapman and Hall, London).

[110] Glasko, V. B. and Kriksin, Y. A. (1984). The principle of quasioptimality for linear ill-posed problems in Hilbert space, *Zh. Vychisl. Mat. i Mat. Fiz.* **24**, 11, pp. 1603–1613, 1758.

[111] Glowinski, R. (1984). *Numerical Methods for Nonlinear Variational Problems* (Springer-Verlag, New York).

[112] Glowinski, R., Lions, J.-L. and Trémolières, R. (1981). *Numerical Analysis of Variational Inequalities* (North-Holland Publishing Co., Amsterdam).

[113] Gordon, C., Webb, D. L. and Wolpert, S. (1992). Isospectral plane domains and surfaces via Riemannian orbifolds, *Invent. Math.* **10**, 1, pp. 1–22.

[114] Grasmair, M. (2010). Generalized Bregman distances and convergence rates for non-convex regularization methods, *Inverse Problems* **26**, 11, pp. 115014, 16 pp.

[115] Grasmair, M. (2011). Linear convergence rates for Tikhonov regularization with positively homogeneous functionals, *Inverse Problems* **27**, 7, pp. 075014, 16 pp.

[116] Grasmair, M. (2013). Variational inequalities and higher order convergence rates for Tikhonov regularisation on Banach spaces, *J. Inverse Ill-Posed Probl.* **21**, 3, pp. 379–394.

[117] Grasmair, M., Haltmeier, M. and Scherzer, O. (2008). Sparse regularization with l^q penalty term, *Inverse Problems* **24**, 5, pp. 055020, 13 pp.

[118] Grasmair, M., Haltmeier, M. and Scherzer, O. (2011). Necessary and sufficient conditions for linear convergence of ℓ^1-regularization, *Comm. Pure Appl. Math.* **64**, 2, pp. 161–182.

[119] Green, P. J. (1990). Bayesian reconstructions from emission tomography data using a modified EM algorithm, *IEEE Trans. Med. Imag.* **9**, 1, pp. 84–93.

[120] Griesmaier, R. (2011). Multi-frequency orthogonality sampling, *Inverse Problems* **27**, 8, pp. 085005, 23 pp.

[121] Gulrajani, R. M. (1998). The forward and inverse problems of electrocar-

diography, *IEEE Eng. Med. Biol. Mag.* **17**, 5, pp. 84–101.
[122] Hadamard, J. (1923). *Lectures on Cauchy's Problem in Linear Partial Differential Equations* (Yale University Press, New Haven, CT).
[123] Hall, P., Ormerod, J. T. and Wand, M. P. (2011). Theory of Gaussian variational approximation for a Poisson linear mixed model, *Stat. Sinica* **21**, 1, pp. 369–389.
[124] Hanke, M. and Groetsch, C. W. (1998). Nonstationary iterated Tikhonov regularization, *J. Optim. Theory Appl.* **98**, 1, pp. 37–53.
[125] Hanke, M. and Raus, T. (1996). A general heuristic for choosing the regularization parameter in ill-posed problems, *SIAM J. Sci. Comput.* **17**, 4, pp. 956–972.
[126] Hanke, M. and Rundell, W. (2011). On rational approximation methods for inverse source problems, *Inverse Probl. Imaging* **5**, 1, pp. 185–202.
[127] Hansen, P. C. (1992). Analysis of discrete ill-posed problems by means of the L-curve, *SIAM Rev.* **34**, 4, pp. 561–580.
[128] Hào, D. N. and Quyen, T. N. T. (2010). Convergence rates for Tikhonov regularization of coefficient identification problems in Laplace-type equations, *Inverse Problems* **26**, 12, pp. 125014, 23 pp.
[129] Hastings, W. K. (1970). Monte Carlo sampling methods using Markov chains and their applications, *Biometrika* **57**, 1, pp. 97–109.
[130] Hein, T. (2008). Convergence rates for regularization of ill-posed problems in Banach spaces by approximate source conditions, *Inverse Problems* **24**, 4, pp. 045007, 10 pp.
[131] Hein, T. (2009a). On Tikhonov regularization in Banach spaces optimal convergence rates results, *Appl. Anal.* **88**, 5, pp. 653–667.
[132] Hein, T. (2009b). Tikhonov regularization in Banach spaces improved convergence rates results, *Inverse Problems* **25**, 3, pp. 035002, 18 pp.
[133] Hein, T. and Hofmann, B. (2009). Approximate source conditions for nonlinear ill-posed problems – chances and limitations, *Inverse Problems* **25**, 3, pp. 035003, 16 pp.
[134] Hein, T. and Kazimierski, K. S. (2010a). Accelerated Landweber iteration in Banach spaces, *Inverse Problems* **26**, 5, pp. 055002, 17 pp.
[135] Hein, T. and Kazimierski, K. S. (2010b). Modified Landweber iteration in Banach spaces—convergence and convergence rates, *Numer. Funct. Anal. Optim.* **31**, 10, pp. 1158–1184.
[136] Heinrich, S. (2001). Multilevel Monte Carlo methods, in S. Margenov, J. Waśniewski and P. Yalamov (eds.), *Lecture Notes in Computer Science*, Vol. 2179 (Springer-Verlag, Berlin), pp. 58–67.
[137] Hintermüller, M., Ito, K. and Kunisch, K. (2002). The primal-dual active set strategy as a semismooth Newton method, *SIAM J. Optim.* **13**, 3, pp. 865–888 (2003).
[138] Hochstadt, H. and Lieberman, B. (1978). An inverse Sturm-Liouville problem with mixed given data, *SIAM J. Appl. Math.* **34**, 4, pp. 676–680.
[139] Hoffman, M. D., Blei, D. M., Wang, C. and Paisley, J. (2013). Stochastic variational inference, *J. Mach. Learn. Res.* **14**, pp. 1303–1347.
[140] Hofinger, A. and Pikkarainen, H. K. (2007). Convergence rate for the

Bayesian approach to linear inverse problems, *Inverse Problems* **23**, 6, pp. 2469–2484.

[141] Hofmann, B., Kaltenbacher, B., Pöschl, C. and Scherzer, O. (2007). A convergence rates result for Tikhonov regularization in Banach spaces with non-smooth operators, *Inverse Problems* **23**, 3, pp. 987–1010.

[142] Hofmann, B. and Mathé, P. (2012). Parameter choice in Banach space regularization under variational inequalities, *Inverse Problems* **28**, 10, pp. 104006, 17 pp.

[143] Hofmann, B. and Yamamoto, M. (2010). On the interplay of source conditions and variational inequalities for nonlinear ill-posed problems, *Appl. Anal.* **89**, 11, pp. 1705–1727.

[144] Hohage, T. and Werner, F. (2014). Convergence rates for inverse problems with impulsive noise, *SIAM J. Numer. Anal.* **52**, 3, pp. 1203–1221.

[145] Hohage, T. and Werner, F. (2013). Iteratively regularized Newton-type methods for general data misfit functionals and applications to Poisson data, *Numer. Math.* **123**, 4, pp. 745–779.

[146] Hou, S., Solna, K. and Zhao, H. (2006). A direct imaging algorithm for extended targets, *Inverse Problems* **22**, 4, pp. 1151–1178.

[147] Huber, P. J. (1981). *Robust Statistics* (John Wiley & Sons Inc., New York).

[148] Hyvönen, N. (2004). Complete electrode model of electrical impedance tomography: approximation properties and characterization of inclusions, *SIAM J. Appl. Math.* **64**, 3, pp. 902–931.

[149] Ikehata, M. (2000). Reconstruction of the support function for inclusion from boundary measurements, *J. Inverse Ill-Posed Probl.* **8**, 4, pp. 367–378.

[150] Imanuvilov, O. Y. and Yamamoto, M. (2001). Carleman estimate for a parabolic equation in a Sobolev space of negative order and its applications, in *Control of Nonlinear Distributed Parameter Systems (College Station, TX, 1999)*, *Lecture Notes in Pure and Appl. Math.*, Vol. 218 (Dekker, New York), pp. 113–137.

[151] Inglese, G. (1997). An inverse problem in corrosion detection, *Inverse Problems* **13**, 4, pp. 977–994.

[152] Ito, K. and Jin, B. (2011). A new approach to nonlinear constrained Tikhonov regularization, *Inverse Problems* **27**, 10, pp. 105005, 23 pp.

[153] Ito, K., Jin, B. and Takeuchi, T. (2011a). Multi-parameter Tikhonov regularization, *Methods and Applications of Analysis* **18**, 1, pp. 31–46.

[154] Ito, K., Jin, B. and Takeuchi, T. (2011b). A regularization parameter for nonsmooth Tikhonov regularization, *SIAM J. Sci. Comput.* **33**, 3, pp. 1415–1438.

[155] Ito, K., Jin, B. and Takeuchi, T. (2014). Multi-parameter Tikhonov regularization – an augmented approach, *Chinese Ann. Math. Ser. B* **35**, 3, pp. 383–398.

[156] Ito, K., Jin, B. and Zou, J. (2011c). A new choice rule for regularization parameters in Tikhonov regularization, *Appl. Anal.* **90**, 10, pp. 1521–1544.

[157] Ito, K., Jin, B. and Zou, J. (2012). A direct sampling method to an inverse medium scattering problem, *Inverse Problems* **28**, 2, pp. 025003, 11pp.

[158] Ito, K., Jin, B. and Zou, J. (2013a). A direct sampling method for inverse electromagnetic medium scattering, *Inverse Problems* **29**, 9, pp. 095018, 19 pp.

[159] Ito, K., Jin, B. and Zou, J. (2013b). A two-stage method for inverse medium scattering, *J. Comput. Phys.* **237**, pp. 211–223.

[160] Ito, K. and Kunisch, K. (1992). On the choice of the regularization parameter in nonlinear inverse problems, *SIAM J. Optim.* **2**, 3, pp. 376–404.

[161] Ito, K. and Kunisch, K. (1994). On the injectivity and linearization of the coefficient-to-solution mapping for elliptic boundary value problems, *J. Math. Anal. Appl.* **188**, 3, pp. 1040–1066.

[162] Ito, K. and Kunisch, K. (2000). BV-type regularization methods for convoluted objects with edge, flat and grey scales, *Inverse Problems* **16**, 4, pp. 909–928.

[163] Ito, K. and Kunisch, K. (2008). *Lagrange Multiplier Approach to Variational Problems and Applications* (SIAM, Philadelphia, PA).

[164] Ito, K. and Kunisch, K. (2014a). Optimal control with $L^p(\Omega)$, $p \in [0,1)$, *SIAM J. Control Optim.* **52**, 2, pp. 1251–1275.

[165] Ito, K. and Kunisch, K. (2014b). A variational approach to sparsity optimization based on Lagrange multiplier theory, *Inverse Problems* **30**, 1, pp. 015001, 23pp.

[166] Ito, K. and Ravindran, S. S. (1998). A reduced-order method for simulation and control of fluid flows, *J. Comput. Phys.* **143**, 2, pp. 403–425.

[167] Jin, B. (2008). Fast Bayesian approach for parameter estimation, *Internat. J. Numer. Methods Engrg.* **76**, 2, pp. 230–252.

[168] Jin, B. (2012a). A variational Bayesian method to inverse problems with impulsive noise, *J. Comput. Phys.* **231**, 2, pp. 423–435.

[169] Jin, B. and Lorenz, D. A. (2010). Heuristic parameter-choice rules for convex variational regularization based on error estimates, *SIAM J. Numer. Anal.* **48**, 3, pp. 1208–1229.

[170] Jin, B., Lorenz, D. A. and Schiffler, S. (2009). Elastic-net regularization: error estimates and active set methods, *Inverse Problems* **25**, 11, pp. 115022, 26 pp.

[171] Jin, B. and Maass, P. (2012a). An analysis of electrical impedance tomography with applications to Tikhonov regularization, *ESAIM Control Optim. Calc. Var.* **18**, 4, pp. 1027–1048.

[172] Jin, B. and Maass, P. (2012b). Sparsity regularization for parameter identification problems, *Inverse Problems* **28**, 12, pp. 123001, 70 pp.

[173] Jin, B., Zhao, Y. and Zou, J. (2012). Iterative parameter choice by discrepancy principle, *IMA J. Numer. Anal.* **32**, 4, pp. 1714–1732.

[174] Jin, B. and Zou, J. (2009). Augmented Tikhonov regularization, *Inverse Problems* **25**, 2, pp. 025001, 25 pp.

[175] Jin, B. and Zou, J. (2010). Numerical estimation of the Robin coefficient in a stationary diffusion equation, *IMA J. Numer. Anal.* **30**, 3, pp. 677–701.

[176] Jin, Q. (1999). Applications of the modified discrepancy principle to Tikhonov regularization of nonlinear ill-posed problems, *SIAM J. Numer. Anal.* **36**, 2, pp. 475–490.

[177] Jin, Q. (2012b). Inexact Newton-Landweber iteration for solving nonlinear inverse problems in Banach spaces, *Inverse Problems* **28**, 6, pp. 065002, 15 pp.

[178] Jin, Q. and Stals, L. (2012). Nonstationary iterated Tikhonov regularization for ill-posed problems in Banach spaces, *Inverse Problems* **28**, 10, pp. 104011, 15 pp.

[179] Jin, Q. and Wang, W. (2013). Landweber iteration of Kaczmarz type with general non-smooth convex penalty functionals, *Inverse Problems* **29**, 8, pp. 085011, 22 pp.

[180] Jin, Q. and Zhong, M. (2014). Nonstationary iterated Tikhonov regularization in Banach spaces with uniformly convex penalty terms, *Numer. Math.* **127**, 3, pp. 485–513.

[181] Johnston, P. R. and Gulrajani, R. M. (1997). A new method for regularization parameter determination in the inverse problem of electrocardiography, *IEEE Trans. Biodmed. Eng.* **44**, 1, pp. 19–39.

[182] Johnston, P. R. and Gulrajani, R. M. (2002). An analysis of the zero-crossing method for choosing regularization parameters, *SIAM J. Sci. Comput.* **24**, 2, pp. 428–442.

[183] Jordan, M. I., Ghahramani, Z., Jaakkola, T. S. and Saul, L. K. (1999). An introduction to variational methods for graphical models, *Mach. Learn.* **37**, pp. 183–233.

[184] Kabanikhin, S. I. and Schieck, M. (2008). Impact of conditional stability: convergence rates for general linear regularization methods, *J. Inverse Ill-Posed Probl.* **16**, 3, pp. 267–282.

[185] Kac, M. (1966). Can one hear the shape of a drum? *The Amer. Math. Monthly* **73**, 4, pp. 1–23.

[186] Kaipio, J. and Somersalo, E. (2005). *Statistical and Computational Inverse Problems* (Springer-Verlag, New York).

[187] Kaipio, J. and Somersalo, E. (2007). Statistical inverse problems: discretization, model reduction and inverse crimes, *J. Comput. Appl. Math.* **198**, 2, pp. 493–504.

[188] Kaltenbacher, B. and Hofmann, B. (2010). Convergence rates for the iteratively regularized Gauss-Newton method in Banach spaces, *Inverse Problems* **26**, 3, pp. 035007, 21 pp.

[189] Kaltenbacher, B., Schöpfer, F. and Schuster, T. (2009). Iterative methods for nonlinear ill-posed problems in Banach spaces: convergence and applications to parameter identification problems, *Inverse Problems* **25**, 6, pp. 065003, 19 pp.

[190] Kaltenbacher, B. and Tomba, I. (2013). Convergence rates for an iteratively regularized Newton-Landweber iteration in Banach space, *Inverse Problems* **29**, 2, pp. 025010, 18 pp.

[191] Kass, R. E. and Raftery, A. E. (1995). Bayes factors and model uncertainty, *J. Amer. Stat. Ass.* **90**, pp. 773–795.

[192] Kaup, P. G. and Santosa, F. (1995). Nondestructive evaluation of corrosion damage using electrostatic measurements, *J. Nondestr. Eval.* **14**, 3, pp. 127–136.

[193] Kazimierski, K. S. (2010). *Aspects of Regularization in Banach Spaces* (Logos, Berlin).

[194] Kazimierski, K. S., Maass, P. and Strehlow, R. (2012). Norm sensitivity of sparsity regularization with respect to p, *Inverse Problems* **28**, 10, pp. 104009, 20 pp.

[195] Ketelsen, C., Scheichl, R. and Teckentrup, A. L. (2013). A hierarchical multilevel Markov chain Monte Carlo algorithm with applications to uncertainty quantification in subsurface flow, preprint.

[196] Kindermann, S. (2011). Convergence analysis of minimization-based noise level-free parameter choice rules for linear ill-posed problems, *Electron. Trans. Numer. Anal.* **38**, pp. 233–257.

[197] Kindermann, S. and Neubauer, A. (2008). On the convergence of the quasioptimality criterion for (iterated) Tikhonov regularization, *Inverse Probl. Imaging* **2**, 2, pp. 291–299.

[198] Kirsch, A. (1993). The domain derivative and two applications in inverse scattering theory, *Inverse Problems* **9**, 1, pp. 81–96.

[199] Kirsch, A. (1998). Characterization of the shape of a scattering obstacle using the spectral data of the far field operator, *Inverse Problems* **14**, 6, pp. 1489–1512.

[200] Kirsch, A. (2002). The MUSIC algorithm and the factorization method in inverse scattering theory for inhomogeneous media, *Inverse Problems* **18**, 4, pp. 1025–1040.

[201] Kirsch, A. and Grinberg, N. (2008). *The Factorization Method for Inverse Problems* (Oxford University Press, Oxford).

[202] Klibanov, M. V. and Timonov, A. (2004). *Carleman Estimates for Coefficient Inverse Problems and Numerical Applications* (VSP, Utrecht).

[203] Kokurin, M. Y. (2010). On the convexity of the Tikhonov functional and iteratively regularized methods for solving irregular nonlinear operator equations, *Zh. Vychisl. Mat. Mat. Fiz.* **50**, 4, pp. 651–664.

[204] Kuchment, P. and Kunyansky, L. (2008). Mathematics of thermoacoustic tomography, *Eur. J. Appl. Math.* **19**, pp. 191–224.

[205] Kullback, S. and Leibler, R. A. (1951). On information and sufficiency, *Ann. Math. Statistics* **22**, pp. 79–86.

[206] Kunisch, K. and Zou, J. (1998). Iterative choices of regularization parameters in linear inverse problems, *Inverse Problems* **14**, 5, pp. 1247–1264.

[207] Lange, K. L., Little, R. J. A. and Taylor, J. M. G. (1989). Robust statistical modeling using the t distribution, *J. Amer. Statist. Assoc.* **84**, 408, pp. 881–896.

[208] Lazarov, R. D., Lu, S. and Pereverzev, S. V. (2007). On the balancing principle for some problems of numerical analysis, *Numer. Math.* **106**, 4, pp. 659–689.

[209] Le, T., Chartrand, R. and Asaki, T. J. (2007). A variational approach to reconstructing images corrupted by Poisson noise, *J. Math. Imaging Vision* **27**, 3, pp. 257–263.

[210] Lechleiter, A., Kazimierski, K. S. and Karamehmedovic, M. (2013). Tikhonov regularization in L^p applied to inverse medium scattering, *In-*

verse Problems **29**, 7, pp. 075003, 19 pp.

[211] Leonov, A. S. (1978a). The choice of the regularization parameter from quasioptimality and relation tests, *Dokl. Akad. Nauk SSSR* **240**, 1, pp. 18–20.

[212] Leonov, A. S. (1978b). On the justification of the choice of the regularization parameter based on quasi-optimality and relation tests, *Zh. Vychisl. Mat. i Mat. Fiz.* **18**, 6, pp. 1363–1376, 1627.

[213] Lepskiĭ, O. V. (1990). A problem of adaptive estimation in Gaussian white noise, *Teor. Veroyatnost. i Primenen.* **35**, 3, pp. 459–470.

[214] Levinson, N. (1949). The inverse Sturm-Liouville problem, *Mat. Tidsskr. B.* **25**, pp. 25–30.

[215] Li, C. and Wang, L. V. (2009). Photoacoustic tomography and sensing in biomedicine, *Phys. Med. Biol.* **54**, 19, pp. R59–R97.

[216] Li, J., Yamamoto, M. and Zou, J. (2009). Conditional stability and numerical reconstruction of initial temperature, *Commun. Pure Appl. Anal.* **8**, 1, pp. 361–382.

[217] Liu, J. (2008). *Monte Carlo Strategies in Scientific Computing* (Springer, Berlin).

[218] Lorenz, D. A. (2008). Convergence rates and source conditions for Tikhonov regularization with sparsity constraints, *J. Inverse Ill-Posed Probl.* **16**, 5, pp. 463–478.

[219] Lorenz, D. A., Schiffler, S. and Trede, D. (2011). Beyond convergence rates: exact recovery with the Tikhonov regularization with sparsity constraints, *Inverse Problems* **27**, 8, pp. 085009, 17 pp.

[220] Lu, S. and Mathé, P. (2013). Heuristic parameter selection based on functional minimization: optimality and model function approach, *Math. Comp.* **82**, 283, pp. 1609–1630.

[221] Lu, S. and Pereverzev, S. V. (2011). Multi-parameter regularization and its numerical regularization, *Numer. Math.* **118**, 1, pp. 1–31.

[222] Lu, S., Pereverzev, S. V. and Ramlau, R. (2007). An analysis of Tikhonov regularization for nonlinear ill-posed problems under a general smoothness assumption, *Inverse Problems* **23**, 1, pp. 217–230.

[223] Lucka, F. (2012). Fast Markov chain Monte Carlo sampling for sparse Bayesian inference in high-dimensional inverse problems using L1-type priors, *Inverse Problems* **28**, 12, pp. 125012, 31 pp.

[224] MacKay, D. J. C. (1992). Bayesian interpolation, *Neur. Comput.* **4**, 3, pp. 415–447.

[225] MacLeod, R. S. and Brooks, D. H. (1998). Recent progress in inverse problems in electrocardiology, *IEEE Eng. Med. Biol. Magazine* **17**, 1, pp. 73–83.

[226] Marčenko, V. A. (1950). Concerning the theory of a differential operator of the second order, *Doklady Akad. Nauk SSSR. (N.S.)* **72**, pp. 457–460.

[227] Margotti, F., Rieder, A. and Leitao, A. (2014). A Kaczmarz version of the REGINN-Landweber iteration for ill-posed problems in Banach spaces, *SIAM J. Numer. Anal.* **52**, 3, pp. 1439–1465.

[228] Martin, J., Wilcox, L. C., Burstedde, C. and Ghattas, O. (2012). A stochastic Newton MCMC method for large-scale statistical inverse problems with

application to seismic inversion, *SIAM J. Sci. Comput.* **34**, 3, pp. A1460–A1487.

[229] Marzouk, Y. M., Najm, H. N. and Rahn, L. A. (2007). Stochastic spectral methods for efficient Bayesian solution of inverse problems, *J. Comput. Phys.* **224**, 2, pp. 560–586.

[230] Mathé, P. (2006). The Lepskiĭ principle revisited, *Inverse Problems* **22**, 3, pp. L11–L15.

[231] Maurer, H. and Zowe, J. (1979). First and second-order necessary and sufficient optimality conditions for infinite-dimensional programming problems, *Math. Progr.* **16**, 1, pp. 98–110.

[232] McLaughlin, J. R. (1988). Inverse spectral theory using nodal points as data – a uniqueness result, *J. Diff. Eq.* **73**, 2, pp. 354–362.

[233] Metropolis, N., Rosenbluth, A. W., Rosenbluth, M. N., Teller, A. H. and Teller, E. (1953). Equations of state calculations by fast computing machines, *J. Chem. Phys.* **21**, 6, pp. 1087–1092.

[234] Meyers, N. G. (1963). An L^p-estimate for the gradient of solutions of second order elliptic divergence equations, *Ann. Scuola Norm. Sup. Pisa (3)* **17**, pp. 189–206.

[235] Mifflin, R. (1977). Semismooth and semiconvex functions in constrained optimization, *SIAM J. Control Optimization* **15**, 6, pp. 959–972.

[236] Minka, T. (2001a). Expectation propagation for approximate Bayesian inference, in J. Breese and D. Koller (eds.), *Uncertainty in Artificial Intelligence 17* (Morgan Kaufmann).

[237] Minka, T. (2001b). *A Family of Algorithms for Approximate Bayesian inference*, Ph.D. thesis, Massachusetts Institute of Technology.

[238] Morozov, V. A. (1966). On the solution of functional equations by the method of regularization, *Soviet Math. Dokl.* **7**, pp. 414–417.

[239] Mueller, J. L. and Siltanen, S. (2003). Direct reconstructions of conductivities from boundary measurements, *SIAM J. Sci. Comput.* **24**, 4, pp. 1232–1266.

[240] Natterer, F. (2001). *The Mathematics of Computerized Tomography* (SIAM, Philadelphia, PA).

[241] Naumova, V. and Pereverzyev, S. V. (2013). Multi-penalty regularization with a component-wise penalization, *Inverse Problems* **29**, 7, pp. 075002, 15 pp.

[242] Neubauer, A. (1987). Finite-dimensional approximation of constrained Tikhonov-regularized solutions of ill-posed linear operator equations, *Math. Comput.* **48**, 178, pp. 565–583.

[243] Neubauer, A. (1988). Tikhonov-regularization of ill-posed linear operator equations on closed convex sets, *J. Approx. Theory* **53**, 3, pp. 304–320.

[244] Neubauer, A. (2008). The convergence of a new heuristic parameter selection criterion for general regularization methods, *Inverse Problems* **24**, 5, pp. 055005, 10 pp.

[245] Neubauer, A. (2009). On enhanced convergence rates for Tikhonov regularization of nonlinear ill-posed problems in Banach spaces, *Inverse Problems* **25**, 6, pp. 065009, 10 pp.

[246] Neubauer, A., Hein, T., Hofmann, B., Kindermann, S. and Tautenhahn, U. (2010). Improved and extended results for enhanced convergence rates of Tikhonov regularization in Banach spaces, *Appl. Anal.* **89**, 11, pp. 1729–1743.

[247] Neubauer, A. and Pikkarainen, H. K. (2008). Convergence results for the Bayesian inversion theory, *J. Inverse Ill-Posed Probl.* **16**, 6, pp. 601–613.

[248] Park, T. and Casella, G. (2008). The Bayesian lasso, *J. Amer. Statist. Assoc.* **103**, 482, pp. 681–686.

[249] Pazy, A. (1983). *Semigroups of Linear Operators and Applications to Partial Differential Equations* (Springer-Verlag, New York).

[250] Pidcock, M. K., Kuzuoglu, M. and Leblebicioglu, K. (1995). Analytic and semianalytic solutions in electrical impedance tomography: I. Two-dimensional problems, *Physiol. Meas.* **16**, pp. 77—90.

[251] Potthas, R. (1994). Fréchet differentiability of boundary integral operators in inverse acoustic scattering, *Inverse Problems* **10**, 2, pp. 431–447.

[252] Potthast, R. (2006). A survey on sampling and probe methods for inverse problems, *Inverse Problems* **22**, 2, pp. R1–R47.

[253] Potthast, R. (2010). A study on orthogonality sampling, *Inverse Problems* **26**, 7, pp. 074015, 17 pp.

[254] Qi, L. Q. (1993). Convergence analysis of some algorithms for solving nonsmooth equations, *Math. Oper. Res.* **18**, 1, pp. 227–244.

[255] Qi, L. Q. and Sun, J. (1993). A nonsmooth version of Newton's method, *Math. Programming* **58**, 3, Ser. A, pp. 353–367.

[256] Ramlau, R. and Resmerita, E. (2010). Convergence rates for regularization with sparsity constraints, *Electron. Trans. Numer. Anal.* **37**, pp. 87–104.

[257] Regińska, T. (1996). A regularization parameter in discrete ill-posed problems, *SIAM J. Sci. Comput.* **17**, 3, pp. 740–749.

[258] Resmerita, E. (2005). Regularization of ill-posed problems in Banach spaces: convergence rates, *Inverse Problems* **21**, 4, pp. 1303–1314.

[259] Resmerita, E. and Anderssen, R. S. (2007). Joint additive Kullback-Leibler residual minimization and regularization for linear inverse problems, *Math. Methods Appl. Sci.* **30**, 13, pp. 1527–1544.

[260] Resmerita, E. and Scherzer, O. (2006). Error estimates for non-quadratic regularization and the relation to enhancement, *Inverse Problems* **22**, 3, pp. 801–814.

[261] Richter, G. R. (1981). An inverse problem for the steady state diffusion equation, *SIAM J. Appl. Math.* **41**, 2, pp. 210–221.

[262] Robert, C. P. and Casella, G. (1999). *Monte Carlo Statistical Methods* (Springer-Verlag, New York).

[263] Rondi, L. (2008). On the regularization of the inverse conductivity problem with discontinuous conductivities, *Inverse Probl. Imaging* **2**, 3, pp. 397–409.

[264] Rondi, L. and Santosa, F. (2001). Enhanced electrical impedance tomography via the Mumford-Shah functional, *ESAIM Control Optim. Calc. Var.* **6**, pp. 517–538.

[265] Rudin, L. I., Osher, S. and Fatemi, E. (1992). Nonlinear total variation based noise removal algorithms, *Phys. D* **60**, 1-4, pp. 259–268.

[266] Rundell, W. and Sacks, P. E. (1992). Reconstruction techniques for classical inverse Sturm-Liouville problems, *Math. Comp.* **58**, 197, pp. 161–183.

[267] Sato, M., Yoshioka, T., Kajihara, S., Toyama, K., Goda, N., Doya, K. and Kawato, M. (2004). Hierarchical Bayesian estimation for MEG inverse problem, *NeuroImage* **23**, 3, pp. 806–826.

[268] Scherzer, O. (1993). Convergence rates of iterated Tikhonov regularized solutions of nonlinear ill-posed problems, *Numer. Math.* **66**, 2, pp. 259–279.

[269] Scherzer, O., Engl, H. W. and Kunisch, K. (1993). Optimal a posteriori parameter choice for Tikhonov regularization for solving nonlinear ill-posed problems, *SIAM J. Numer. Anal.* **30**, 6, pp. 1796–1838.

[270] Schmidt, R. O. (1986). Multiple emitter location and signal parameter estimation, *IEEE Trans. Ant. Prop.* **34**, 3, pp. 276–280.

[271] Schöpfer, F., Louis, A. K. and Schuster, T. (2006). Nonlinear iterative methods for linear ill-posed problems in Banach spaces, *Inverse Problems* **22**, 1, pp. 311–329.

[272] Schuster, T., Kaltenbacher, B., Hofmann, B. and Kazimierski, K. S. (2012). *Regularization Methods in Banach Spaces* (Walter de Gruyter, Berlin).

[273] Seeger, M. W. (2008). Bayesian inference and optimal design for the sparse linear model, *J. Mach. Learn. Res.* **9**, pp. 759–813.

[274] Seidman, T. I. and Vogel, C. R. (1989). Well posedness and convergence of some regularisation methods for non-linear ill posed problems, *Inverse Problems* **5**, 2, pp. 227–238.

[275] Semenov, S. (2009). Microwave tomography: review of the progress towards clinical applications, *Phil. Trans. R. Soc. A* **367**, 1900, pp. 3021–3042.

[276] Shepp, L. A. and Vardi, Y. (1982). Maximum likelihood reconstruction for emission tomography, *IEEE Trans. Med. Imag.* **1**, 2, pp. 113–122.

[277] Siltanen, S., Mueller, J. and Isaacson, D. (2000). An implementation of the reconstruction algorithm of A. Nachman for the 2D inverse conductivity problem, *Inverse Problems* **16**, 3, pp. 681–699.

[278] Somersalo, E., Cheney, M. and Isaacson, D. (1992). Existence and uniqueness for electrode models for electric current computed tomography, *SIAM J. Appl. Math.* **52**, 4, pp. 1023–1040.

[279] Somersalo, E., Cheney, M., Isaacson, D. and Isaacson, E. (1991). Layer stripping: a direct numerical method for impedance imaging, *Inverse Problems* **7**, 6, pp. 899–926.

[280] Stuart, A. M. (2010). Inverse problems: a Bayesian perspective, *Acta Numer.* **19**, pp. 451–559.

[281] Tang, J., Han, W. and Han, B. (2013). A theoretical study for RTE-based parameter identification problems, *Inverse Problems* **29**, 9, pp. 095002, 18 pp.

[282] Tarantola, A. (2005). *Inverse Problem Theory and Methods for Model Parameter Estimation* (SIAM, Philadelphia, PA).

[283] Tautenhahn, U. and Hämarik, U. (1999). The use of monotonicity for choosing the regularization parameter in ill-posed problems, *Inverse Problems* **15**, 6, pp. 1487–1505.

[284] Tautenhahn, U. and Jin, Q.-N. (2003). Tikhonov regularization and a posteriori rules for solving nonlinear ill posed problems, *Inverse Problems* **19**, 1, pp. 1–21.

[285] Tibshirani, R. (1996). Regression shrinkage and selection via the lasso, *J. Roy. Statist. Soc. Ser. B* **58**, 1, pp. 267–288.

[286] Tikhonov, A. N. (1943). On the stability of inverse problems, *C. R. (Doklady) Acad. Sci. URSS (N.S.)* **39**, pp. 176–179.

[287] Tikhonov, A. N. (1963a). On the regularization of ill-posed problems, *Dokl. Akad. Nauk SSSR* **153**, pp. 49–52.

[288] Tikhonov, A. N. (1963b). On the solution of ill-posed problems and the method of regularization, *Dokl. Akad. Nauk SSSR* **151**, pp. 501–504.

[289] Tikhonov, A. N. and Arsenin, V. Y. (1977). *Solutions of Ill-posed Problems* (John Wiley & Sons, New York).

[290] Tikhonov, A. N. and Glasko, V. B. (1965). Use of the regularization method in non-linear problems, *USSR Comput. Math. Math. Phys.* **5**, 3, pp. 93–107.

[291] Tikhonov, A. N., Glasko, V. B. and Kriksin, J. A. (1979). On the question of quasi-optimal choice of a regularized approximation, *Dokl. Akad. Nauk SSSR* **248**, 3, pp. 531–535.

[292] Tikhonov, A. N., Leonov, A. S. and Yagola, A. G. (1998). *Nonlinear Ill-Posed Problems. Vol. 1, 2* (Chapman & Hall, London).

[293] Tipping, M. E. (2001). Sparse bayesian learning and the relevance vector machine, *J. Mach. Learn Res.* **1**, pp. 211–244.

[294] Troianiello, G. M. (1987). *Elliptic Differential Equations and Ostacle Problems* (Plenum Press, New York).

[295] Tur, M., Chin, K. C. and Goodman, J. W. (1982). When is speckle noise multiplicative? *Appl. Optics* **21**, 7, pp. 1157–1159.

[296] Ulbrich, M. (2002). Semismooth Newton methods for operator equations in function spaces, *SIAM J. Optim.* **13**, 3, pp. 805–842.

[297] Vaiter, S., Peyré, G., Dossal, C. and Fadili, J. (2013). Robust sparse analysis regularization, *IEEE Trans. Inform. Theory* **59**, 4, pp. 2001–2016.

[298] Vardi, Y., Shepp, L. A. and Kaufman, L. (1985). A statistical model for positron emission tomography, *J. Amer. Stat. Ass.* **80**, 389, pp. 8–20.

[299] Vinokurov, V. A. (1972). Two remarks on the choice of the regularization parameter, *Ž. Vyčisl. Mat. i Mat. Fiz.* **12**, pp. 481–483.

[300] Vinokurov, V. A. and Shatov, A. K. (1974). Estimation of the reuglarization parameter in non-linear problems, *Zh. v̄ychisl Mat. mat. Fiz.* **14**, 6, pp. 1386–1392.

[301] Wainwright, M. J. and Jordan, M. I. (2008). Graphical Models, Exponential Families, and Variational Inference, *Foundations and Trends in Machine Learning* , pp. 1–305.

[302] Wang, B. and Titterington, D. M. (2005). Inadequacy of interval estimates corresponding to variational Bayesian approximations, in R. G. Cowell and Z. Ghahramani (eds.), *Proceedings of the Tenth International Workshop on Artificial Intelligence and Statistics* (Society for Artificial Intelligence and Statistics, Barbados), pp. 373–380.

[303] Wang, J. and Zabaras, N. (2005). Hierarchical Bayesian models for inverse

problems in heat conduction, *Inverse Problems* **21**, 1, pp. 183–206.

[304] Wang, K. and Anastasio, M. A. (2011). Photoacoustic and thermoacoustic tomography: image formation and principles, in O. Scherzer (ed.), *Handbook of Mathematical Methods in Imaging* (Springer), pp. 781–815.

[305] Wasserman, L. (2000). Bayesian model selection and model averaging, *J. Math. Psychol.* **44**, 1, pp. 92–107.

[306] Werner, F. and Hohage, T. (2012). Convergence rates in expectation for Tikhonov-type regularization of inverse problems with Poisson data, *Inverse Problems* **28**, 10, pp. 104004, 15 pp.

[307] White, F. M. (1988). *Heat and Mass Transfer* (Addison-Wesley, Reading, MA).

[308] Xie, J. and Zou, J. (2002). An improved model function method for choosing regularization parameters in linear inverse problems, *Inverse Problems* **18**, 3, pp. 631–643.

[309] Xu, Z. B. and Roach, G. F. (1991). Characteristic inequalities of uniformly convex and uniformly smooth Banach spaces, *J. Math. Anal. Appl.* **157**, 1, pp. 189–210.

[310] Yamamoto, M. (2009). Carleman estimates for parabolic equations and applications, *Inverse Problems* **25**, 12, pp. 123013, 75 pp.

[311] Zarzer, C. A. (2009). On Tikhonov regularization with non-convex sparsity constraints, *Inverse Problems* **25**, 2, pp. 025006, 13 pp.

[312] Zhou, Z., Leahy, R. M. and Qi, J. (1997). Approximate maximum likelihood hyperparameter estimation for Gibbs priors, *IEEE Trans. Imag. Proc.* **6**, 6, pp. 844–861.

[313] Zou, H. and Hastie, T. (2005). Regularization and variable selection via the elastic net, *J. R. Stat. Soc. Ser. B Stat. Methodol.* **67**, 2, pp. 301–320.

[314] Zou, H., Hastie, T. and Tibshirani, R. (2007). On the "degrees of freedom" of the lasso, *Ann. Statist.* **35**, 5, pp. 2173–2192.

Index

ψ-minimizing solution, 35, 110

analytic continuation, 226
approximate inference, 267
  expectation propagation, 277
  Gaussian approximation, 275
  mean field approximation, 271
  variational Bayes, 271
Arcangeli's rule, 59
augmented Lagrangian method, 177
augmented Tikhonov regularization, 77, 92
  balancing principle, 79
  Bayesian inversion, 242
  variational characterization, 78

Bakushinskiĭ's veto, 66
balanced discrepancy principle, 95
  Broyden's method, 100
balancing principle, 79, 94, 137, 250
  Bayesian inversion, 243
  consistency, 139
  fixed point algorithm, 101
  sensitivity analysis, 85
  value function, 81
Bayesian inference, 235
  augmented Tikhonov regularization, 241
  Bayes factor, 244
  Bayes' formula, 235
  hierarchical model, 240
  likelihood, 235
  maximum a posteriori, 241
  Poisson inverse problem, 275
  prior distribution, 235
  sparsity, 258, 273
  Tikhonov regularization, 240
Bayesian information criterion, 246
Bregman distance, 46
  Pythagoras identity, 49
  symmetric Bregman distance, 51

Cauchy problem, 6
complete electrode model, 21
conditional stability, 160

direct sampling method, 218
discrepancy principle, 55, 135
  convergence rate, 57
  damped, 56
  secant method, 65
  sensitivity analysis, 58
  two parameter algorithm, 62
  upper bound, 64, 127
  value function representation, 55
  variational inequalities, 159

electrical impedance tomography, 20, 175, 281
exact penalty method, 184
exponential family, 295

factorization method, 217
fidelity functional, 289

first-order augmented Lagrangian method, 179
fixed point algorithm, 88

Gauss-Newton method, 188
Gel'fand-Levitan-Marchenko transformation, 228
Gibbs sampler, 256
  scale mixture, 258
  sparsity, 258

Hanke-Raus rule, 66, 137, 140

inverse scattering problem, 10, 147, 212, 218
  Born's approximation, 13
inverse source problem, 8, 223
inverse Sturm-Liouville problem, 13, 232

Kullback-Leibler divergence, 268

Lagrange calculus, 175
Lagrange multiplier theory
  Yoshida-Moreau approximation, 181
Laplace's method, 245
linear sampling method, 215

Markov chain Monte Carlo, 250
  approximation error approach, 266
  convergence, 258
  convergence diagnostics, 256
  Gibbs sampler, 256
  Metropolis-Hastings algorithm, 253
  multilevel, 262
  preconditioning, 261
  reduced-order model, 264
maximum likelihood method, 248
  EM algorithm, 249
Metropolis-Hastings algorithm
  acceptance probability, 254
  independence sampler, 255
  random walk proposal, 254
model function, 59
  balancing principle, 80
Monte Carlo method, 250
  error, 251
  importance sampling, 252
multi-parameter Tikhonov regularization, 91
MUSIC, 210, 212, 248

necessary optimality, 173
Newton differentiability, 189
noise model, 238, 289
  additive noise, 289
  Gaussian noise, 239, 289
  Laplace noise, 290
  multiplicative noise, 291
  Poisson noise, 20, 293
  salt-and-pepper noise, 292
  speckle noise, 292
nonlinearity condition, 131
  parameter identification, 142

parameter identification, 141
posterior probability density function, 242
Prékopa-Leindler inequality, 297
primal-dual active set method, 180
prior distribution
  Gamma distribution, 240
  Markov random field, 239

reciprocity gap functional, 223, 226
Robin inverse problem, 7, 145

second-order necessary condition, 130
second-order sufficient condition, 131
semismooth Newton method, 180, 189
singular value decomposition, 265, 286
singular value system, 285
source condition, 45, 131, 132
  higher order, 52
  necessary optimality condition, 128
  parameter identification, 142
  variational inequalities, 152, 156
sparsity, 154
  Bayesian inference, 258, 273
  Bayesian information criterion, 247
sparsity optimization, 193

Tikhonov functional
local convexity, 124
uniqueness, 126
Tikhonov regularization, 29, 240
L^1-TV model, 32, 40
fidelity, 29
multi-parameter regularization, 91
regularization, 29
regularization parameter, 30
sparsity regularization, 33
Tikhonov functional, 29

value function, 39, 55, 93
differentiability, 42
Radon-Nikodym derivative, 43

Printed in the United States
By Bookmasters